The Emergent Metaphysics in Plato's Theory of Disorder

The Emergent Metaphysics in Plato's Theory of Disorder

S. R. Charles

LEXINGTON BOOKS

A Division of
ROWMAN & LITTLEFIELD PUBLISHERS, INC.
Lanham • Boulder • New York • Toronto • Oxford

LEXINGTON BOOKS

A division of Rowman & Littlefield Publishers, Inc.
A wholly owned subsidiary of The Rowman & Littlefield Publishing Group, Inc.
4501 Forbes Boulevard, Suite 200
Lanham, MD 20706

PO Box 317
Oxford
OX2 9RU, UK

British Library Cataloguing in Publication Information Available

Library of Congress Cataloging-in-Publication Data

Charles, S. R. (Sarai Robin), 1951-
The emergent metaphysics in Plato's theory of disorder / S.R. Charles.
p. cm.
Includes bibliographical references and indexes.
ISBN 978-0-7391-0911-3
1. Plato. 2. Metaphysics. 3. Cosmology. 4. Plato. Timaeus. 5. Plato. Laws X. I. Title.
B398.M4C43 2005
113—dc22 2005026381

Printed in the United States of America

™ The paper used in this publication meets the minimum requirements of American National Standard for Information Sciences—Permanence of Paper for Printed Library Materials, ANSI/NISO Z39.48–1992.

To Plato

He had all the highest conditions for making the most of his passage through earthly life. Of noble ancestry, he inherited a bodily strength and power enabling him to sustain the efforts necessary in order to acquire all the knowledge of his times and to increase it; he was not compelled in any way to struggle for material existence, being a wealthy citizen in the wealthiest city of his times; he was born after a generation which included some of the greatest poets of mankind, and had himself an exceptional poetical talent, which he reserved entirely for the purposes of his philosophical teaching. He did not live in isolation, like Descartes or Spinoza, nor in a whirl of worldly interests, like Leibniz, nor in humiliating dependence upon an absolute government, like Kant or Hegel, his freedom of speech and teaching . . . secured by the crime committed against Socrates. . . . What limits can be set to the intellectual progress of such a philosopher? He stands far above his great teacher, far above his great pupil, alone in his incomparable greatness.

—W. Lutoslawski,
The Origin and Growth of Plato's Logic (526–27)

CONTENTS

PART II

COMMENTARY ON *LAWS X*
The 'Ανάγκη/'Ananke' (Necessity) and 'Demiurge' Dichotomy and a Journey into 'Soul'

Acknowledgments

I would like to thank the many scholars and students who have contributed over the years through their writings, dialogue and questioning to the ideas contained in this book. At the same time, I wish to take full responsibility for any errors of thought, a certain boldness in tone and at times, the uncompromising manner in which I have chosen to tackle this very difficult area of philosophy.

Specifically I would like to thank T. M. Robinson, Professor Emeritus, University of Toronto; the late T. J. Lang, Professor, St. Michael's University, Toronto; Dr. I. A. Massey, Toronto; Paolo G. Crivelli, Oxford; J. P. Mackey, Divinity, University of Edinburgh; as well as Dory Scaltsas, Timothy Sprigge, Stephen Priest, A. S. Mason, V. M. Hope, University of Edinburgh; and Andros Loizou, University of Central Lancashire. I wish also to thank Mr. D. B. Robinson, Hon. Fellow in Classics, University of Edinburgh, for reviewing my translation of the *Timaeus* and for his insight and input into its final version. It cannot be overstated, moreover, that without the passionate philosophical writings of A. E. Taylor, F. M. Cornford and R. D. Archer-Hind, upon whose research and findings I comment and upon whose genius I am indebted, this book would founder.

Finally, I would to thank Lexington Books and their staff for publishing this work and in particular, with deep appreciation, Acquisitions Editor, Jason Hallman and Associate Editor, Production, Sheila-Katherine Zwiebel, and in completing the project, Acquisitions Editor, Rob Carley, as well as Editorial Assistants, Kathryn Funk and Ryan Quick, and last but not least, the ever-so-patient proof readers.

S. R. Charles
October 31, 2005

Abbreviations

(see Bibliography for full details)

A-H	Archer-Hind (whose work is *The Timaeus of Plato*)
DK	Democritus – Fragments, H. Diels and W. Kranz, *Die Fragmente der Vorsokratiker*
DP	Jowett, *The Dialogues of Plato*
DPM	Teloh, *The Development of Plato's Metaphysics*
EGP	J. Barnes, *Early Greek Philosophy*
LG	*Laws* (Plato)
LSL	Liddell & Scott, *Greek-English Lexicon*
OCT	Burnet, *Oxford Classical Texts* (*Timaeus*)
PC	Cornford, *Plato's Cosmology*
PT	Taylor, *Commentary on Plato's "Timaeus"*
PU	Vlastos, *Plato's Universe*
TI	*Timaeus* (Plato)
TP	Archer-Hind (A-H), *The Timaeus of Plato*

References to translations of the *Timaeus* other than that of the author are cited as *Ti.* preceded by the translator's name or are simply referred to by the translator's name, followed by the relevant page number(s) or line within the Greek text. This also applies to Plato's *Laws*, abbreviated *Lg.*

INTRODUCTION

"*Not even the gods can war against Ananke*"
(from a poem by Simonides, as quoted by Plato in the *Protagoras* 345d)

This book is an exploration of Plato's understanding of disorder as it is found in his later works: the *Timaeus* and *Laws X*. Additionally, it is also a study of the emergent metaphysics that follows upon such an investigation. Although a position that claims Plato has a discrete theory of disorder, which this is, is open to dispute, given that for the most part it is couched in mythic terms within his cosmology (the *Timaeus*) and in religious terms within *Laws X*, it is my hope to challenge, or at least diminish, some of the doubts which may be raised concerning such an endeavor. It is my hope, moreover, to demonstrate that when a fuller account of Plato's understanding of disorder is considered, both within its own terms and as can be demonstrated within core works touching upon his metaphysics at large, a deeper and more substantial understanding of Plato's overall metaphysics becomes possible, whereby age-old philosophical problems become less formidable.

I will now summarize some of the main points made and conclusions arrived at, necessarily leaving out most of the argumentation and support for these to the reader's own judgment following a fuller reading of this work. Firstly, this book is not in any way a compilation or critique of the huge body of secondary literature that has been written on the *Timaeus* and *Laws X* concerning 'necessity' or 'evil.' Although 'necessity,' or more properly 'ananke,' and 'evil' are discussed, these ideas form part of a larger discussion concerning Plato's treatment of 'disorder' as it appears in each of these two works. For practical and scholastic reasons, I have narrowed the commentators whose writings I discuss within the body of the text to F. M. Cornford, A. E. Taylor, and R. D. Archer-Hind, since the groundwork of these scholars forms the tone of modern-day scholastic opinion concerning Plato's understanding of 'disorder.' More recent writers, although regrettably only a few, I confine to the footnotes, whose opinions, for the most part, form a refinement or amendment to the conclusions of these earlier writers. It is, I suggest, Cornford's view (which heavily followed upon that of

Archer-Hind) that has persisted as the one predominantly held by most scholars today concerning 'ananke,' although many twists and turns of nuance have invariably refined it. Notably, Luc Brisson is a contemporary champion of the Cornfordian position. It is a view, broadly put, that affords no metaphysical or ontological status to 'ananke,' understood by Cornford as mechanical motion within an ungenerated cosmos and by Brisson, as the secondary causes outlined by Plato. A. E. Taylor, on the other hand, would have been uncomfortable with a view which so opposed Plato's own story of the cosmos that it reduces his myth to mere myth, thereby challenging almost every aspect of it. As a result, Taylor leaves much of Plato's cosmology unchallenged, but also uninvestigated. In this book I attempt to move on from Taylor's position and investigate that part of the cosmological myth from the *Timaeus* wherein we find 'ananke' and which, I argue, persists to some degree as a causal force in *Laws X* as the 'Non-beneficent Soul.' Namely, therefore, this is an investigation of Plato's understanding of 'disorder' and its cause. However, rather than dismantle the myth, notably that concerning 'ananke,' from the *Timaeus* or dismiss the metaphysical content from Plato's theological discourse in *Laws X*, I look at them as written to see if what they represent or explain can stand on their own terms and if they can, whether they afford elucidation to the broader spectrum of Plato's metaphysics. My conclusion is that they can and do, and that a softening of any reinterpretation of Plato concerning what have become metaphysical cast-offs, to wit, 'ananke' and the 'non-beneficent soul,'[1] is not only warranted, but demanded by a serious investigator of Plato's thought.

I begin with a study of the *Timaeus*, in which seven passages, roughly divided up into five areas of investigation, form Part I of this book. These passages are translated anew and presented in both the Greek and English. Chapter 1 opens with a core passage relating to 'ananke' (*Ti.* 47e3-48d7), which most commentators have translated as 'necessity,' but which I argue is wrong. By looking afresh at this passage it can be shown that a close parallel is drawn by Plato between 'nous' (translated here as 'mind') and 'ananke' (translated as 'might'), in which the suggestive imagery of a race, soon enlarged to a battle for victory, emerges. Notably, from this vantage these two powers are presented as parallel,[2] although unequal, powers. This opening image of a 'race,' although soon dropped, is ultimately won by 'nous,' which power is described as "impelling" 'might' to bring about the greatest good during the Demiurge's activity (the source of 'nous') through sensible persuasion. I argue against the position that 'ananke' is at any time a power for good or that it was somehow "persuaded" by 'nous' to

1. T. M. Robinson (*Plato's Psychology*, 163) denies in Plato any claim to what can be termed two souls, one 'good' and one 'bad,' dismissing this as a result of Plato expressing himself "most unfortunately."

2. These powers are presented much in the same way as a racer is paired at the start of a race with another racer of unequal ability.

act intelligibly thereafter. I argue, rather, that 'nous' was *de facto* a stronger primary power than 'ananke,' with the consequence that the Cosmos (the capital here pointing to Plato's notion of a *divine* universe, having been wrought by the Demiurge) came into being, further arguing that the idea of "persuasion" forms part of Plato's mythic language. The idea, I suggest, of friendly collusion is used by Plato solely to turn the reader's attention towards the formation of the cosmos and the intelligent work of the Demiurge (the stated purpose of the *Timaeus*) and away from further considerations of 'ananke' other than how it served the Demiurge at the outset of generation. Chapter 1, therefore, centers chiefly on considerations of 'ananke' as understood from this passage.

I also discuss the possible development of ἀνάγκη ('ananke') as a technical term for Plato, appropriated from the Greek words ἀνά (meaning 'up') and ἀγκή (meaning 'arm') with its associated meaning of physical force, which does not appear to have been previously considered by other commentators. In light of this, however, I do not then interpret 'ananke' in terms of physical force and subsequent to this, reinterpret it as the collective term for secondary causes. I argue, rather, that Ananke is first of all to be understood, along with the Demiurge, as one of two powerful and primal metaphysical powers existing outwith the cosmos before its generation into an organized universe and therefore outside of all time and change. It is only later, in the final third of the *Timaeus*, where Plato, with his concerns now totally earthward (the cosmos having been fully generated), relinquishes all metaphysical references to Ananke, long subdued (for the most part) by the Demiurge. As a technical term serving his metaphysics and this part of his cosmology over, 'Ananke' I argue is then dropped by Plato and this word is returned to its ordinary use or understanding as physical force or 'necessity' ('ananke') as found in the common prepositional phrase ἐξ ἀνάγκης, 'necessarily.' The mistake that I argue Cornford and his followers make is in understanding 'Ananke' for the whole of the *Timaeus* according to this later development.

In chapter 2 of Part I, I continue an earlier discussion concerning Plato's understanding of what it is to be a 'cause' and demonstrate from this key passage (*Ti.* 46c7-e6) that Ananke cannot possibly be a secondary cause and that the 'two kinds of causes' referred to here are represented by the Demiurge and Ananke as their source and not by the συναίτια, which, as its name suggests, is not an αἰτία or true cause.

I also undertake a listing and study of the phrases in the *Timaeus* relating to the notion of the English translation 'as far as he could' to see how they are used in this work, detailing several meanings, namely, the ideas of 'likeliness' or a 'copy' as it relates to its 'Form,' 'curtailment of perfection,' as well as 'perfection' (i.e. used as a superlative). My conclusion is that these various Greek phrases offer little support to the idea that the *Timaeus*, although containing mythic material, is itself a myth.

Finally, I present Plato's understanding of 'cause' as it relates to that reported of Anaxagoras, Empedocles, and Thucydides. Additionally, an in-depth comparison between the physical theories of Democritus and Plato is also drawn up, where I conclude that Plato borrowed or accepted many of the same ideas as Democritus, differing essentially only in his understanding of causality. Thus, I argue that Plato had developed a theory of causation, namely, that of the 'soul' with regard to ends and motion within the cosmos, and of the Demiurge and Ananke with regard to primary causation, which Democritus had failed to do.

Chapter 3 is a study of the passage at *Ti.* 29d7-30c1. I begin by looking at the phrase 'as far as possible' with its meaning here of 'the very best.' I argue that it is only *before* its generation or *after* its generation that Plato will allow imperfection or disorder to exist in the universe, but not *during* it, when only the Demiurge's hand is at work and Ananke is left with no active role. The so-called 'persuasion' of Ananke in the sense of anthropomorphically 'assisting' the Demiurge during this process is therefore denied both figuratively and mythically.

In this chapter I also discern two *kinds* of disorder, namely, 'natural' disorder and 'unnatural,' with only the latter relating to Ananke and 'evil,' just as I discern two *kinds* of order, 'natural' and 'unnatural' (as in 'chance' order), with again, the latter relating to Ananke. In both cases that which is 'natural' relates to 'nous,' pointing to its generation as an object or end by the Demiurge.

I highlight in this chapter the phrase 'all that was visible' with its meaning of the naturally visible, as in 'not too small to be seen.' I further note the possibility latent within Plato's cosmology not only of a 'non-visible' substratum (too small to be seen), but also an 'invisible' substratum, discussing in some detail Plato's theory of triangles and its relevance to this notion. I conclude that Plato, without ever compromising his metaphysical position regarding the primal powers, the Demiurge and Ananke, proves himself to be a profound realist, gifted with great scientific imagination, as regards his understanding of the physical world.

Chapter 4 and the three passages from the *Timaeus* presented here (41a3-b6, 42e7-43a3, and 68c7-d7) highlight the sharp divide that Plato draws between the creative ability of the Demiurge and the creative element within the natural order, which latter, while it includes 'the divine,' does not include the Demiurge. This division has given both cause and rise, perhaps more than any other feature of the *Timaeus*, to the widely held position that this work is essentially mythic and its metaphysical players pure myth. The Demiurge is then understood as 'reason' within the cosmos and Ananke, if discussed, as 'unreason' in its various guises. Even should the notion of the Demiurge be understood as meaning 'God,' retaining for Plato a theological perspective, inevitably and with no exceptions to my knowledge, Ananke is still understood as being entirely a mythic figure and is invariably understood to represent either secondary causes or physical motion within the cosmos. I argue, however, that while Plato does make use of myth to explain what cannot be seen or the metaphysical realm generally, neither the Demiurge nor Ananke are mythic figures in the sense of being less real than portrayed, but rather, are *at least* as real. They are mythic only in the sense of being essentially unknowable and therefore unknown and because of this, necessarily imagined. To this end I present Plato's axiology with regard to his cosmology, his adherence to the notion of 'like generating like,' and his use of descriptive names as opposed to mythic names when referring to the Demiurge and Ananke.

Chapter 5 forms the closing chapter of Part I and brings the reader's attention to the passage at *Ti.* 68e1-69c6. This chapter focuses on Plato's move in the final third of the *Timaeus* from the metaphysical realm and the generation of the cosmos to a cosmos now long generated and fully operational. As a consequence of this change of aspect from a metaphysical perspective to that solely of the physical world, I argue that the primary causes, the Demiurge and Ananke, existing for Plato entirely outwith the cosmos which has been generated, give way to 'kinds' of causes, 'the divine' and 'the necessary,' active within it. Thus, dropped is Plato's interest in the primary causes, Ananke and the Demiurge, bringing his vision full forward from that of chaotic beginnings and powerful causal players to a finished and reasonable cosmos, viewed and understood entirely from within and in its own terms. However, while I argue that there is a change, it is not a material or essential change with regard to Plato's previous understanding. Rather, what guides Timaeus's speech now is his earlier promise to his friends to end with the story of man. Therefore, what becomes important to Plato and what he now stresses to the end of this work are the secondary causes ('the necessary') and their ends ('the divine'). These causal echoes, so-to-speak, of primal beginnings, the necessary and the divine, serve in turn an end far removed from that of the generation of the cosmos, namely, the happiness of man, thereby bringing Plato's cosmology and his 'likely' account to its envisaged completion.

Whereas an investigation of the relevant passages from the *Timaeus* opens my research on Plato's understanding of disorder, it is a study of eight passages from *Laws X*, sometimes in smaller sections and roughly divided up into four areas of investigation, which completes it and forms Part II of this book. These passages are not translated, the Greek presenting few problems, but are presented in both the Greek and English, with the English translation by A. E. Taylor. Chapter 1 focuses on the passage at *Lg.* 887c5-888a2, which sets the tone for the subsequent chapters, presenting unequivocally Plato's belief in the gods, namely, that they are the "most certain of realities" and "their being beyond the remotest shadow of a doubt," which he will set out to prove in *Laws X* in the face of those who with contempt "force us into our present argument." I argue against the position, however, that what follows is simply a polemic centered round a philosophical discussion or that it has little to offer of philosophical value to Plato's metaphysical scheme from the *Timaeus*. It proves much more than this. Not only does he offer several arguments for the existence of the gods, whose existence in the *Timaeus* was assumed, Plato is now prepared in this work to account for disorder and evil within the world of change, and in particular, within man. This is a major step forward for Plato, where earlier in the *Timaeus* his philosophical gaze rested almost exclusively on the creator (the Demiurge) and the created (the generated cosmos), leaving Ananke, the disordering power, "persuaded" and its works of chaos as vestiges only.

Chapter 2 presents the passage at *Lg.* 895e10-897b5 (which, for the purposes of explication, I have divided into ten smaller sections), wherein Plato outlines his main argument for the existence of the gods, namely, the primacy of soul and its identity with primal becoming and movement, ultimately arguing that it is the universal cause of all change and motion. I argue that whereas in the *Timaeus* 'soul' was understood as generated, in *Laws X* Plato has substantially developed his notion of 'soul,' so that it has now become 'that which generates.' Thus, except for the World Soul, which is not discussed, 'soul' is presented in *Laws X* as though it were eternal or 'begotten,' but never generated and appears to take over the role of the primary causal powers outlined by Plato in the *Timaeus*.

I point out that what is crucial to Plato's understanding of causality and movement in *Laws X* is his earlier understanding from the *Timaeus* that there is no beginning to physical movement, rather, there is only a beginning to *organized* physical movement, preceded by chaos, a state untouched by 'nous' or its ends. This leads to the question of dualism, concerning which I argue that Plato is an incontrovertible dualist. This dualism, however, which I describe and detail as a 'parallel account,' does not relate to the Forms, but rather, to Plato's developed notion of 'soul' and the ability which these psychic powers (there are more than one) have to impact the physical world with either order or disorder, whilst remaining themselves entirely psychical. I argue that the Forms, on the other hand, are not featured in *Laws X,* having been subsumed under the noetic singularity of 'nous' as understood and now identified with 'soul' and as identified earlier with the Demiurge. Their metaphysics, moreover, was never discussed in the *Timaeus*, the Forms appearing to have been given the role by Plato of stabilizing reality for the observer so as to allow a measure for discerning truth over true opinion and thus, in a sense, acting for Plato as an early analytic tool. I argue, however, that he does not provide support in his ontology for the separate existence of the Forms apart from the thinker reflecting upon them, with the consequence that their existence is entirely dependent upon 'nous.' Thus, if retained, the Forms become redundant with respect to 'nous' as it is understood within Plato's developed notion of 'soul.' It is for this reason, I suggest, that he does not in the end retain the Forms as a dynamic part of his metaphysical scheme and is why they do not feature in *Laws X*.

In chapter 2, I also argue that whereas 'soul' appears to be 'begotten' rather than generated, God (the Demiurge) for Plato is not begotten and remains the ultimate source of begotten 'soul.' Equally, I argue that the primal souls in *Laws X*, the Beneficent Soul and the Non-beneficent Soul, are not begotten, but represent for Plato a further development in his understanding of the primal powers, the Demiurge and Ananke. Where formerly they were drawn solely as transcendent primary causes in the generation of the cosmos, they are now redrawn and consequently understood from within the Cosmos as powerful primal souls, immanent in a world long generated.

Chapter three focuses exclusively on 'soul,' not as part of an argument for the existence of the gods, but rather, with questions regarding its nature and role in the world. The passages chosen are *Lg.* 897b7-d1 and 898a3-c5, divided into seven segments. Plato asks what manner of soul controls the heaven and earth and their whole circuit and answers that it can only be the "supremely good soul." This also raises for Plato the question of the possibility of *another* kind of soul, namely, the hypothetical existence of that [soul] which is the cause of procedures which are distraught or without order, answering this time with the "evil [soul]." I note that in the *Timaeus*, questions concerning the *kind* of soul would never have been raised, since in that work 'soul' was always good precisely because it had been generated by the Demiurge. This is not the case in *Laws X*, however, where 'soul' is not described as generated. Thus, questions concerning the 'kind' of soul can now be asked, since there is no guarantee that a soul will be good. This in turn raises the possibility of the existence of a 'non-virtuous soul,' which the Athenian in this dialogue proposes would account for motion that is never regular or uniform, has no order, plan or law, and has "kinship with folly of every kind."

In the fourth chapter of Part II and the final one of this work, I note the change of perspective in *Laws X* over that of the *Timaeus*, where timeless eternity and the 'unmoved mover' and its assistants (meaning the 'heavenly gods' and not the accessory causes or Ananke) of generation give way to timely events and the 'self-moved movers' or primal souls of regeneration, in particular, the 'Beneficent' and 'Non-beneficent' Soul(s). The passages chosen are *Lg.* 898d9-e3, 899b3-10, 905d8-e3, and 906a2-c6. By the time Plato wrote *Laws X* the reasonableness of the world is not in question and the World Soul, I argue, is assumed. What Plato attempts to account for in this work, therefore, is the unreasonable behavior of man and what for Plato were irrational beliefs, particularly those concerning the gods. For this he must draw upon an all but forgotten notion from the *Timaeus*, namely, an irrational primal power, Ananke, infusing it with new power and presence in its disordering role and re-introducing it *within* the world as the 'Non-beneficent Soul.' For Plato, mankind can never be the source of its own evil, having been generated by a perfect God. The human soul, however, can will wrong ends and thus make wrong choices concerning its own good, which is due, Plato argues, to both its human condition and the counter influence of irrational soul(s). Thus, adhering to the principle of 'like with like,' it is 'soul' or specifically, the 'Non-beneficent Soul,' that Plato suggests can move the human soul into making wrong choices. How it does this Plato is not prepared to say, again arguing for a purely dualistic understanding of 'soul.'

Finally, it is the notion of the human 'will' in its ability to make right and wrong choices and therefore of choosing 'good' or 'evil' upon which I end this book. Such choice is never for Plato an entirely human issue, nor is it a deterministic one, which principle, I argue, is foreign to Plato's thought, given the freedom of the human soul to choose not only well, making right choices, but also poorly, making wrong ones. In the end Plato leaves us with what I call his 'theory of disorder,' which theory traces a long anoetic trail, from that of Ananke, to the Non-beneficent Soul, and finally, to its emergence in the human soul. What Plato's theory encompasses, I argue, are a bold epistemology, an ancient ontology, and ultimately, a persistent metaphysics of disorder and evil as found in man and the world.

PART I

COMMENTARY ON THE *TIMAEUS*

'Ανάγκη / 'Ananke' (Necessity) and the Original Text

Chapter 1

1 *TI.* 47E3-48D7

1.1 Original Greek

The above passage from the *Timaeus* is presented in full, first in the most authoritative critical edition of the Greek[3] text, and then with a new English translation, followed by comments and comparative notes on other translations.

Τὰ μὲν οὖν παρεληλυθότα τῶν εἰρημένων πλὴν βραχέων, ἐπιδέδεικται τὰ διὰ νοῦ δεδημιουργημένα· δεῖ δὲ καὶ τὰ δι' **ἀνάγκης** γιγνόμενα τῷ λόγῳ παραθέσθαι. μεμειγμένη γὰρ οὖν ἡ τοῦδε τοῦ κόσμου γένεσις ἐξ **ἀνάγκης** τε καὶ νοῦ συστάσεως ἐγεννήθη· νοῦ δὲ **ἀνάγκης** ἄρχοντος τῷ πείθειν αὐτὴν τῶν γιγνομένων τὰ πλεῖστα ἐπὶ τὸ βέλτιστον ἄγειν, ταύτῃ κατὰ ταὐτά τε δι' **ἀνάγκης** ἡττωμένης ὑπὸ πειθοῦς ἔμφρονος οὕτω κατ' ἀρχὰς συνίστατο τόδε τὸ πᾶν. εἴ τις οὖν ᾗ γέγονεν κατὰ ταῦτα ὄντως ἐρεῖ, μικτέον καὶ τὸ τῆς πλανωμένης εἶδος αἰτίας, ᾗ φέρειν πέφυκεν· ὧδε οὖν πάλιν ἀναχωρητέον, καὶ λαβοῦσιν αὐτῶν τούτων προσήκουσαν ἑτέραν ἀρχὴν αὖθις αὖ, καθάπερ περὶ τῶν τότε νῦν οὕτω περὶ τούτων πάλιν ἀρκτέον ἀπ' ἀρχῆς. τὴν δὴ πρὸ τῆς οὐρανοῦ

3. Burnet, *Timaeus*, OCT, 47e4-48e7. All future references to the *Timaeus* unless otherwise indicated will be to this edition. The lineage follows very closely with that of the Oxford edition.

γενέσεως πυρὸς ὕδατός τε καὶ ἀέρος καὶ γῆς φύσιν θεατέον αὐτὴν καὶ τὰ πρὸ τούτου πάθη· νῦν γὰρ οὐδείς πω γένεσιν αὐτῶν μεμήνυκεν, ἀλλ' ὡς εἰδόσιν πῦρ ὅτι ποτέ ἐστιν καὶ ἕκαστον αὐτῶν λέγομεν ἀρχὰς αὐτὰ τιθέμενοι στοιχεῖα τοῦ παντός, προσῆκον αὐτοῖς οὐδ' ἂν ὡς ἐν συλλαβῆς εἴδεσιν μόνον εἰκότως ὑπὸ τοῦ καὶ βραχὺ φρονοῦντος ἀπεικασθῆναι. νῦν δὲ οὖν τό γε παρ' ἡμῶν ὧδε ἐχέτω· τὴν μὲν περὶ ἁπάντων εἴτε ἀρχὴν εἴτε ἀρχὰς εἴτε ὅπῃ δοκεῖ τούτων πέρι τὸ νῦν οὐ ῥητέον, δι' ἄλλο μὲν οὐδέν, διὰ δὲ τὸ χαλεπὸν εἶναι κατὰ τὸν παρόντα τρόπον τῆς διεξόδου δηλῶσαι τὰ δοκοῦντα, μήτ' οὖν ὑμεῖς οἴεσθε δεῖν ἐμὲ λέγειν, οὔτ' αὐτὸς αὖ πείθειν ἐμαυτὸν εἴην ἂν δυνατὸς ὡς ὀρθῶς ἐγχειροῖμ' ἂν τοσοῦτον ἐπιβαλλόμενος ἔργον· τὸ δὲ κατ' ἀρχὰς ῥηθὲν διαφυλάττων, τὴν τῶν εἰκότων λόγων δύναμιν, πειράσομαι μηδενὸς ἧττον εἰκότα, μᾶλλον δέ, καὶ ἔμπροσθεν ἀπ' ἀρχῆς περὶ ἑκάστων καὶ συμπάντων λέγειν. θεὸν δὴ καὶ νῦν ἐπ' ἀρχῇ τῶν λεγομένων σωτῆρα ἐξ ἀτόπου καὶ ἀήθους διηγήσεως πρὸς τὸ τῶν εἰκότων δόγμα διασῴζειν ἡμᾶς ἐπικαλεσάμενοι πάλιν ἀρχώμεθα λέγειν.

1.2 Translation

The ideas that have guided our discussion,[4] except for a small portion, demonstrate the workings of Mind. But one must also set aside for inquiry[5] the crea-

4. See R. G. Bury, trans. (Plato) *Timaeus,* 109: "The foregoing part of our discourse"; H. D. P. Lee, trans. (Plato) *Timaeus*, 66: "In almost all we have said"; Cornford, *Plato's Cosmology* (hereafter abbreviated as *PC*), 160: "our foregoing discourse." With reference to Τὰ . . . παρεληλυθότα τῶν εἰρημένων, by translating παρεληλυθότα as 'ideas which have guided,' I have not mistakenly taken παρεληλυθότα from the verb παρελούνω ('to drive by'), which is incorrect, but rather, have reflected in my translation the accepted primary meaning given in *Liddell & Scott's Greek-English Lexicon* (henceforth abbreviated *LSL*) for παρέρχομαι, a deponent, meaning 'to go by' or 'pass by,' while at the same time taking into consideration the form which this verb has taken. Since παρεληλυθότα is both a perfect and neuter plural participle, the nuance which must be retained is not only the idea of a present state resulting from a past action, but it must be

tions of Might. For, in fact, the origin of this Cosmos[6] is a mixture of both Might and Mind battling together. However, Mind was the ruler of Might, impelling it to bring about the greatest good for most of the things generated. Thus, in this manner, by sensible persuasion of inferior Might, this universe was wrought together.

Consequently, if, of whatever came into being, one inquires truly after these matters, then the inquiry must also bring into view this particular cause which leads astray and how it grew to be nurturing. Therefore, it is necessary to go back again and grasp all that belongs to these matters, starting anew, once more, from another origin. So thus, concerning these same matters now, just as those earlier, one must start again from the beginning.

It is necessary to look at the nature itself both of fire and water, air and earth then, before the birth of the heavens, and at the παθή or chaos before this. For up to now no one has explained their birth and yet, as one who knows what fire possibly is and each one of these, we talk of origins, rendering them as building

---

shown in relation to the tense of the leading verb ἐπιδέδεικται from ἐπι- δείκνυμι, 'to show forth,' ἐπιδέδεικται itself a past, 3rd person singular, indicative. Hence, I have translated παρεληλυθότα as 'ideas' (since words said, τῶν εἰρημένων, become ideas) which 'have guided' (a slightly weaker and more suitable notion for this passage than to 'have driven') the discussion, since such a translation properly acknowledges the subtlety of the tense relationship of the past participle to the past main verb. To say 'have guided' indicates a certain priority in time, yet a present state, of the 'ideas' to the action of the main verb. In other words, these ideas, by 'guiding' the discussion, exist before they demonstrate. In the other translations the participle and the leading verb are represented as taking place simultaneously.

5. A possible alternative reading for τῷ λόγῳ is "in fair game." The primary meaning given for λόγῳ in both the abridged and intermediate versions of the *LSL* for Attic Greek is 'talk' or 'pretence,' with τῷ λόγῳ meaning 'in pretence,' opposite to ἔργῳ, 'in reality.' One of the meanings of 'pretend' is to play a game. If "in fair game" were chosen as the translation for τῷ λόγῳ this would work well, since not only is the notion of a game in keeping with the imagery Plato has in place, 'fair' at the same time strengthens the idea that attention is now to be given to the other contestant. Moreover, when τῷ λόγῳ is translated as a dative of respect using 'fair' ("in fair game"), this retains the adverbial sense of 'in pretence,' which "in game" alone (without the 'fair') would not have.

6. The word 'Cosmos,' with the specific meaning of the organized universe, is written for the most part with a capital letter throughout to reflect Plato's belief that the Cosmos or generated (organized) universe is a single, living, and divine being. See *Ti.*30d1-31a1, 34b8-9, 55d4-6, and 92c4-9. However, when referred to at the beginning of the discussion or in a general way, as found in the *Introduction*, 'cosmos' is not capitalized. Similarly, the word 'universe,' which may or may not refer to the organized universe (i.e. generated Cosmos), is not normally capitalized, especially when used in reference to *Laws X*, where Plato is arguing against those who claim that any of its four elements were themselves causal or 'divine.'

blocks of the universe. Nor is it appropriate for them, even by someone of little understanding, to be considered as reasonably represented, as it were, by mere 'syllables.'

But now then, at least with us, let the following hold. Whether explained as a first principle, principles, or whatever way it seems to be concerning all things, it must not be presented now and not for any other reason but due to the difficulty in giving a full account of our views after the manner of our present discussion. You must not, therefore, suppose that I want to expound. Nor, moreover, would I be able to persuade myself that it was right should I attempt in attending to so great a task. While preserving the meaning of the likely account that was laid down from the beginning concerning all things individually and as a whole, I will undertake nothing of less likelihood, rather more, recounting an even earlier beginning.

And now then, let us begin again our discourse, invoking God at the outset of what is being said, as Protector, to guide us safely through unnatural and strange territory to grounds of probability.

1.3 Commentary and Notes

1.3.1 Discussion of Ti. 47e3-48a5 and the imagery of a race/battle

In this first selection from the *Timaeus* the language which Plato employs is suggestive of the imagery of a race, enlarging it to a battle for victory, between νοῦς and ἀνάγκη. To this end, he uses παρεληλυθότα from the verb παρέρχομαι, 'to go by' or 'to pass by'; παραθεσθαι from παρατίθημι, 'to set before oneself' or 'to have set before one'; and τῷ λόγῳ, translated 'for inquiry,' but also possibly a dative of respect with the sense of 'talk' or 'in pretence' and translatable as 'in fair game.'[7] At least initially, the contenders of this race and battle are presented as equals, which idea is supported by the use of exact literary parallelism between the phrases presenting the two principal actors or subjects, νοῦς and ἀνάγκη:

> τὰ διὰ νοῦ δεδημιουργημένα and
> τὰ δι᾽ ἀνάγκης γιγνόμενα

Thus, Plato, at the outset of his portrayal of the beginning of the Cosmos presents νοῦς and ἀνάγκη, translated here respectively as 'Mind' and 'Might,' on an equal footing, with both operating as causal agents[8] in the creation of the mix

---

7. See my note 3.

8. A. Vallejo, in his article "No, it's not a Fiction" (in *Interpreting the Timaeus-Critias*, eds. T. Calvo and L. Brisson, 141–48) distinguishes two kinds of causes within this passage, but denies that the cause(s) which works with intelligence (nous) has exis-

or μεμειγμένη . . . τοῦ κόσμου γένεσις. With respect to their ontological status, moreover, both are understood as having existence or being before the γένεσις takes place and as being eternal.

Neither Cornford nor Taylor comment on this opening sentence or on what appears to be the imagery of a race or contest between two equal contestants. From a philosophical perspective, however, I would suggest that by using this imagery Plato is making an important claim. Not only is he giving ἀνάγκη similar, if not the same, ontological status as νοῦς, he is also placing it within his scheme of things before the generation of the κόσμος and thus, is endowing ἀνάγκη with some of the same essential features as νοῦς, namely, of being uncreated and eternal with respect to the cosmos. It is from this perspective, moreover, in sharing a degree of equality, that one begins to understand the open disdain which Plato later feels towards the 'non-benevolent soul' in *Laws X*, which in Part II I argue is a further development of his understanding of ἀνάγκη from the *Timaeus*.

Even more crucial to our discussion here, however, is the rationale behind Plato having set the stage in this way. The main question with which the *Timaeus* is concerned is 'how the universe began.' At 27a2 Plato writes:

> Σκόπει δὴ τὴν τῶν ξενίων σοι διάθεσιν, ὦ Σώκρατες, ᾗ διέθεμεν. ἔδοξεν γὰρ ἡμῖν Τίμαιον μέν, ἅτε ὄντα

---

tence before the generation of soul, thereby not identifying this cause with either the Demiurge or with 'nous,' but with 'soul.' With respect to the cause(s), which, destitute of reason (ananke), "produce their sundry effects at random and without order," he interprets this cause as an 'event' (as do those who hold that 'ananke' refers to secondary causes or mechanical motion) or more properly as a "chain of events" without purposeful causality existing prior to generation in an eternal precosmical chaos. This interpretation of these causes, however, breaks down when pitted against Plato's developed notion of 'soul' introduced in *Laws X* as Beneficent and Non-beneficent Souls, akin to the major causal players of the *Timaeus* (see Part II). Whereas Vallejo's claim is that what Plato means by chaos is the "absence of teleological causality," whereas 'necessity' is to be understood as the "co-operative agent of the demiurge" always present in the world (143) and the two are not to be equated, he does nothing to solve the problem of why, if these events or this "co-operative agent" needed the Demiurge in order to become organized, that the earlier *disorganized* events did not equally require a cause to sustain their *dis*organization. In saying, moreover, that the "precosmical world as a whole does not have a cause, because γένεσις is eternal" (142), not even to sustain that chaos, Vallejo is stating more than what Plato was prepared to say. Although the γένεσις is argued to have existed "before the Heavens came into being," this is not sufficient to say that it is eternal, but only that the triangles had to have taken form prior to the generation of the Heavens. Just as Plato claims that the ordered cosmos had a cause (the Demiurge), so I will argue throughout the remainder of Part I (as well as in Part II), that Plato was fully prepared to admit to a disordering cause (Ananke), although with reticence due to the nature of this cause and the purpose of the *Timaeus*.

ἀστρονομικώτατον ἡμῶν καὶ περὶ φύσεως τοῦ παντὸς εἰδέναι μάλιστα ἔργον πεποιημένον, πρῶτον λέγειν ἀρχόμενον ἀπὸ τῆς τοῦ κόσμου γενέσεως, τελευτᾶν δὲ εἰς ἀνθρώπων φύσιν·

Consider now the order of the feast we have arranged for you, Socrates. Seeing that Timaeus is the best astronomer among us and knows the most concerning the nature of the universe, having made it his expertise, he shall speak first, beginning with the generation of the cosmos and ending with the nature of man.

The order of this "feast" is to begin with the origin of the universe and end with the nature of man. In keeping with the celebratory tone of this original opening, Plato ends his discussion on the "workings of Mind" and begins his next main section on the "creations of Might" by alluding to the imagery of a race, yet another social event or continuation of such an event. It is essential to both the backdrop which Plato now has in place (a cosmology) and the direction in which his philosophical understanding of these metaphysical powers, νοῦς and ἀνάγκη, is moving (as *opposing* forces) that this race be between two equal contestants. To this end his writing at this juncture becomes a synthesis of both his literary acumen and philosophical understanding.[9] Accordingly, Plato places these two clues within the prose, one through literary device, i.e. the parallelism noted above and the other through literary style, i.e. the imagery of a race, subtly informing the reader of his intention. This imagery, I suggest, becomes an important part of the evidence that Plato wants considered and will ultimately help in forming the fuller answer he wishes to give in regard to the origin of the cosmos.

In this instance Plato is grappling with the difficult philosophical problem that 'soul' cannot be the cause of inharmonious and disordered motion. He builds slowly the solution to this problem. Already at *Ti.* 28a4 Plato states that everything which becomes and thus changes must do so owing to some cause:

9. Plato's literary and philosophical expertise was gained over a long period, both before and during his tenure of over twenty years as a founding and leading member of the Academy. Along with the *Laws*, the *Timaeus* likely belongs to his last main group of writings written in the remaining ten to twelve years of his life (e.g. see Lee, 22). In addition to being well honed in dialogue-writing technique when he wrote the *Timaeus*, having already composed over fifteen dialogues, the fact that Plato chose the dialogue format to present his philosophical treatises is of itself enough to warrant a careful look at the prose settings. With regard to the *Timaeus*, although after the introductory section the dialogue form is dropped and the main body of the work then becomes a continuous exposition by Timaeus, the prose style nonetheless persists, which Plato will use to present his philosophical views contained therein.

> πᾶν δὲ αὖ τὸ γιγνόμενον ὑπ' αἰτίου τινὸς ἐξ ἀνάγκης γίγνεσθαι παντὶ γὰρ ἀδύνατον χωρὶς αἰτίου γένεσιν σχεῖν.
>
> And again, everything which comes to be must[10] come into being by some cause. For nothing is able to have come into being without a cause.

He repeats this again at 28c2:

> τῷ δ' αὖ γενομένῳ φαμὲν ὑπ' αἰτίου τινὸς ἀνάγκην εἶναι γενέσθαι.
>
> But again, with respect to that which has come into being we say it must have come to be by some cause.

I have used 'must' here in the translation, rather than 'might,' to distinguish a pending change in Plato's focus, namely, his introduction of an axis of comparison between ἀνάγκη and νοῦς at *Ti.* 47e-48a. At neither *Ti.* 28a nor 28c has Plato yet distinguished ἀνάγκη as a cause itself, merely acknowledging causal force generally, i.e. implied in 'must' or 'of necessity,' in all acts of γένεσις. At this juncture the causal power behind generation is still under consideration by Plato. He is aware, however, that the physical forces within the κόσμος, like the Cosmos itself, are subject to change, forming part of the process of 'becoming' and therefore point to another cause or causes as the source of generation.[11] It is this idea, I suggest, which drives the change of focus Plato introduces later at *Ti.* 47e-48a, where, with νοῦς, ἀνάγκη is first presented as a substantive and potential cause in its own right. Thus, although at *Ti.* 28a and 28c ἀνάγκη is very closely tied with the idea of the γένεσις, by being part of the common prepositional phrase ἐξ ἀνάγκης and treated as a modal (must) or adverbial (of ne-

10. I have chosen 'must' rather than the stronger sense of 'necessarily' or 'of necessity' in order to point to the possibility that the original meaning of ἐξ ἀνάγκης as a prepositional phrase, deriving etymologically from a combination of the preposition ἀνά = 'up' combined with ἀγκή = 'arm,' has a direct bearing on its later usage and may have a specialized use with respect to physical force when used by Plato in discussing the generation of the cosmos. See *Dictionnaire Étymologique de la Langue Grecque*, ed. P. Chantraine, Klincksieck, Paris, 1968, 83. Further, whereas 'must' still retains a possible connection to the idea of 'physical force' and can be understood in this sense, 'of necessity' implies a logical or practical necessity in which no reminiscence of an earlier provenance needs be discerned. For example, one might say, "you must have twisted your ankle" on proposing a diagnosis, but would not say, "you necessarily twisted your ankle." Lee has also chosen "must" for this passage (40). At both *Ti.* 28a and 28c Taylor and Cornford have translated ἀνάγκη in the stronger sense of 'necessity.'

11. See *Ti.* 27d5-28a6; see also *Ti.* 46c7-e6.

cessity), it is still subordinated to the idea of finding the cause behind the γένεσις and is not yet considered as a causal contender itself. However, the idea of ἀνάγκη remains a close linking concept between the γένεσις and the αἰτία, which proximity Plato makes full use in further developing his causal ideas involving ἀνάγκη.

At *Ti.* 28c there are two possibilities for Plato's treatment of ἀνάγκη. First, it would appear odd that Plato should repeat within such close proximity two almost identical statements, the other being at 28a. The first possibility and the one which I propose, is that in this second passage ἀνάγκην is the subject of the infinitive εἶναι forming part of the indirect discourse introduced by the verb φαμὲν. By the introduction of ἀνάγκη as a substantive, therefore, but still following very closely upon its prior meaning in reference to ὑπ' αἰτίου τινὸς and its use at 28a, Plato has now fully prepared his reader in a two-stage fashion, at 28a and 28c, for what appears to be a decisive move at *Ti.* 48a1, where in combination with νοῦς, ἀνάγκη is named as the general cause of the universe. Thus, at 48a ἀνάγκη is now directly contrasted with νοῦς in which instance I have translated this not as 'must' or 'of necessity,' but as 'might' to note the development which ἀνάγκη is undergoing. This historical movement of ἀνάγκη, therefore, begins with the idea of 'force,' moves to that of 'cause' and specifically, the primal cause, Ananke, and finally rests with the idea of 'necessity,' where it has now been separated off from both the source and effect of any causal agency, i.e. that of (brute) force. My further suggestion, which is shown in my translations of 28a and 28c by the use of the word 'must,' is that Plato has signaled this change of meaning in ἀνάγκη at 48a, from that of 'force' to 'cause,' earlier at 28c. The following diagram illustrates this possible development of ἀνάγκη as it was first understood apart from its later use within the prepositional phrase ἐξ ἀνάγκης ('necessarily') or in its close association with the idea of an αἰτίον or 'cause':

1.3.2 Diagram A – Proposed historical development of ἀνάγκη/ananke as 'force' in Plato's *Timaeus*

The meaning of ἀνάγκη as it came to be understood:

Things created	come/came to be	through 'necessity' (i.e. of logical conditions or physical force)
		'necessarily'/'must'
(28a: πᾶν . . . τὸ γιγνόμενον)	(γίγνεσθαι)	(ἐξ ἀνάγκης)
(28c: τῷ γενομένῳ)	(εἶναι γενέσθαι)	(ἀνάγκην)
		through/of 'Might' and 'Mind'
(47e4-5: γιγνόμενά)		(δι' ἀνάκης)
(48a1: μεμειγμένη)	(ἐγεννήθη)	(ἐξ ἀνάκης τε καὶ νοῦ)

The idea of ἀνάγκη as 'force' in turn points to an even earlier development:

FORCE	→	MODAL FORCE/NECESSITY
An early etymology:		A later, derived meaning:
ἀνά + ἀγκή	→	ἀνάγκη
(up + arm)		(must, necessarily, of necessity)
= physical force		(28a: ἀνάγκης)
		(28c: ἀνάγκην)

Thus, it is with regard to the above historical background of ἀνάγκη (rich for the purposes of myth and metaphor) or one similar to it that I suggest Plato gives free play in his use of this word when describing the generation of the cosmos.

The second possibility is that this mirroring of 28a by 28c is unimportant, with the introduction of ἀνάγκη, in its substantive form ἀνάγκην, likewise carrying no importance. In this scenario ἀνάγκη as ἀνάγκην can simply be glossed over as though it were in its prepositional form ἐξ ἀνάγκης and then translated as 'necessarily.' I reject this second possibility on the grounds that not only is it inaccurate, but it fails to capture the meaning of Plato's richly nuanced prose, which when taken at face value is fully comprehensible. The first reading also better aids the understanding of what follows, with less having to be interpreted as metaphor.

1.3.3 A new meaning for ἀνάγκη

Beyond its historical development within the Greek language, however, Plato also imposes on ἀνάγκη a technical meaning specifically for use in his cosmology.[12] By drawing upon the idea of physical force from its original etymology and combining this with the idea of 'necessity' from its later usage, Plato allows ἀνάγκη to take on substantially more meaning and importance than it previously had, becoming in its own right a 'cause.' Accordingly, at 47e5-48a of the *Timaeus* ἀνάγκη has now become a cause of the κόσμος, although its nature is still to be articulated and initially is described only metaphorically as a cause. This can be illustrated as follows:

12. I agree with Teloh that "we cannot assume that Plato uses a rigorous, regimented technical vocabulary; odds are, and in several cases we have strong evidence, that the same word has different uses in different contexts" (*The Development of Plato's Metaphysics*, 15, hereafter abbreviated *DPM*), which, in this case, I suggest, is substantiated by Plato's highly skilled use of the word 'ananke.'

1.3.4 Diagram B – Development of ἀνάγκη/ananke as a primal cause in Plato's *Timaeus*

ἀνάγκη

(physical force + modal/logical necessity)

↓

now becomes a causal agent in the origin of the Cosmos, which origin is metaphorically described as a mixture of both ἀνάγκη and νοῦς battling together:

= one of two primal causes

αἴτια

↓

47e5, 48a1	ἀνάγκη(ς)	↔	47e4, 48a1	νοῦ
	Might			Mind

↓

soon specifically referred to as a cause

48a7 τῆς . . . εἶδος αἰτίας[13]

Thus, by 48a7 Plato refers to ἀνάγκη as a particular cause. Moreover, as a cause which 'leads astray' but 'grew to be nurturing,' Plato announces his intention to inquire further into its singular nature and role in his cosmology. The development of ἀνάγκη from being the object of a somewhat innocuous, albeit important, prepositional phrase at 28a, namely, that everything which comes to be *must* have a cause, to something which at 48a, along with νοῦς, has being of a grand stature, i.e. is uncreated and therefore eternal, and is a being, moreover, which, possessed of a will, is capable of leading astray, presents a huge development of this concept and in a very short space. The extent of the development of ἀνάγκη, combined with the brevity in which it takes place and the explicitness by which Plato now states his concerns and intention to explicate ἀνάγκη further, preclude an analysis which understands this development as mere hyperbole, metaphorical speech, or symbolism of something else. Such brevity and explicitness, I would argue, by minimizing confusion, is in keeping with the

13. Plato changes from using the neuter noun αἴτιον at *Ti.* 28a and 28c to using the feminine noun αἰτία here.

borrowing of a well-known term, but now used in a new and technical way. Not only has Plato prepared the reader for this move at *Ti.* 28a and 28c, but he explicitly acknowledges it at 47e3-48a5.

Furthermore, just as a careful reading of the Greek text shows Plato to have a dynamic sense of ἀνάγκη, taking on new meaning as the context changes, likewise different words need to be used when translating it in order to avoid not only falsification by oversimplification, but also ambiguity. The above, I suggest, is compelling evidence not to retain for every instance of ἀνάγκη the adverbial sense 'of necessity' or 'necessarily,' a meaning derived from the common, but later usage of the prepositional phrase ἐξ ἀνάγκης. With respect to the latter, Plato clearly acknowledges its later use by employing this prepositional phrase exclusively in the final third of the *Timaeus*, where the cosmos as fully operational is being described. Moreover, the issue of whether this 'necessity' is modal, practical, or logical is easily glossed over, yet would impact the overall meaning of a passage. Additionally, the possible early etymology from ἀνά and ἀγκή with the associated meaning of physical force is too quickly passed over or simply forgotten. Finally, in explicitly stating at 48a7 that ἀνάγκη is a cause in its own right, Plato discourages subsequent readings in the *Timaeus* which would indicate it functioned merely as a modifier, which readings, should they arise, would have to be carefully substantiated.

Thus, to use the adverbial sense of 'necessarily' or any other similar single meaning throughout the *Timaeus* or to build upon the idea of 'necessity' a derivative meaning, e.g. mechanical motion, when trying to understand or translate ἀνάγκη is, I suggest, erroneous.[14] The reasons can be summarized as follows. First, no compelling evidence is given for retaining one meaning throughout, on the contrary, a careful reading of the text would indicate otherwise. Second, when the development of ἀνάγκη is mapped in the *Timaeus* a subtle but dynamic movement with regard to its meaning is revealed. Third, the idea of 'necessity' is itself ambiguous, there being several different kinds of 'necessity.' Should this ambiguity or dynamism go unrecognized, it commits the reader to

14. Leucippus's remark that "Nothing happens at random, but everything from reason and necessity" (B2), which 'reason,' Vlastos explains, is the reason of necessity or "the reasoning of mechanistic explanation which excludes rigorously and systematically considerations of value from the determination of matter of fact" (*Plato's Universe*, 30, hereafter abbreviated *PU*) is clearly at odds with Plato's understanding of νοῦς and ἀνάγκη discussed here. Value is intrinsic to Plato's metaphysics and is perhaps why, I suggest, Plato chose to highlight these two primal powers in just the way he did in the *Timaeus* and later in *Laws X* (transformed in this later work as 'Souls'; see Part II), expressing in this way his opposition to the atomist position regarding these causal powers and their relation to the world. The Leucippus fragment is in H. Diels and W. Kranz, eds., *Die Fragmente der Vorsokratiker*.

the fallacy of equivocation or to misunderstanding in general.[15] The modern philosophical notion that the correctness of an argument depends upon a word or expression maintaining a constant meaning throughout must be very carefully applied when drawing conclusions concerning ἀνάγκη. I would argue that in choosing the dialogue format, consistency of this kind was of less concern to Plato than the dynamic quality and possibilities of conversation and the ideas which would consequently emanate, and that his treatment of ἀνάγκη attests to this. Thus, to apply rules of consistency to the whole of the *Timaeus* becomes an anachronism if applied too strictly, since evidence for argumentation dependent on such consistency is almost entirely missing. Rather, Plato presents his cosmology as an observer making reasonable or what for him were likely deductions about the universe and how it came about. As he builds his cosmology he develops and presents his own understanding of it, incorporating not only his knowledge of physical causes and certain phenomena (e.g. movement of the heavenly bodies), but also his knowledge of metaphysical causes, the latter based upon certain *a priori* ideas, assumptions that flow from these, as well as his own beliefs. In so doing, Plato builds a vocabulary capable of supporting this cosmology and one that allows for fluidity of usage and changes in meaning.

1.3.5 Discussion of *Ti.* 48a1 (μεμειγμένη γὰρ οὖν . . . ἐγεννήθη·) and the meaning of συστάσεως/συνίστημι

In the sentence that immediately follows, Plato, by employing συστάσεως, follows through with the two central ideas he had already introduced in his prior opening sentence, namely, of a contest between two equals and of the force that such a match would create. What is essential to note here is that it is neither ἀνάγκη on its own nor νοῦς which is considered to be the cause of the universe, rather, it is the force created by their 'striving together,'[16] with the consequence

15. Understanding ἀνάγκη variously as modal, practical, or logical necessity is an example of the possible equivocation which might occur, causing at the very least a lack of clarity and at worst, confusion and erroneous conclusions concerning certain passages. Further, by translating ἀνάγκη by the word 'must' or 'necessity' in the course of drawing a particular conclusion but where certain of the passages containing ἀνάγκη upon which the conclusion rests have actually reverted to its earlier and more basic meaning of 'physical force' or alternatively, to Plato's newly derived meaning of metaphysical 'cause,' would also result in general misunderstanding. The subsequent interpretation would assume the meaning of 'must' or 'necessity' (or some notion developed from these, e.g. mechanical motion) in all instances used in drawing its conclusions, whereas in fact the meaning of ἀνάγκη would, I argue, have borne different meanings depending upon where it occurred in Plato's cosmology.

16. The meaning of 'strive' used here is that of 'to struggle in opposition' or 'to contend.' I argue that Plato's understanding is that the physical universe was wrought out of the force created by this 'striving' between the two powers, Ananke and the Demiurge,

that it is this force which brings about the universe. To understand the meaning of συστάσεως here in the weaker sense of 'combination' as Bury,[17] Lee,[18] and Cornford[19] have all done belies the imagery which Plato is carefully creating, namely, that of a battle between two combatants, who between them create enough force to shake the Cosmos into existence. What is clearly being emphasized by Plato is the idea of the force created by this union and not merely the fact that ἀνάγκη and νοῦς were combined. Strengthening this idea, moreover, is the fact that συστάσεως translated as 'combination' adds nothing new to the idea already stated by μεμιγμένη or 'mixture,' with the consequence that it appears superfluous. Conversely, except in this one instance συστάσεως is not understood anywhere else in the *Timaeus* either by Bury, Lee, Cornford, or Taylor to mean 'combination' in the sense of joining two disparate, unordered physical or metaphysical objects. Rather, once Plato has moved on from this short opening metaphor the essential meaning of συστάσεως in each of the four remaining instances is that of a completed 'structure,' which instances are invariably those of something that has already been ordered and do not involve either νοῦς or ἀνάγκη.[20] Thus, to accept 'combination' as the meaning of συστάσεως at *Ti.* 48a2, where references elsewhere are consistently to a completed structure, is awkward and therefore, I suggest, unlikely.

There are two further points worth considering. First, the verb συνίστημι while used in a general sense throughout the *Timaeus*, does not appear to be used in a technical way, i.e. reserved for a specific context. It presents as only one among several verbs or verb formations used in the *Timaeus* by Plato to mean 'combine' or 'construct.' Thus, for example, whereas συνιστάς is used for the 'Constructor' of the Cosmos (*Ti.* 37c7), so is δημιουργὸς (29a3) and frequently (e.g. 40c1-2; 41c5; 42e8; 69c2; 75b8[21]). Likewise, while συνίστημι is used to mean 'construct' (in the sense of 'combine') when referring to the Cosmos (i.e. 29e1), so is the verb δημιουργέω (29a7-b1; 69c4), its use preceding that of συνίστημι. Moreover, not only δημιουργέω, but several other verbs appear throughout the *Timaeus* carrying similar meanings of blending,

---

and not to any conjugal striving between these powers, which its other meaning 'to make effort' might imply. The fact that the universe originated out of this contentious striving, however, makes 'strive' uniquely suitable, since the organized universe was also generated as a result of their combined, albeit acrimonious, effort.

17. Bury, *Ti.*, 109.

18. Lee, *Ti.*, 66.

19. Cornford, *PC*, 160.

20. Συστάσεως is translated variously as a 'compound,' 'construction,' or 'combination' (of a fifth regular solid) at *Ti.* 55c4, as a 'structure,' 'texture,' or 'cubes' (of earth) at 60e4, as the 'structure' (of the belly) at 78b2, and finally, as the 'structure,' 'course,' or 'constitution' (of diseases) at 89c5.

21. This last example is in the plural, referring to the 'Constructors of our being,' also called the 'συστήσαντες' at *Ti.* 71d5-6.

mixing, compounding (sometimes translated as 'fashion' or 'fabricate,'[22] but with the idea of 'compounding') or combining, e.g. τεκταίνομαι (28c6; 33b1; 36e1), ποιέω (34b3), συγκεράννυμι (37a3-4), ἁρμόζω (41b1), προσυφαίνω (41d1-2), συγκολλάω (43a2), συνέχω + δεσμός (43a2), συντήκω (43a3), ειμί + συμπαγές (56e7), συμμίγνυμι (57d4; 60b8; 74c7; 76d5), μίγνυμι (59d4; 59e8), συμπήγνυμι (60e3), κεράννυμι (68c1; 68c3), συγκεράννυμι (68c2), συναρμόζω (74c7), συντάσσω (83b7), and συμπλάσσω (83d6). An interesting example is the verb συνδέω at 32b1, meaning 'to bind,' for while δέω is joined by καί with συνίστημι (συνέδησε καὶ συνεστήσατο) and has the meaning of 'joined' (i.e. the four elements), the latter (συνεστήσατο) is clearly referring to what has just been 'constructed' or the οὐρανόν. Thus, here, Plato is purposely not choosing the verb συνίστημι, but rather, another verb, συνδέω, to carry the meaning of 'to join together,' retaining for συνίστημι the more general sense of 'constructing' or a 'construction.'

A second and important observation is that the opening structure of the *Timaeus* as a whole supports what can be taken as a metaphorical opening to Plato's description of its second beginning, while at the same time giving evidence for the dynamic use of συνίστημι. To wit, in the opening of the *Timaeus* Plato moves from a mythical setting at 25a6 (συνέστη), passing quickly to a description of the physical Cosmos with its metaphysical implications at 29e1 (συνιστὰς). Likewise, when introducing his new description of the universe, Plato moves from what I argue is an imaginative battle at 48a2 (συστάσεως), at the same time re-introducing his two key metaphysical players or forces, but then moves immediately to a description of the physical universe (συνίστατο) where, once again, he begins anew a working out of the metaphysical implications. I would proffer the opinion that Plato's reason for staging his entries in this way, i.e. imaginatively, when attempting to give description to the beginning of universe is to emphasize that what he is about to offer is a 'likely' or possible account, rather than an exact or compelling one. Moreover, the use of metaphorical language at 48a2 in contrast to the myth employed at the beginning of the *Timaeus* is in keeping with both the considerably smaller metaphysical stature of ἀνάγκη in comparison to νοῦς and the importance of the opening of the work over secondary considerations.

Taylor[23] in his commentary, along with Bury and Lee, translates συστάσεως as 'combination.' Thus, he comments as follows:

> i.e. ἐκ συστάσεως ἀνάγκης καὶ νοῦ 'from a combination of ἀνάγκη and νοῦς.'

---

22. E.g. Bury.
23. A. E. Taylor, *A Commentary on Plato's Timaeus* (henceforth abbreviated *PT*), 303.

This is not what the text states, but rather:

ἐξ ἀνάγκης τε καὶ νοῦ συστάσεως

If Taylor is merely clarifying the syntactical structure by his re-arrangement of this short Greek passage I am curious as to why, for whereas he presents the syntax of this passage as being fairly obvious, and likewise its meaning, he offers no reason for the need to clarify the syntax or support for his reading. I would like to offer some observations with respect to this. First, 'ἐκ' with 'συστάσεως' is never cited by *LSL* as meaning 'from a combination,' rather, this phrase is quoted once with the meaning of 'a standing together, close conflict, battle,' modifying μάχη (*Hdn.* 4.15.3). Plato is also cited as using συστάσεως with this same meaning, but not with ἐκ and again, closely joined to μάχη by καί (*Lg.* 833a). Moreover, σύστασις from the root συνίστημι, meaning a 'bringing together,' is not mentioned at all as occurring in the *Timaeus*, whereas σύστασις from the root συνίσταμαι meaning a 'combination,' although cited, is never cited with respect to *Ti.* 48a2, but rather, only to other passages relating to the ongoing explication of the cosmology (i.e. 32c6, 36d8, 75b3). Σύστασις is also understood in the sense of 'formation' at 89b5, but other than these few instances, neither σύστασις nor συστάσεως are again cited from the *Timaeus*. It should be noted, moreover, that *LSL* does not cite συστάσεως at *Ti.* 48a2 as meaning 'combination,' even though it would have presented a clear example from a key passage, if in fact that were its meaning.

It is also possible that συστάσεως is in fact an adverb and not, as Taylor suggests, a noun in the genitive case.[24] There are two points in favor of this reading. First, its form is in keeping with that of an adverb, i.e. ending in -εως and second, so is its placement, i.e. placed directly before the verb. I would further argue that the dynamic sense of the interaction between ἀνάγκη and νοῦς, whether of 'combining' or of 'battling,' is more precisely caught by adverbial modification than by the more static nominal.

With respect to a nominal reading, however, prepositions such as ἐκ do not strictly govern cases, but rather, serve to bring out more clearly or emphasize the meaning of the cases with which they stand. Thus, should the text at 48a1-2 actually read ἐκ συστάσεως, which it does not, with συστάσεως remaining a genitive noun, this argues just as favorably for the idea of 'separation' or 'combating' (i.e. the genitive of separation) as it does for 'combining,' whereby the genitive of origin, relation, or causal means are argued to be in play, all of which are indicated by the genitive case.

I queried above the reason for Taylor wishing to clarify the syntax of this passage, while being satisfied, as it were, with merely asserting the meaning of

24. There is no evidence presented in the *LSL* for this reading.

the passage after alteration of its syntax. Reasons were clearly lacking. In response I argued that this passage should indeed be commented upon, but without alteration. Rather, it needs to be looked at first and foremost as Plato wrote it. I would like to further suggest that the key to Taylor's lack of argumentation is evidenced in his very next comment with regard to 48a3. Here he quotes Hume and the "growing discovery of rational connexion, as science advances."[25] Although Taylor is explicating the reason why νοῦς is successful in 'persuading' ἀνάγκη in "the majority of cases," his method is to refer back to the καὶ joining ἀνάγκη with νοῦς at 48a1, where he states "at first we began by seeing only 'conjunction'," quoting as he puts it "Hume's famous antithesis." We cannot read or assume to understand Plato according to Hume's eighteenth-century understanding of science. It is Hume, however, whom Taylor is quoting in support of his understanding of these passages and not Plato. As such, Taylor's presentation of Hume's scientific acumen, while interesting, does not offer tenable support to his alteration and consequent reading of this passage.

Thus, what is presented in Taylor's commentary as a simple and at first glance, harmless switch of syntax, supposedly to clarify the syntactical structure of the passage, is questionable upon closer scrutiny. It becomes indicative of the further distortion which will follow throughout Taylor's commentary in his reading of ἀνάγκη and νοῦς. In understanding συστάσεως as 'combination,' an abstract, non-active noun, the causal powers ἀνάγκη and νοῦς henceforth become secondary or mere explanatory features of the cosmology, with references to either as key players mitigated to various degrees of allegory. Συστάσεως, on the other hand, by losing any active sense it might have had from its original verbal root presents a devitalized and ultimately positivist depiction of the origin of the Cosmos, which, while a fitting testimony to the appreciation of the advances of scientific method from Hume's day on, is oddly out of place for Plato and is a depiction, I would argue, Plato neither intended nor described.

In like manner and with the same consequence, Cornford[26] translates συστάσεως as 'combination,' feeling neither the need to open up this passage for discussion nor, like Taylor, to clarify its syntax. However, the consequence of this choice, becomes, as it did for Taylor, a crucial move away from a deeper understanding of ἀνάγκη, an understanding which would take into consideration any literary clues that were in place, not just the larger scientific picture, and one which would also take into consideration any possible original use by Plato of the Greek language in describing what for him was the likely origin of the Cosmos, wherein fresh use of the language would at times not only have been naturally forthcoming, but also necessary.

25. Taylor, *PT*, 303.
26. Cornford, *PC*, 160–77.

1.3.6 The meaning of γένεσις

The meaning of γένεσις at *Ti.* 48a1, I suggest, is singular, found within the context of a unique opening metaphor re-introducing the beginning of the Cosmos. I would argue, however, that in these first few lines its meaning is closer to that of 'origin' than 'generation.'[27] What is at issue here is the positing of a pre-generative viewpoint *prior to* the generation of the universe as distinct from a reference to the universe *during* its physical generation, thus supporting Plato's stated view that the Cosmos had a beginning and rejecting interpretations arguing otherwise. I offer the following reasons. First, 'generation' in the sense of the production or manner, in other words, the 'how' in which the Cosmos came into being, is yet to be discussed and will not be discussed until after the invocation to the protective deity at 48d4. Second, although, like ἀνάγκη and συστάσεως, γένεσις will also shift in meaning once it is outside its opening metaphor, i.e. from 'origin' to 'generation,' it does so specifically to accommodate a change of focus, namely, from an opening reference to the primary event, i.e. the 'battling together' of νοῦς and ἀνάγκη, which event creates the force driving the γένεσις, to a description of how the individual items of the universe were actually made and in what order they were made, which focus will occupy Timaeus to the end of his speech. Third, nowhere in the *Timaeus* is an actual theory of mechanical motion ever made explicit, but if one were to assume such a theory as its basis, it could then be argued that at 48a1 'generation' is a more apt translation for γένεσις, rejecting 'origin' as being more diffuse in meaning and therefore open to unnecessary metaphysical speculation. This I would argue, for all of the above reasons, is wrong. Choosing an ontological and event-based reading of 47e3-48a5 requires neither a reductive nor figurative understanding of this passage, but rather, a metaphorical one, where the big metaphysical questions are envisioned but not asked. They are put aside to address the physical ones, the metaphysical questions having been thought too difficult to answer or the listener not able to understand.[28] Granted, therefore, that in the opening metaphor, a metaphysical basis and not merely a physical one is being assumed, should the more difficult metaphysical questions be articulated, then religious answers cannot be overlooked and in fact, may be required, which is the position Plato argues passionately for in *Laws X*.

A description of the manner in which this mixture brought about the Cosmos is what follows next. Plato, however, gives little detail at this juncture except to highlight that "Mind was the ruler of Might," impelling the greatest good for most of the things generated. Nonetheless, from this short description we do get a glimpse of the respective powers of both νοῦς and ἀνάγκη. Mind, which always brings about the greatest good, seems to have achieved this for most

27. See *Ti.* 28b.
28. See *Ti.* 28c and 48c.

things generated, but not for all, leaving us with the clear implication that Might, not being rendered totally powerless, was able to bring about a lesser good for certain, but far fewer things. In other words, the greatest good was not achieved consistently or completely for all things generated in the Cosmos due to residual power remaining with and supposedly having been used by ἀνάγκη or Might, the disordering cause, during the generative process. Plato does not give further details or clarification of what he means here. This apparent lack of detail, however, is understandable if Timaeus's speech is understood primarily as that of an observer. Thus, it was observed that "Mind was the ruler of Might" and upon further consideration a theory followed. In other words, Timaeus is not presented as basing his initial understanding on *a priori* ideas and then positing a world according to these ideas, but quite the opposite. He observes the world and from this, extrapolates a theory of its origin. There is no evidence here of either Plato's theory of forms[29] or a theory of mechanical motion. Rather, Plato, through Timaeus' opening remarks, notes that the Cosmos appears for the most part to be perfectly good, i.e. "the greatest good" having been achieved for most things generated, but not completely so, crediting to Mind what is wholly good and to Might what is not. It is from this juncture, I suggest, that Plato turns to *a priori* ideas in order to understand and explain these causes, and in so doing also turns to mythic language.

1.3.7 Νοῦς and ἀνάγκη

For Plato the workings of νοῦς and ἀνάγκη capture the very essence of generation, both the act (their 'striving together') and the result (the organized universe), and yet, of themselves, they are neither. He has already declared that to discover the Maker and Father of this Universe (that is, the generated or organized universe), the agent, is "a hard task indeed," deeming it impossible to tell all men about Him.[30] Therefore, it is not surprising that Plato does not spell out fully what νοῦς and ἀνάγκη are. He cannot, for, like the 'Maker of the Universe,' they remain behind the scenes with only their effect being felt and hence, articulated. For this reason to ask who and what they are remain largely unanswerable and perhaps inappropriate questions, since their being is prior to all generation. Plato addresses this problem by either leaving νοῦς and ἀνάγκη alone within a context with very little explanation or by using metaphor or similar language, which to some degree is anthropomorphic, when describing and using them within his cosmology. However, in so doing he is not reducing their metaphysical status to mere symbolism or to some abstraction about the Cosmos, for nowhere in the *Timaeus* does he intimate this and in *Laws X* he argues

29. Although a clear reference is made to the Forms beginning at *Ti.*28c5, no such reference is made in the later passage beginning at 47e3.

30. *Ti.* 28c3-5.

against such views.[31] Rather, with the assistance of myth and metaphor—the language of the unknown—and likely storytelling, Plato enlarges the context of his cosmology to include what existed before the generation of the Cosmos. He does this, moreover, in such a way so as to also allow a certain precision of description (i.e. both νοῦς and ἀνάγκη have qualities akin to the physical world later described and as causal forces are understood by Plato to be the source of its generation), as well as passion, especially in *Laws X*, for what ultimately remained unseen and unknown to ordinary man.

1.3.7.1 A discussion of νοῦς

I have chosen to translate νοῦς as 'Mind' in order to incorporate from the text what I believe is the best reading. Taylor also understands νοῦς as 'Mind,' offering several points in support of this, some of which are historical (see Appendix).[32] 'Mind' generally does not refer to just reason, the will, the subconscious, the conscious, the soul, intelligence, identity, thought, or personality alone, but it can and often does refer to any or all of these aspects of mental or spiritual phenomena, depending on what ideas are being demonstrated or what act is being carried out. At *Ti.* 47e4 Plato specifically states that the ideas presented up to this point by Timaeus demonstrate the workings of νοῦς or as I have translated it, 'Mind.' However, the demonstration of such workings tell us little about νοῦς itself other than that there is rationality involved. To be able to figuratively race alongside ἀνάγκη, to battle with it, and ultimately to impel it to bring about the greatest good for most things generated, does not imply rationality or intelligence alone, but is, I suggest, much more akin to action which requires the totality of mind. In this regard, I follow Taylor's understanding, in which he has translated νοῦς within the *Timaeus* as 'Mind' up to and including this passage at 47e3-48d7. Occasionally Taylor restricts his understanding to a certain aspect of mind, e.g. 'thought' (at *Ti.*29b6), but not to the degree that he is willing to incorporate, even intermittently, a substantially different understanding and hence, translation.

Plato's use of νοῦς in describing the generation of the universe is one of the more difficult complexities of the *Timaeus*. A telling question to help unravel this, however, is to ask why Plato refers not only to 'the God'[33] in the generation of the world, but also to a 'Demiurge,' in addition to 'nous.' Moreover, if one were to concede that the 'God' mentioned at *Ti.* 30a2 is in fact the Demiurge

31. See *Lg.* 886b-888d.

32. See Taylor, *PT*, 74, 82, 117, 123, 153, 175, 182–83, 247, 292–93 as set out in the Appendix.

33. I follow the practice of A. S. Mason (*Reason and Necessity in Plato's Timaeus*, D. Phil. thesis, Braesnose College, Oxford, 1990, 1, fn. 2) and others in generally keeping the article when referring to the primary deity, so as to avoid potential confusion with the Christian concept of 'God.'

who generated the universe, what is Plato's purpose in introducing 'nous' as well? Pointedly, what is 'nous'? With regard to the first question, I find no evidence within the *Timaeus* in support of the argument that the Demiurge referred to at 68e2 or 69c3 is anything other than 'the God' referred to at 30a2. These references are used interchangeably and synonymously, the Demiurge referring to the God's activity in generating the world much in the same way that a man may be referred to as a father in relation to his children and an engineer in relation to his work or even a gardener in relation to a hobby, but nonetheless remaining one and the same man. Elsewhere[34] I have argued that the Demiurge does not represent the whole of what the God is—Plato denies that it is possible to truly know the 'Father of the Universe' (he is "past finding out," *Ti.* 28c3-5)—but rather, bears witness to and names a specific activity of the God, namely, his role as a craftsman in the generation the world. The fluidity with which Plato uses these and other names for God bears witness to this. Even later, in *Laws X*, when Plato's understanding concerning the God has developed still further, his intermingling of names for this primary god, that of 'King' (*Lg.* 904a6) or the 'supremely good Soul' (*Lg.* 897c7), remains the same. In each case Plato is highlighting an aspect of the God's activity in relation to the generated world.

With the regard to the second question or Plato's reason for introducing 'nous' as an adjunct to 'the God' as a causal source in generating the world, the answer proves more complex. However, if one also remembers that Plato does not allow the creator God to generate the world *ex nihilo* or 'out of nothing,'[35] rather, it was generated out of chaotic and mostly unorganized material (and never out of *purposely* organized material, but by chance organization) and additionally, incorporates the assistance of ideal Forms to aid the Demiurge in constructing the world, then it comes to no surprise that Plato should isolate the reasoning function of this god in the form of 'nous.'[36] What all of this suggests is that as the causal scheme is disclosed in order to accord with human understanding, so Plato increases the primary god's attributes (e.g. 'Mind') and the tools (e.g. chaotic material and the Forms) he uses. I argue elsewhere concerning the primal powers that the Demiurge is all 'nous' or rationality to the degree that Ananke is not. However, I do not argue that 'nous' for Plato is *only* rationality and in this regard I differ substantially from Cornford.[37] Part of this disagree-

---

34. For example, see §3.3.3.1 and §4.3.7 of Part I and §4.3.6 of Part II.

35. The idea that 'nothing comes from nothing' is a negative statement of the principle of sufficient reason.

36. Gregory Vlastos captures the richness of the Demiurge's character when he states that "the high god of the *Timaeus* is not so much a governor as a philosopher, a mathematician, an engineer, and, above all, an artist. That he is Reason personified is taken for granted; Plato alludes to this from time to time, but feels no need to say so formally. He does not call his deity 'Nous' or 'Logos,' but 'Demiourgos' – literally, 'Craftsman' [*Plato's Universe*, 26]."

37. See Cornford, *PC*, 33–37.

ment lies in Cornford's rejection of the idea that the myth which Plato has constructed refers to anything ontologically real in the metaphysical realm, allowing it only to refer to aspects of the physical world. However, as I later argue, Plato's burgeoning metaphysical retinue in the opening scenes of his cosmology is for the sake of understanding the divine, however imperfect our knowledge of it is, in its role in the world's generation. To understand the divine at all we must bring it into human terms and that is exactly what Plato has done, but to say that that is *all* that the divine is goes much too far. Mind or 'nous' for Plato represents the best of man disembodied, in other words, his 'mind,' but which is now free to act in its every capacity. In this form it becomes the highest vision Plato can approximate to in intimating the active power of the God and as regards the final question of what 'nous' is, provides his closest answer to the divine identity.

Cornford's understanding of νοῦς is articulated as early as page 27 of his commentary where he is discussing 27c-29d of the *Timaeus*. Thus, he writes:

> He [Plato][38] was certain that the visible world exhibited the working of a divine intelligence aiming at what is good, and he held it to be of the utmost importance for the conduct of human life that this should be believed. The truth is best conveyed by the image of the divine maker, pictured as distinct (like the human craftsman) from its model, his materials, and his work. But he here warns us not to imagine that in using this image, he has declared the true nature of the cause. It is to be taken, not literally, but as a poetical figure. The whole subsequent account of the world is cast in a mould which this figure dictates. What is really an analysis of the elements of rational order in the visible universe and of those other elements on which order is imposed, is presented in mythical form as the story of a creation in time. . . . He does not mean that any actual state ever came into existence by these stages.[39]

Cornford's exegesis of the *Timaeus* at this point is strongly interpretative and does not, I suggest, provide an accurate analysis of the text as it stands. First, he fails to give sufficient grounds for why Plato, who by Cornford's own admission believed that the world exhibited the working of a divine intelligence aiming at what is good, should suddenly drop the 'divine' and now only be concerned with the 'intelligence,' i.e. "what is really an analysis of the elements of rational order in the visible universe" or why Plato would devote most of the *Timaeus* to describing a state that never came into existence. Second, arguing from *The Republic*, depicted here as an analysis of the ideal State cast into the form of a history, Cornford suggests that Plato used a "similar device" in the *Timaeus*. Thus, he argues that whereas in *The Republic* Plato did not mean that any actual state ever came into existence by these stages, by analogy Plato does not tell us

38. Square brackets are my own.
39. Cornford, *PC*, 27.

ever came into existence by these stages, by analogy Plato does not tell us in the *Timaeus* what the sustaining cause is and could not tell us "without stepping outside the framework of the very myth he is constructing."[40] However, Plato *does* attempt to tell us something about the ultimate cause of the world in his account, which argues against the idea that he is constructing a myth in the strong sense that Cornford suggests. The only analogy which Cornford can accurately make here with *The Republic* would be that in the *Timaeus* no actual state ever came into existence by the stages Plato describes. This, however, is only partially true in that Plato's description of how the universe came into being could only ever be, by his own admission, a 'likely' account and not an exact one, with the so-called "stages" necessarily being imagined and therefore, while inexact, nonetheless representing a real process of generation involving real stages. What *is* true is that the actual state, in this case the universe, did indeed come into existence. Moreover, as stated above, Plato does speak of a cause whereby it came into existence, namely, the 'Demiurge' and gives some detail about this cause even if he has to claim difficulty in doing so.[41] Third, in arguing that Plato could not tell us what the sustaining cause is, not only has Cornford conflated two issues, namely that of a cause and that of a sustaining cause,[42] he also begs the question, for in presupposing in his analysis that Plato is constructing a myth, he must forgo the possibility that Plato can state the cause. Finally, although Cornford's query is directed specifically at a 'sustaining' cause and what this cause might be, these considerations follow upon presuming that this cause is the one in question. He does not consider the possibility of a non-sustaining primary cause.[43] Plato's description, however, would indicate other than a sustaining cause, setting out as he does the generated universe as being

40. Ibid.

41. See *Ti.* 28c, 29a, 29d-30c, 31b, 32b, and further examples throughout the *Timaeus*, where, as Bury has noted (7), various references to "God" (in the singular), "the Divine" Power, as well as "the Gods" taking an active part in operations marked for rational purpose, are made. Bury writes: "The Role of the World-Artificer, the 'Demiurge,' is similar to that of the Anaxagorean 'Nous' (Reason). He is not a Creator, in the strict sense; that is to say, he does not make things *ex nihilo* but only imposes order and system on pre-existing Chaos. Nor does he continue to act directly, *in propria persona*, throughout the process of world-building, but, at a certain stage, hands over his task to the created star-gods."

42. Cornford, *PC*, 26. Here Cornford states: "It follows that the 'cause' of this becoming must be a perpetually sustaining cause," but offers no argument for this assertion.

43. Unfortunately, Cornford has fallen into the same trap, I suggest, as that which he accuses the Christian apologists (e.g. see *PC*, 35), namely, of confusing a particular Christian doctrine with Plato's understanding of God. To wit, God understood as a 'sustaining' creator or cause is prevalent in many Christian theological circles, but is notably absent (as Cornford points out) in the *Timaeus*. This is not to say, however, that Plato's primary cause could not sustain if it so wished.

entirely self-sufficient and self-sustaining.[44] Moreover, the fact that the Demiurge could choose to dissolve anything which it has generated (although according to Plato would not destroy something which was good and in possessing a 'will' which has "a stronger and more sovereign bond" than those to which the heavenly Gods were previously bound, would prevent dissolution by another 'will'),[45] does not mean that the universe requires this primary cause ('the God') to sustain it.

1.3.8 The 'likely' tale/myth

In trying to match the *Timaeus* with Cornford's analogy of *The Republic*, what could be stated as being comparable would be the stages of the construction of the Cosmos, not the actual final state. However, Plato speaks early on of a "likely" tale (29d), not of an unlikely tale, and although this point has been widely debated and still is, it is not a closed issue, tending to form two camps. There are those who understand myth or a 'likely tale' as Taylor sees it:

> He [Plato][46] wishes to make it clear from the outset that the whole of his cosmology makes no claim to be regarded as 'exact science'. Properly speaking it is not 'science' but 'myth', not in the sense that it is baseless fiction, but in the sense that it is the nearest approximation which can 'provisionally' be made to exact truth.[47]

Taylor argues for this view from Plato's own theory of forms. Whereas in pure mathematics "you get absolute finality and exactitude, just because there is no change or movement or life in the objects you are studying, e.g. integers, triangles, ellipses, and the like," time not needing to be taken into account in studying them because there is no movement or change, equally so, "Cosmology and biology, and 'pure' physics itself" or in Timaeus's own words 'that which is becoming only and never is existent,'[48] " . . . can never in his [Timaeus's] or Plato's opinion be rigorously 'exact',"[49] because such things are incessantly undergoing change and variation, requiring perpetual revising and improving on the results, making 'passage' the fundamental fact about 'Nature.'[50]

There are also those, however, who, like Cornford, understand 'myth' from the perspective of being poetic or symbolic, and thus representational of something else, which in this case, at least for Cornford, becomes "an analysis of the

---

44. See *Ti.* 32c5-34b9.
45. See *Ti.* 41a7-b6.
46. Square brackets my own.
47. Taylor, *PT,* 59.
48. *Ti.* 28a.
49. Taylor, *PT,* 59.
50. Ibid., 60.

elements of rational order in the visible universe," with 'Reason' becoming the preferred choice of meaning for νοῦς.[51] Although the capital 'R' does appear to give some voice to the 'divine' of "divine intelligence," in effect the opposite happens. Due to a complete rejection by Cornford of all religious import and hence, of any relevance it might have in a commentary on and translation of the *Timaeus*, in his analysis of νοῦς what ultimately results from such a modus operandi is a divinization of human reason, rather than a position such as Taylor's where divine intelligence remains as an object of human understanding and not an attribute of it. With regard to νοῦς, however, Cornford's position, being at once both revisionist, in that he redirects Plato's imagery to coincide with his own highly interpretative exegesis ("divine Reason")[52] and reductionist, in claiming that the text does not really mean what it says ("It must be said of the Demiurge that, as a mythical symbol, . . . he is not really a creator god, distinct from the universe he is represented as making"),[53] cannot be what Plato had in mind when he stated:[54]

> ἀγαθὸς ἦν, ἀγαθῷ δὲ οὐδεὶς περὶ οὐδενὸς οὐδέποτε ἐγγίγνεται φθόνος· τούτου δ' ἐκτὸς ὢν πάντα ὅτι μάλιστα ἐβουλήθη γενέσθαι[55] παραπλήσια ἑαυτῷ. ταύτην δὴ[56] γενέσεως καὶ κόσμου μάλιστ' ἄν τις ἀρχὴν κυριωτάτην παρ' ἀνδρῶν φρονίμων ἀποδεχόμενος ὀρθότατα ἀποδέχιοτ' ἄν.
>
> He was good, and in which good no envy whatsoever concerning anything is possible. Being therefore free from this he wished all things to be as much as possible like himself. Accepting this principle, then, from men of wisdom as being above all one of supreme authority, it can be accepted as absolutely true.

In this passage the ἀγαθὸς is clearly prior to and separate from the πάντα. Cornford's understanding and translation of νοῦς as 'Reason,' therefore, is entirely derived from his interpretation of the *Timaeus* as purely mythical in nature and of his explication of the Demiurge specifically as a mythic symbol, to which he feels no need to add further arguments.

51. See *Ti.* 47e; Cornford, *PC*, 160; Bury, 109.
52. Cornford, *PC*, 41.
53. Cornford, *PC*, 38.
54. *Ti.* 29e-30a.
55. Bury, at *Ti.* 29e4, has a reversed word order: γενέσθαι ἐβουλήθη.
56. Bury, *Ti.* 29e4. This passage has δὲ instead of δή.

1.3.9 One cause versus more than one

Archer-Hind, like Cornford, translates ἀνάγκη as 'necessity' and νοῦς as 'reason' (but with a small 'r'), providing similar, but in some ways fuller reasons for his translation and possibly the basis for Cornford's understanding written fifty years later. Because of this richer explication it is worthwhile to discuss Archer-Hind's argument in full. According to him it is completely senseless to speak of the notion of ἀνάγκη as an independent force external to νοῦς and for this reason ἀνάγκη will be considered here along with his discussion on νοῦς.[57]

His argument is as follows: The one cardinal doctrine of Platonism is that the only ἀρχὴ κινήσεως is ψυχή.[58] From this he concludes that:

> Plato does not mean that there is a blind force existing in nature, acting at random and producing hap-hazard effects. Such a conception is totally foreign to his system, in which the one cause, the one ἀρχὴ κινήσεως, is ψυχή.[59]

I do not dispute that for Plato ψυχή is the only cause of ordered motion within the Cosmos. This is assumed to be the case throughout the *Timaeus* and is clearly articulated in *Laws X.* What is disputed, however, is how narrowly Archer-Hind has taken this doctrine and what he concludes from it, namely, that there is only one cause. Plato does not specify at any point in the *Timaeus* that there is only "one cause," asserting, rather, that everything that becomes or changes must do so owing to *some* cause.[60] The closest he comes to this idea of one cause is when he acknowledges that the Demiurge or 'Father of the Universe' is a superior cause or 'the best of causes.'[61] Moreover, in his telling of the ordering of the Cosmos, after the World Soul has been fashioned, followed by the generation of the material body of the world and heavens of the universe by this God or 'Best of Causes,' Plato articulates two more causes, namely, Ananke (Necessity)[62] and the Accessory Cause(s),[63] specifically noting the latter as subordinate. A reason for positing these two additional sources of causation beyond providing a fuller description of the physical story is suggested at *Ti.* 28a6-b1.

---

57. Archer-Hind, *The Timaeus of Plato* (hereafter abbreviated as A-H, *TP*), 166 ff.

58. A-H, *TP*, 162.

59. Ibid.

60. *Ti.* 28a4-5.

61. *Ti.* 29a5-6: ὁ δ' ἄριστος τῶν αἰτίων.

62. *Ti.* 48a6-7: "this particular (kind of) wandering cause (τὸ τῆς πλανωμένης εἶδος αἰτίας)."

63. *Ti.* 46c7-46d7: "Accordingly, these are all assistant causes (πάντα ἔστιν τῶν συναιτίων)."

Here Plato proposes the idea that whenever the maker of anything keeps his mind's eye on the eternally unchanging and uses it for the pattern after which he realizes the form and capacities[64] of his product, then the result must be good. He can only hold the Demiurge responsible for results that are completely good, since the Demiurge himself is described as being wholly good. Consequently, Plato must look to other causes to account for all aspects of the Cosmos both before and after its ordering, causes which are not completely good, which come to be, or which change. For this he turns to Ananke, both before and after the world has been generated, and also to the World Soul and the accessory causes after it has been generated.[65]

1.3.10 The accessory cause or συναίτια

With regard to the latter, Plato clarifies what he means by the accessory causes.[66] Although he refers to these as causes, he makes it clear that it is not because they are what he would consider to be *true* causes. For Plato true causes within the generated Cosmos must possess ψυχή or soul capable of self-motion, conceding only that the accessory causes are sometimes mistaken as possessing such a power. Rather, he calls all causes accessory or secondary which do not move of their own accord, but when moved by a true cause or ψυχή, themselves become objects of movement by being moved and thereby become capable of moving other objects upon coming into contact with them. Thus, they *assist* the original cause in bringing about a particular end either by being moved or by acting as intermediaries moving others to bring about such an end through their auxiliary motion.[67] It is in explaining the origin of such ends in his description of the Cosmos that requires Plato to bring in the notion of the accessory cause.

---

64. The function of such an object would follow upon its form and capacities, and like these and for the same reason, would also be good.

65. The fact that Cornford takes the Demiurge to be a completely mythic character or non-being in the normal sense of the word will be discussed later. His view is that both the Demiurge and chaos are symbols for "real elements in the world as it exists" (*PC*, 37). I will argue that although there is myth in the *Timaeus* it is given at the beginning of the dialogue in the form of the Atlantis myth (*Ti.* 20d7-25d6). This is a second-hand account related by Critias as passed on to him through his relatives by Solon, the famous ancient poet. Plato's account of the Cosmos, on the other hand, is through the voice of Timaeus, a scientist and is not a second-hand account, but is his own understanding with no indication that it is to be taken as myth. On the contrary, it is explicitly stated that this account is to be understood as a likely account, although it does contain metaphor and due to the nature of metaphysics, cannot be exact.

66. *Ti.* 46e1-3. See §2.3, Part I, for a 'Commentary and Notes' on *Ti.* 46c7-46e6.

67. In Plato's account, the end brought about by an accessory cause is always good, pointing to the need to clarify exactly how Ananke can be a cause in any efficient sense, if at all.

1.3.11 Discussion of αἰτία / 'cause'

Plato is articulate in his understanding and explanation in the *Timaeus* of what constitutes a 'cause' or αἰτία and no more so than when he clarifies why an accessory cause or συναιτία cannot be a primary cause or αἰτία. He leaves no room for equivocation. In Plato's metaphysical scheme, the term 'αἰτία' or 'cause' is used exclusively for the work which his key metaphysical players, the Demiurge, Ananke, and the World Soul, must carry out in order for the physical world to be or have the existence it does.[68] Accordingly, these αἰτίαι account for its subsequent stability, change, perfection, and lack thereof.[69]

In reviewing the actual use of αἰτία or 'cause' as this word is found in the *Timaeus*, the following can be ascertained.[70] First, it or a close derivative appears in the text at least 39 times and presents as having three distinct, clearly defined uses. Its first appearance is at *Ti.* 18e3 and is best translated as 'was due to.' In instances such as these, there are no references to either the Demiurge or Ananke (Necessity), and the idea of metaphysical or physical causation is entirely lacking. What remains is an allusion to deductive reasoning (i.e. Premise 1: X has been allotted a spouse with the same intelligence as himself. Premise 2: There has been no pre-arranged specially designed drawing up of lots for spouses. Conclusion: therefore, this has happen entirely—'was due to'—by chance, rightly or wrongly deduced, and in this initial case, wrongly, as we are told that the allotments have been 'fixed,' so that like partners end up with like in each case. The next two instances, at 22c1 and 22e4, suitably follow with the same translation, referring to aspects of history, and form the preamble to the *Timaeus* proper and thus occur before Plato sets about to describe the generation of the Cosmos.

From *Ti.* 28a4 down to 48a7, however, instances of αἰτία refer almost exclusively to ontological causation[71] or to the 'coming into being' of the Cosmos and to the metaphysical causes bringing about such an event. Plato clearly states that such a depiction is his express purpose for this extended passage regardless of the difficulties which this engenders or more precisely, *because* of those difficulties and of sometimes missing the mark. An exception appears at 29d7, where the meaning of αἰτίαν is clearly the abstract noun 'reason,' i.e. "let us therefore

---

68. Causality for the modern thinker often denotes efficient causality and less often material or formal causality, whereas in religious circles causal interest often gets referred to final and first causes, but in all cases the reference and importance of 'cause' remains grounded in the 'stuff' of the world. A shift to Plato's thinking would see the relative importance of the latter diminish and a rejection of all but metaphysical causes as true causes. See *Ti.* 46d7-e6.

69. Outside of metaphysical discussions, a less rigorous and broader meaning for 'cause' is occasionally sought, e.g. *Ti.* 18e3.

70. See Index II.

71. In the sense of 'primary' causation, without which the Cosmos would have no being.

state the reason," but which cannot be confused with the Demiurge as 'cause,' since in this instance a separate word is used to describe the Demiurge, namely, ὁ συνιστάς. Similar passages where the abstract noun 'reason' (αἰτίαν) or 'reasons' (αἰτίας) is again found, but while still on the subject of the generation of the Cosmos, are at *Ti.* 33a6 and 38d7 respectively. At 40b4 αἰτία is best translated as '(from this) origin,' but again, there is little room for mistaking its meaning here. Whereas Bury translates this passage literally "from this cause,"[72] plausibly referring to the Demiurge who did in fact (we are told) generate the stars, the passage which immediately follows and where Plato subsequently dismisses a discussion of the origin (τὴν γένεσιν) of the other gods as too great a task (*Ti.* 40d6-7), would argue for a similar meaning for these two passages, i.e. that of 'origin.' *Ti.*42e3, on the other hand, refers to 'cause' in the sense of 'source,' i.e. "a cause of evil to itself, " whereas 44c7 refers to 'causes' generally, whether physical or metaphysical. The passages at 46c7, d1, e6, and 47a2 all refer to accessory causes.

Finally, from the instance of αἰτία at *Ti.* 57c8 down to its last instance at 88a7, Plato, for the most part, is using αἰτία or 'cause' in the sense of 'process,' meaning the physical reasons for an event occurring or alternatively, as 'cause' generally, which might include the idea of process. Once again, the exceptions are clearly defined. Two distinct places are at 68e4 and 68e6, and at 76d6 and 76d7. At 68e4 Plato is making a distinction between a true cause and a subservient cause, and at 68e6 he is referring explicitly to metaphysical causes, the ἀναγκαῖον and the θεῖον, the 'necessary' and the 'divine.' Similarly, at 76d6 he is referring to the συναιτίοις or accessory causes, explicitly not to first causes, and at 76d7, by using the superlative, to 'the greatest of causes' or as best understood here, to 'divine Purpose.' Further, at 67e4 the meaning of αἰτίας from its context, as at 22e4, is best understood as 'reasons,' whereas at 69a7, αἰτίων refers to various kinds of 'causes,' and thus, to 'causes' generally. Thus, in this final section, except where noted, Plato is dealing exclusively with physical processes, e.g. the growth of hair, the causes of respiration, etc. and the use of αἰτία is consequently borrowed and adapted with this specific purpose in mind. There is no equivocation. The material surrounding the *in situ* use of αἰτία as occasioned in all three sections of the *Timaeus*, the historical preamble, the metaphysical description of the origins of the Cosmos, and finally, its physical description, in each case clearly sets the parameters with respect to the various meanings of αἰτία, which in turn emerge distinctive and well-defined.

My concern is with Plato's understanding of 'cause' as a metaphysical entity and, specifically, as 'Cause' in its primary sense.[73] Understood in this way, the

72. Bury, 85.

73. The use of capitals is avoided except where a distinction is required or being made. In this case 'Cause' with initial uppercase 'C' refers specifically to a primary cause.

idea of Cause is one of prime mover and is entirely an *a priori* consideration for Plato and is not an *a posteriori* one following upon judgments from sense data. Thus, with regard to metaphysical causation he does not begin with an outcome and then work backwards to an explanation until at one point a logically economic universe emerges.[74] Rather, Plato begins, quite literally, both his metaphysical explanation and his cosmology with an uncompromising commitment to a core belief in a God who is perfectly good,[75] but not alone,[76] and to philosophical premises that support this belief, particularly those involving his theory of knowledge and which incorporate his theory of forms or ideas.[77] To ask whether or not the "unfelicity" or non-perfection of an outcome (i.e. of the generated universe) should be explained in terms of some cause, e.g. Ananke (Necessity), is a non sequitur in the case of the *Timaeus*, as it is no more valid a question for its writer than asking whether the felicity or perfection should be explained in such terms. Plato does neither. After Timaeus's opening remarks,[78] Plato does not continue his inquiry with the world and then posit by way of explanation, causes; rather, based on his (Timaeus's) initial observations he affirms now what he considers to be the primal causes of the universe, the Demiurge and Ananke, and then draws his understanding of the world from this. Thus, once the Cosmos has been observed as a completed entity and its goodness noted, and apart from any further observations, Plato's template for *understanding* its generation proceeds first from these Primal Causes and other proposed metaphysical features, and then to the world itself and not the other way round. These in turn are based on Plato's core religious beliefs, as well as his metaphysical and philosophical assumptions concerning goodness or lack thereof, and not on empirical considerations.

Although the generated Cosmos does figure in Plato's decision to posit the various metaphysical components of his cosmology, they are never limited to it

74. My position is opposite to that of Vlastos (although the universe remains the same in both accounts), who, in describing Plato's method and purpose in writing his cosmology states that "if you cannot expunge the supernatural, you can rationalize it, turning it paradoxically into the very source of the natural order, restricting its operation to a single primordial creative act which insures that the physical world would be not chaos but cosmos forever after. Plato accomplished this by vesting all supernatural power in a Creator who was informed by intelligence and was moved to create our world by his love of beauty and by his pure, unenvying, goodness" (*PU*, 97), thereby granting to Plato a "perversely original way" of giving rational men a pious faith to live by for two millennia. My suggestion is that Plato found (and argued) a higher power to be the source of that rationality, which both the *Timaeus* and *Laws X* strongly support, and not that he *made* it the source to serve some pragmatic end, i.e. the conscious expunging of the supernatural by rationalizing it to ensure that a rational explanation of the cosmos is furthered.

75. *Ti.* 29d7-30a7.

76. See *Ti.* 27c1-d4.

77. See *Ti.* 27d5-29d3.

78. See §1.3.6.

and for the most part supersede the dictates of scientific, but never rational, explanation and necessarily so—Being preceded Becoming. Rationality for Plato involved 'nous,' the contemplation of Being, and the belief in its inherent intelligibility. The explanations provided in his inquiry move constantly forward and never backwards, although restated anew several times. Moreover, the world of Being and that of Becoming are not, I suggest, dependent worlds. Although the fact of their separate existences is deductively realized or understood (a maker of something must have a model in mind), the existence itself of Being and of Becoming do not require that they have a relationship. These worlds simply exist and their contents are discoverable. A wrong idea of one will not necessarily obtain the other world wrong. Once the heavens have been generated and 'Becoming' or the Cosmos is operative, 'Becoming,' like 'Being,' is also explained in its own terms or those of 'becoming' and secondary causes. Equally, the Primal Causes, like the world of Being, lie outside the generated world or that of 'Becoming' and arguably also lie outside of 'Being' as it is described.[79]

Thus, the felicity or unfelicity of the outcome (the outcome being the Cosmos itself) is not explained by Plato in the positing of Primal Causes which are either good (the Demiurge) or evil (Ananke)—explanations for the workings of the universe remain within it and are discoverable therein—rather, the outcome of Plato's inquiry begins with the Primal Causes themselves and moves forward through the world of Being to include Becoming or the Cosmos. The metaphysical movements of the Demiurge and Ananke in bringing about order or disorder respectively are in turn given as the reason for the existence of the Cosmos. It is unlikely that Plato envisioned that the details of the physical world should be examined in the light of these metaphysical assumptions (e.g. the Primal Causes), as they did not extend for the most part[80] to such detail.[81] As such, the accountability of the metaphysical grounds he sets out is very restricted. In Plato's own words, "the world the Demiurge constructed was a single, complete whole, designed to supply its own nourishment from its own decay and to comprise and cause all of its own processes, its Framer having thought it better to be self-sufficient than dependent on anything else":

---

79. I later argue, however, that the concept of 'Being' is closely associated in Plato's understanding with the causal operations of 'Nous' ('Nous' existing as the causal aspect of the Demiurge or God) and that these two metaphysical players are intimately connected.

80. See the discussion immediately following on the World Soul.

81. The fact that Plato's philosophical system of Forms has been held up to the scrutiny of the greatest minds and has been judged and found wanting from its very outset, rejected as untenable at worst and unnecessary at best from the time of his most gifted pupil, Aristotle, to the present day as regards the empirical world and its details, bears witness to this.

> ( ) αὐτὸ γὰρ ἑαυτῷ τροφὴν τὴν ἑαυτοῦ φθίσιν παρέχον καὶ πάντα ἐν ἑαυτῷ καὶ ὑφ ἑαυτοῦ πάσχον καὶ δρῶν ἐκ τέχνης γέγονεν· ἡγήσατο γὰρ αὐτὸ ὁ ξυνθεὶς αὔταρκες ὂν ἄμεινον ἔσεσθαι μᾶλλον ἢ προσδεὲς ἄλλων. (*Ti.* 33c7-d3)

Accordingly, the very basis on which the universe was organized, namely, a principle of self-sufficiency in which it would comprise and cause all of its own processes, precludes attempts to connect causal factors[82] in the world with those existing outside of it or before its generation. Indeed, such attempts appear foolish. Whatever took place between the Demiurge and Ananke to bring about the generation of the world happened before[83] this averred independence of the world from everything that preceded it.

1.3.12 Denial of the four elements as causes

Furthermore, Plato patently denies that any of the four elements, fire, water, air, or earth are causal factors in themselves. He describes them, rather, as the four constituents of the universe.[84] These bodies form the whole by their proportion, at which point the Cosmos came into existence, and are only causal by way of their being assistant or auxiliary to a true cause.[85] In *Laws X* he repeats this assertion:

> πῦρ καὶ ὕδωρ καὶ γῆν καὶ ἀέρα φύσει πάντα εἶναι καὶ τύχῃ φασίν, τέχνῃ δὲ οὐδὲν τούτων, καὶ τὰ μετὰ ταῦτα αὖ σώματα, γῆς τε καὶ ἡλίου καὶ σελήνης ἄστρων τε πέρι, διὰ τούτων γεγονέναι παντελῶς ὄντων ἀψύχων· (*Lg.* 889b1-5)

> Fire and water, earth and air – so they say – all owe their being to nature and chance, none of them to art; they, in turn, are the agents, and the absolutely soulless agents, in the production of the bodies of the next rank, the earth, sun, moon.[86]

82. Except for the noetic movement in the world produced by the World Soul, having been formed precisely for this purpose (being a true αἰτία possessed of ψυχή or 'soul'), all other causes in the world are accessory only.

83. At the very least ontological priority is implied here, the heavens having not yet been made.

84. *Ti.* 32b3-c8.

85. *Ti.* 46c7-d7.

86. Taylor, *Laws X*, 279.

While Plato is vehemently arguing against the concluding portion of the above assertion, namely, that the earth, sun, and moon are simply in the next rank, and are thus equally soulless, in no way does he depart from his original premise contained in the first part and as previously set out in the *Timaeus*, namely, that fire, water, earth, and air are absolutely soulless agents. His argument, which clarifies this, begins a little further on in *Laws X*, at 891c7-d4 where he questions the validity of the reasoning of those who would make soul derivative of the soulless agents, fire, water, earth, and air. Soul, he will claim, is elder-born than all bodies and the prime source of all their changes and transformations.[87] He then moves on to show that soul is more primitive than fire or air, or the like, and that its defining factor is its capacity to cause its own movement, as well as the movement of others, which is shown, Plato argues, in the movement of the sun, the moon, and the stars.[88] This in turn is what ranks them (the souls that drive the heavenly bodies, and to the extent that they do, the heavenly bodies themselves) as gods, which self-motion is clearly superior to the motion of any of the elements, which latter motion is always caused from without by another agent. From this he will argue:

1. Motion, however it arises, which is induced by something else and which never confers self-motion on anything else, is secondary, being change in a truly soulless body.[89]

2. Thus, soul is prior to body, governing the real order of things, and body being subject to governance.[90]

3. Moreover, the characteristics[91] of soul, i.e. "mood and habits of mind, wishes, calculations, and true judgments, purposes, and memories, will also be prior to physical lengths, breadths, and depths, in virtue of the priority of soul itself to body."[92]

---

87. *Lg.* 892a2-b1.

88. For Plato, the identity of the soul(s), whose self-movement drives the movement of the heavenly bodies, is so closely linked to the identity of those celestial bodies themselves that he asserts, taking the sun as an example, it is this *soul* (not this body) which provides us all with light and that "every single one of us is bound to regard it as a god" (*Lg.* 899a7-10).

89. *Lg.* 896b4-8.

90. *Lg.* 896b10-c3.

91. Taylor has translated this as "characters." See his translation, *PT*, 288.

92. *Lg.* 896c9-d3.

4. Plato sums up his argument by stating that given the soul has been shown to be the cause of all the contraries, then we are bound to assert it as the *universal* cause.[93]

5. Indwelling soul, then, controls all things universally that move anywhere and thus, heaven itself,[94] including individual celestial bodies.[95]

6. It is shortly after this juncture that Plato explains the manner in which he considers the sun or moon or stars to be a god or gods, namely, by being moved by soul, either by being driven from within or from without, or "in what way soever."[96] He comes to this conclusion from the conjecture that the source of their self-movement is invisible, comparing this with the self-movement of a man whose soul is also invisible and can only be discerned by an act of pure understanding.[97]

7. After this Plato is able to exclaim: "Will any man who shares this belief bear to hear it said that all things are not 'full of gods'?"[98]

Thus, Plato asserts, not only in the *Timaeus*, but again in *Laws X*, supported to some extent by argument, that the elements are entirely soulless, having no capacity of self-movement. As a consequence, moving back to the argument of the *Timaeus*, they are not true causal factors.

Plato has preempted the possibility of the Demiurge and Ananke[99] playing a further causal role in the physical universe once it has been generated by the priority of their ontology.[100] However, the one metaphysical causal factor,[101]

---

93. *Lg.* 896d5-8.
94. *Lg.* 896d10-e2.
95. *Lg.* 898d3-7.
96. *Lg.* 898e8-899a10.
97. *Lg.* 898d9-e3.
98. *Lg.* 899b8-10.

99. Although 'Necessity' is a common translation for the causal force ἀνάγκη, where possible I use the transliteration 'Ananke' to avoid a misunderstanding which can arise from this reading. However, I do use 'Necessity' where this particular reading is being discussed.

100. The negative impact of Ananke as a causal factor in the soul of man after his generation will be taken up by Plato in *Laws X*. See the discussion in Part II, §3.3.3, where the counterpart of Ananke within the generated world is now understood by Plato as the 'Non-beneficent Soul.'

101. The World Soul, having been created by the Demiurge for the precise purpose of generating noetic motion in and for the world, is not too far removed from the modern concept of a 'factor,' since although the World Soul is described here as being 'the best creation of the best of intelligent and eternal things' (*Ti.* 37a1-2), endowed as it is with reason and harmony from an unending and divine source (*Ti.* 36e4-5), apart from its par-

unlike the Demiurge and Ananke, which does imbue the world with its ongoing presence in material matters is the World Soul, which was constructed by the Demiurge for the sake of the world to provide it with its several movements.[102] Although the World Soul is eternal, unlike the Demiurge and Ananke its immortality, by way of it being constructed (which is not the case with either the Demiurge or Ananke), extends only *a parte post* into infinity, being described here as existing 'for all time.'

1.3.13 From timely indissolubility to eternity: An argument from Ananke

Prior to the section on the World Soul, Plato also makes an important move regarding the infinity of the body of the world or Cosmos. It is deemed infinite, not, as in the case of the Demiurge and Ananke, because it is ungenerated, since clearly it has been generated, but rather, because of the providence of God, whose strength and sovereignty prevailed and continues to prevail over that of Ananke. This is an important claim Plato makes and is part of a larger deductive argument involving Ananke and its apparent insurgent presence as a metaphysical force. He states in the *Timaeus*, at 32c2-4, that the body of the Cosmos is indissoluble by anything except its compounder, but not without explaining a little further on at 41a7-b6 that its compounder, the Demiurge, described earlier as being perfectly good,[103] because of this goodness, would not do such a thing in any event. Similar indissolubility is also attributed to the heavenly bodies at 38b6-c3. Thus, the Cosmos, theoretically at least, becomes eternal from the point of its generation, with Plato explicitly making the move from it being indissoluble to being eternal at 40b4-6 when discussing the fixed stars.[104]

Two initial observations can be made with regard to these passages and the making of the Cosmos, its Soul, and the Living Creatures within it. First, at no point is Ananke given any role in either the making of the body of the organized

---

ticipation in the specific pragmatic end for which it was made, all of which, its being, its participation, and the end, are necessarily good having been engineered by the Demiurge who is wholly good, the World Soul is not judged of itself to be either good or bad apart from its role in the world for which it was fashioned. This is patently *not* the case for Plato with regard to either the Demiurge or Ananke, neither of which is similarly generated.

102. Plato makes it abundantly clear in the *Timaeus* that the World Soul was generated *by* the Demiurge and is not to be identified with this primary causal power and as Lutoslawski correctly states: "This demiurge is outside the world, and different from the world's soul," imparting to the world "its shape and present nature," from *The Origin and Growth of Plato's Logic* (London: Longmans, Green, and Co., 1897), 475.

103. *Ti.* 29e1-2.

104. In this passage Plato uses the word αἴδια from ἀίδιος (ἀεί), meaning 'everlasting,' for ἀιέδιος (*LSL*). The non-temporal eternity of the primary causes is prior to any 'everlastingness' attributable to the physical universe, which latter necessarily remains contingently everlasting due to its dependent and generated nature.

world, i.e. the generated universe, in part or in whole, or its Soul, materially or mechanically, e.g. from the perspective of process. Materially, the world was formed by the Demiurge alone, who achieved this by introducing measurable relations, internal and external, to the elementary particles of what was otherwise a chaotic, albeit visible, world.[105] Although Plato describes the elementary particles of the pre-cosmic chaos as sometimes being in a certain order, this was only by chance. The order produced by the Demiurge, on the other hand, was never random, but was achieved in and through the use of the Demiurge's power of 'nous.' According to Plato's cosmology, however, there is also a power behind all disorder and randomness, which metaphysical power he names 'Ananke.' Consequently, Ananke would also be the cause, inadvertently, of any random order.

Implicit to Plato's theory is the assumption that these elementary particles were comprised minimally of the two basic triangles later ordered by the Demiurge,[106] since it is these which are then described as having been ordered (by the Demiurge) into more complex shapes and which surfaces upon further ordering became the basic elements, fire, earth, water, and air, of the ordered universe. Prior to their further ordering by the Demiurge, Ananke would have held these basic triangles in their non-ordered state, being a metaphysical power without reason and thus, having no capacity to order them.[107] These basic triangles, therefore, would not only form the basis of the consequent ordering of the universe described by Plato, but would also form the basis of its non-ordered state or particular level of it at which Plato begins his cosmology. I would concur with Cornford that:

> There is really no warrant for attributing to Plato this atomistic picture of irregular particles moving at random in a void.[108] Atoms were

105. *Ti.* 69b2-c3.

106. *Ti.* 53c4-d4.

107. This is vastly different from Teloh's understanding of 'ananke' (understood by him as 'necessity'). According to his interpretation 'necessity' is closely linked with the receptacle or 'nurse of becoming,' which receptacle had two diverse but not inconsistent functions (*DPM*, 214–15): He writes, first, that it was 'space' which received all generated things, and second, it molded the stuff in it and in fact "is that out of which becoming is formed." It is precisely this molding stuff "made up of rudimentary physical powers (52e), with these powers operating according to what we would now call physical laws" and "organized according to the mechanistic principles of the *physiologoi* . . . [which] constitute the domain of necessity." This complex and somewhat ingenious idea of 'necessity' being the "molding stuff" of the cosmos, which in turn is made up of "rudimentary physical powers" remains perplexing and in need of further explanation. Plato's account does not provide this explanation or make this connection between space and 'ananke,' nor does he suggest it.

108. See my discussion of 'void' within the comparison between Democritus and Plato at §2.3.5.2, Part I.

> completely determined particles of solid substance, separated by intervals of nothingness, which gave them room to move about. Plato's Space is not a void which remains completely distinct from particles moving in it; it is a Recipient which affords a basis for images reflected in it, as in a mirror—a comparison that could not be applied to atoms and a void.[109]

I would further propose that Plato deduced these basic particles to be triangles due to the fact that according to him only triangles were capable of being formed into the surfaces of which solids were composed, and whereas they could be so formed, they need not be and thus could remain in a disordered (non-ordered) state. Plato does not address the issue of the possibility of a greater degree of disorder achievable for these random particles, i.e. whether the triangles or particles themselves were ordered into their basic shape from something else even more basic.[110] His main concern in the *Timaeus* was the ordering of the Cosmos and such order could be achieved by the ordering of these two kinds of, previously random, triangles. He regresses no further than is necessary. Thus, while I would agree with Cornford that Plato stops short at the positing of triangles "without pursuing the analysis of the triangles themselves into simpler principles,"[111] I do not agree that "Plato indicates that there is something arbitrary in starting from this assumption" or that "Plato's reason for stopping short at triangles was perhaps the need to keep his exposition within reasonable bounds."[112] Rather, Plato states "the principles which are still higher than these are known only to God and to the man who is dear to God."[113] When further explanation would take too long or diverge too far off topic, Plato acknowledges this.[114] He does not do so here. Cornford's identity of 'the man who is dear to God' or as he translates this phrase, "men favoured by heaven," with the philosophic mathematicians whose analysis would stretch "back to the ultimate premisses of the science,"[115] if a proper analysis was fully borne out, is not a position suggested here or supported by Plato. Ultimately, *Laws X* will attest to Plato's rejection of any hypothesis omitting the deity as first principle ('grand

---

109. Cornford, *PC*, 200.

110. See *Ti.* 48b6-c2. In this passage Plato is stating that the four elements, fire, water, earth, and air are far from being the basic (not even comparable to syllables in an alphabet) building blocks of the universe, with the further implication that the basic triangles themselves might have a greater story, which Plato will not even attempt to tell. As this passage continues (48c2–d4), Timaeus asserts that to tell a story of the basic principles would in fact be too difficult to explain in the context of the discussion he was undertaking.

111. See Cornford, *PC*, 212.

112. Ibid., 213.

113. *Ti.* 53d6-7.

114. See footnote 110 with reference to *Ti.* 48b6-c2.

115. Cornford, *PC*, 213.

and primary') or built on the notion that the works of nature themselves can offer up to the scientists their own ultimate premises, mathematical or otherwise, upon inquiry. Cornford's footnoted conclusion to his discussion of this passage (*Ti.* 53c-d) that "the *Timaeus* is a myth of the physical world, and therefore has no need to go further back than the surface,"[116] must methodically omit all record of Plato's repeated affirmations of the god or Demiurge as first principle and alone privy to ultimate origins, to affirm this. As such, Cornford's conclusion remains highly interpretative and substantially removed from what Plato originally sets forth.[117]

The World Soul in the *Timaeus* was likewise formed by the Demiurge, before the body of the world, to be its dominating and controlling (as the source of

---

116. Ibid., fn. 4.

117. Similar to Cornford in his approach to the *Timaeus*, seen purely as myth, R. D. Mohr reads the creation story of the *Timaeus* as though "not intended to be read literally" (see *The Platonic Cosmology*, 1). Accordingly, he suggests " 'all the furniture' of Platonism—gods, souls, Ideas, matter, space, properties, natural and artificial kinds—are seen related each to all within a single frame," namely, "the major branches of his speculative thought—epistemology, metaphysics, theology, physics, and to an extent logic and ethics," and in so doing "casts light on all of Platonism." What neither Cornford nor Mohr acknowledge, however, is that this approach to the *Timaeus* is dependent on extensive interpretation of not just a few, but of every piece of "furniture" within that story or myth. This is to ignore Plato's repeated claim that what he is telling is a *likely* story, not an *un*likely story, but which an understanding of this work as pure 'myth' or a fanciful tale essentially assumes. It may be argued, however, that if the causal players or other metaphysical features within Plato's cosmology are *not* viewed as essential features of a myth, e.g. are *not* interpreted as 'reason,' 'motion,' or 'disorder' in the world as it exists now (Cornford) or in the case of Plato's primary ordering cause, that it is *not* viewed as a "necessary existent whose essence is rationality . . . to serve as a crafting agent who constantly works to bring that which falls away from a paradigmatic Form into accord with it" (Mohr, *The Platonic Cosmology*, 183), thereby introducing "standards and measures into the phenomenal realm," (Mohr, *The Platonic Cosmology*, 3), then the *Timaeus* as it stands is too fantastic to be meaningful or 'likely.' Such an assumption, I suggest, needs to be challenged in light of certain of the features of Plato's cosmology having become completely dismantled (e.g. 'Ananke') or made completely subservient to this understanding of an already existent world (with or without a pre-existing 'chaos') in the process of carrying out such an interpretation. Nothing of the myth is left requiring it in the first place. Should, on the other hand, the *Timaeus* be approached as a likely cosmological account, which contains mythic features only (but is not itself a myth) while explaining the supra-physical or metaphysical features of the cosmos before and after its generation, and in so doing is found to be sufficiently cogent so as to satisfy Plato's reasons for writing it (i.e. provides a meaningful explanation for the generation of the cosmos according to his perspective and understanding, and which understanding is not fanciful in terms of that explanation), then this, I suggest, is the preferred reading since it is closer to Plato's own account. In contrast to a 'literal' account, this reading of the *Timaeus* is best described as a 'literate' account, where the understanding of each passage requires to be assessed and re-assessed according to the ongoing dynamics of Plato's thought, and never in spite of it, nor conversely, restricted to the words by which this thought is conveyed.

its noetic movement) power.[118] According to Plato's 'likely' account, it was compounded of various mixtures of Existence, Sameness, and Difference as its constituent parts, and was then divided up and bound together in proportion. The Demiurge then arranged for it to encompass the completed Cosmos and act as its source of unending rational life (partaking in reasoning and harmony).[119] This precludes Ananke from ever being the cause of any of the Cosmos's several motions within time, since all ordered movement in the universe, except that initiated by mankind (having its own source of rational movement by way of its less pure soul, in that the human soul was generated by the Demiurge, assisted by the young gods),[120] was caused by the World Soul. Moreover, whatever ordered movement was not *directly* caused by the World Soul or human souls was *in*directly caused by them.[121] Such indirect or secondary movement was the result of ongoing processes realized by auxiliary causes carrying through the original momentum initiated by its causal source (the World Soul or human soul), but which auxiliary causes, in any event, were never direct causes of movement themselves. Thus, Ananke, by virtue of its ontology—a primal cause and thus ungenerated, but never the source of ordered movement—is excluded from being the source of either movement or process within the ordered world.

Second, and decisive for understanding the extent of the metaphysical stature given to Ananke and the implications that follow thereon, is the role Ananke plays in Plato's structuring of his argument for timely indissolubility. Embedded here within Plato's theology is a deductive argument for the eternity of the generated universe. This not only allows, but also requires Plato to make a move from dissolubility to indissolubility of the Cosmos within time, and ultimately to immortality. This move, however, is subtle, presented by Plato as a dismissive remark made against Ananke and remains almost hidden within the exception which Timaeus grants to the Cosmos's 'compounder,' the Demiurge.

With regard to the first observation, some clarification concerning 'time' and the generation of the Cosmos is required. Although the meaning of Plato's famous passage[122] describing time as a "moving image of eternity" is profoundly compact and carries with it the impact of poetry, it is not poetry. Plato has already carefully explained in prior passages leading up to this one exactly what

118. *Ti.* 34c4-35a1.

119. *Ti.* 36d7-37a5.

120. *Ti.* 41d4-42e4.

121. Plato is never clear in the *Timaeus* on exactly *how* 'soul' is the cause of rational movement ("unceasing and intelligent life") in the universe or what this actually means, only that it *is* the cause of such movement. His depiction of the generation of the soul as strips permeating the universe is clearly metaphoric, since according to Plato we cannot see the soul. In Part II, however, dealing with *Laws X*, where Plato further develops his notion of 'soul,' this relationship between the soul and movement is again discussed and explicated in greater detail. See Part II, §2.3 ff.

122. *Ti.* 37d5.

his understanding of movement is, both causal and non-causal, how 'image' figures in the process of generated being, and what the relationship of eternity to the generated Cosmos is. Germane to our discussion of Ananke, however, is Plato's averment here that the copy, i.e. the Cosmos, "has been, is, and will be throughout the whole of time."[123] For Plato 'the whole of time' is something quite different from the parallel notion of 'eternity.' Time only comes into being with the generation of the heavens, whereas eternity carries the notion of having always existed, thus existing before the heavens came into being, during their depiction of time, and forever after should they be dissolved.[124] Moreover, time has movement, a beginning, and an end, whereas eternity has none, remaining forever at one with itself.[125] It is upon the equivocation of this passage concerning 'the whole of time,' however, or similar passages, that Ananke (when understood in this way, referred to as 'Necessity') is argued to be a mere physical factor. Thus, Necessity (Ananke) becomes associated with mechanical motion, a moving metaphor so-to-speak, in a physical world which according to this interpretation has always existed (i.e. 'throughout the whole of time') and whereby the notion of eternity is dropped altogether. Against such a position is the fact that much of what Plato states, except where he is describing physical phenomena, has either to be abandoned, ignored, or passed off as metaphor in order to achieve this. A severe abatement of this sort would have to be a last resort upon finding no rational basis or coherence in the metaphysical ideas Plato presents in the *Timaeus*. I argue against taking such a view.

On the other hand, it could be argued that the face-value facility with which Plato's scheme is presented and is so easily understood approximates too closely to storytelling to be anything but. As earlier suggested, however, and as Plato repeatedly reminds his readers, the 'storytelling' quality of the *Timaeus* reflects the "likelihood" of his account whereby he is attempting to describe certain metaphysical realities. This requirement of metaphor or 'likeliness' is due to both the priority[126] and perfection of these *a priori* entities, existing as they do before the generated universe and, except for the 'Nurse of Becoming,'[127] outside of all change. A description of this sort, therefore, cannot be exact. It is patently absurd to ascribe to Plato the need to invent a Demiurge and Ananke in order to describe the ordering process of mental and physical phenomena within

123. *Ti.* 38c1-3: τὸ μὲν γὰρ δὴ παράδειγμα πάντα αἰῶνά ἐστιν ὄν, ὁ δ' αὖ διὰ τέλους τον̀ ἄπαντα χρόνον γεγονώς τε καὶ ὢν καὶ ἐσόμενος.

124. *Ti.* 38b6-c3.

125. *Ti.* 37d6: ἐν ἑνὶ 'in unity' or 'singular.'

126. That is, before the generated heavens and thus, before time.

127. The 'Nurse of Becoming' is a difficult concept when considering the congruence of Plato's cosmic scheme and is rife with potential problems in that it appears to exist prior to the generated Cosmos and yet is subject to change. This and associated difficulties are discussed at §1.3.19, Part I.

the universe, as he has proven himself quite capable of discussing each of these independently from his separate discussions on the two metaphysical causes,[128] making such a need redundant and completely pointless. Moreover, the 'whole of time' described by Plato is clearly not inclusive of the eternity wherein his key metaphysical causal players, the Demiurge and Ananke, had earlier striven to bring about the Cosmos. This would require a logical impossibility, namely, for the Cosmos to have existence before being generated. Not only does the generation of the Cosmos account for the beginning of its existence, but it also accounts for the periodic motion of the heavenly bodies within the Cosmos, which motion gives us our sense of time, 'time' being the measure of this motion. Furthermore, Plato makes a point of not only declaring at the very outset of the *Timaeus* that the Cosmos came into being,[129] but also of arguing that this *must* be the case. He reasons that since all things which are perceptible by the senses—of which the world in its totality is comprised—are objects of sensation and opinion, and therefore, change, then they must have a cause. If such objects have a cause, then they also must have come into being.

Nonetheless, it is also equally true, and unequivocally so, that prior to the generation of the Cosmos there was for Plato another kind of metaphysical reality, one which was neither causal nor caused, but rather, existed eternally as pure 'being,' after which all things were patterned, Plato's Forms or Ideas.[130] In addition to these metaphysical entities, there was also for Plato a pre-generation physical reality consisting of basic, randomly moving triangles. This chaos, moreover, was contained within 'space.' If something exists, it must exist somewhere or not exist at all.[131] Finally, while remaining mostly unorganized in form and random in movement, this chaos, due to chance movements, was not entirely disorganized, but also consisted of triangles which had become randomly organized into what Plato calls 'becoming' and which 'becoming,' when taken over by the Demiurge, would eventually comprise the whole of the universe.[132] It is thus this three-fold claim of the pre-existence of Being, Space, and Becoming, in addition to the above-noted equivocation regarding time and eternity, which together lead to a general misinterpretation of Plato's cosmology in support of a non-metaphysical

128. See, for example, his discussion in the *Timaeus* of the ordering of regular solids and of their assignation to the four elements at *Ti.* 53c4-56c7 (Ananke is only referred to at the very end by way of reminder of its seemingly cross-purposed metaphysical impact on generation, being one of both willingness to work with the Demiurge and yet at the same time, because of this 'willingness' and its disordering nature, a lessening of perfection, i.e. that which is 'permitted,' on the work of the Demiurge). See also Plato's discussion of the diseases of the soul or mind beginning at *Ti.* 86b1or again, immediately following that, his discussion of the balance of the mind and body beginning at 87c1.

129. *Ti.* 28b7-c2.

130. *Ti.* 51e6-52a4.

131. *Ti.* 52b3-5.

132. *Ti.* 52d2-4.

position with regard to Ananke and the Demiurge, and a positivist, atheistic position overall for the *Timaeus.* A careful examination of the text, however, shows such a position to be untenable. At this juncture it is only necessary to point out that Plato saw the positing of 'being,' 'space,' and incipient 'becoming' prior to the generation of the Cosmos as prerequisite to its generation. Thus, their existence was not simply commensurate with the Cosmos, never having existed apart from it and now portrayed anachronistically as mythic detail. Rather, prior and eternal, their existence was logically required *before* generation could take place. According to Plato, the Cosmos had to have a pattern or idea after which all things are fashioned ('being'), there had to be a place in which it could have existence ('space') or be nothing at all, and it had to be made out of something ('becoming'). Finally, and importantly for Plato, there were powerful primal forces behind both the generated Cosmos and the non-generated Chaos, which forces he distinguishes respectively as the Demiurge and Ananke. Thus, when Plato now marks out in his scheme that the generated Cosmos will exist 'for all time,' as will the celestial gods or heavenly bodies, it has a decidedly peculiar ring to it, since earlier in the *Timaeus* he has not referred to time at all, our perception of it having only just come into being with the generation of the heavenly bodies. Prior to this section on time, Plato simply makes reference to the 'eternal' without seeking to qualify it, since there was no thought when referring to what is eternal of such an eternal object ever having a beginning or an end. As noted above, the Demiurge, Ananke, the Forms (or 'being'), space (the 'nurse'[133] or receptacle of becoming), and incipient becoming ('becoming' at its earliest stage) are all described by Plato as being eternal and therefore, as having existence before the world was generated.[134]

With regard to the second point made above concerning Plato's subsequent argument for timely indissolubility as a hidden rebuff against the eternal and chaotic powers of Ananke, this argument takes the following form:

1. Order is the source of all goodness.

2. The Demiurge is the ordering Cause.

3. The Demiurge is altogether good based on the fact that it is the ordering Cause and on this fact alone.

---

133. *Ti.* 52d4-5 (τὴν δὲ δὴ γενέσεως τιθήνην). Plato's reference to the place of becoming or 'space' as the 'nurse' of becoming is congruent with his previous metaphor of birth at 50c7-d4 (for a discussion see Cornford, 187), where the receptacle is referred to as the 'mother,' the model or forms as 'the father,' and incipient being as the 'offspring.'

134. *Ti.* 52d2-4.

4. The Demiurge generated or 'ordered' the Cosmos.

5. Disorder is the source of all evil.

6. Ananke (a transliteration of the Greek word ἀνάγκη) or 'Anarchy'[135] is the disordering Cause.

7. Ananke is altogether not good or, conversely, is 'bad,' based on the fact that it is the disordering Cause and on this fact alone.

8. Ananke did not prevent the Cosmos from being ordered or generated, or more precisely, could not, and therefore is given a role in its generation based on this fact and on this fact alone.

9. Ananke or more properly, its effect, chaos, remains in the generated world due to the fact that it was not completely subdued or organized by the Demiurge.

10. The Demiurge is eternal.

---

135. Although derived differently, ἀνάγκη (see my proposed etymology at §1.3.2, Part I) understood as a primary cause shares the idea of disorder with the English word 'anarchy' (from the Greek *a(n)-*, expressing want, 'without' + *arkhos* 'chief, ruler') and is, I suggest, a good translation, although not problem free (see fn. 431). In translations of the *Timaeus*, however, ἀνάγκη is most often translated as 'necessity,' but which word I argue becomes its meaning (or a derived meaning) only in the latter third of this work. What actually transpires is that a change in meaning takes place throughout the course of the *Timaeus*. For the most part, therefore, I have left 'Ananke' in its transliterated, untranslated form, except for the above and when translating it as 'Might' upon first being introduced to coincide with Plato's contrast of Ananke with Nous or 'Mind' in the early development of the Cosmos. I argue that for the first two thirds of the *Timaeus* Ananke is understood by Plato as a disordering primary cause. Accordingly, for this part of the *Timaeus* 'Anarchy' provides a better translation and is closer in meaning to ἀνάγκη than 'Necessity,' thereby depicting not only its force, but also its wholly negative or disordering impact. The apparent lapse on the part of Ananke or collusion suggested by Plato between the Demiurge and Ananke which allowed the Demiurge to order the Cosmos (see *Ti.* 48a2-5) need only follow from the observation that the universe was in fact ordered and for the most part remained so. Following this line of reasoning, and keeping the opposing natures of these two powers intact, the Demiurge would then be seen as having the more powerful 'will,' with Ananke having had to either concede or submit to allow this ordering. Mythically, this asymmetry is easy to portray as momentary agreement or collusion between these primal powers. Plato does appear to make the above subsequent claim regarding the respective opposing 'wills' of the Demiurge and Ananke at *Ti.* 41a8-b6. For a discussion of the concept of 'will' in ancient Greece see Albrecht Dihle, *The Theory of Will in Classical Antiquity* (Berkeley, University of California Press, 1982), chaps. II and III.

11. Whatever the Demiurge generates directly is indissoluble except by the Demiurge, i.e. the Cosmos as a whole (which includes the Heavens) (*Ti.* 32c2-4, 41a7-8), the World Soul (36e4-5), and the human soul (41c6-d1), based on this fact and on this fact alone.

12. Whatever the Demiurge does not generate directly[136] (and while excluding the immortal part of the human soul, does include the human body, the mortal parts of the human soul built onto the divine part [69c3-d6], as well as the weaving of these together [41d1-4]), i.e. the three kinds of living creatures, are not eternal or conversely, are dissoluble or mortal (41c2-6), based on this fact and on this fact alone.

13. Thus, anything generated or bonded together, including the Gods (the Heavenly Bodies), is not entirely immortal or indissoluble (41b2-3) and therefore is dissoluble either by the Demiurge (prem. 11) or due to their mortal nature (prem. 12).

14. Only a wicked 'will' would do this (*Ti.*41a8-b6), i.e. dissolve anything which is good and all generated things, in as much as they are ordered, are good.

15. Because the Demiurge is entirely responsible for the ordering of the Cosmos and is altogether good (premises 3 and 4 above), it could never be the Demiurge's 'will' to dissolve the Cosmos or any part of it.

16. Ananke is also eternal.

17. Because Ananke is the disordering Cause and is altogether not good or conversely, is 'bad' (premises 6 and 7 above), it could only be the will of Ananke that would bring about the dissolution of the Cosmos or any part of it.

---

136. How the natural objects of the world, e.g. a maple tree, a gold nugget, etc., apart from the three kinds of living creatures (those inhabiting water, air, and land) fit into Plato's scheme with respect to dissolubility is obscure, not having been discussed. They appear to evolve, rather than to have been generated, out of the 'Nurse of Becoming.' This difficulty will be dealt with later. It would seem, however, that these objects, including the elements out of which they are formed, in not requiring or exhibiting any sort of self-movement or presence of soul, would not have been generated either by the Demiurge or by one of his divine delegates, nor could they have been, given that they are soulless. The presence of the World Soul, moreover, does not seem to affect these individual objects, which objects for Plato remain subject to secondary causes for their movement. They remain linked to generation, nonetheless, due to the fact that they are ordered, thereby exhibiting the presence of 'nous.' These objects, therefore, are properly described by Plato as having been taken over by the Demiurge from "all that was visible" (*Ti.* 30a2-6), rather than having been generated.

18. However, such dissolution is impossible, since the will of the Demiurge is a stronger and more sovereign will than that of Ananke (41b4-6).

19. Thus, as a consequence of the existence of the eternal presence of the disordering power of Ananke or 'Anarchy' in the generated Cosmos (prem. 9) and the fact that only it would dissolve something which was good (prem. 17), but combined also with the fact that the Demiurge is the stronger and more sovereign eternal will in the Cosmos (prem. 18) and therefore would prevent this from ever happening, the Demiurge must now declare all things previously generated (the three kinds of mortal creatures yet to be fashioned [41b7]), beginning with the Heavenly Bodies because of their central role in the ordering of 'time,' to be no longer dissoluble, since this would be impossible given the above premises. Plato asserts this, moreover, in order to also affirm the certainty of his premises, which are all *a priori* or follow therefrom and could be disputed.

    The reason, therefore, that the generated Cosmos is now declared to be *in*dissoluble throughout all time is due to the fact that given the Demiurge and Ananke are both eternal, but with Ananke being the disordering cause, whilst the Demiurge is the sole ordering one and thus, is all good (and given the ordering of the Cosmos, the more powerful cause), the latter must now assert its superiority over Ananke for the *full length* of their existence, which is forever. What otherwise would follow is that whatever has been ordered could be dissolved. The indissolubility of the ordered Cosmos, however, is now a sign of the Demiurge's continued superiority over Ananke in time, just as the ordered Cosmos was a mark of its initial superiority outside of time. Plato does not argue that because there is no 'evil will,' generated things are indissoluble, but rather, because of the goodness, power, and sovereignty of the Demiurge over such an evil will is this the case, with the Demiurge himself declaring that the things he has thus far generated will never be dissolved or taste death (*Ti.* 41b3-4), with the result that they now become eternal *a parte post*.

The importance of this argument is the fact that Plato was logically required to make it and precisely because of the metaphysical, and not physical, existence of Ananke. If Ananke did not actually exist as one of two primary metaphysical causes, the Demiurge being the other, then Plato could have left the generated Cosmos (as a whole) as potentially dissoluble throughout time, since it would have had no opposing force to ever bring about its dissolution, reflecting in this manner its likeness to the Eternal Living Being. Further on (at §3.3.2) I discuss the difference between natural disorder and unnatural disorder in the *Timaeus*. Plato clearly accounts for natural disorder or the natural processes of decay and resurgence that occur in the Cosmos. These are considered good and part of the ordering process. This type of decay, therefore, involving natural death, is not what he is rejecting here. Rather, what Plato is opposing by his declaration that the Cosmos is indissoluble in time, is that the Cosmos *as a whole* is indissoluble. Without Ananke the Demiurge would have been the only metaphysical being in existence and would not have required such a declaration, since it would have been against its nature to dissolve anything that was good, which the Cosmos, in its entirety, was. It is precisely because of the eternal presence of Ananke and its 'will' to return the Cosmos to chaos that the Demiurge is required to assert its own superiority and thus claim indissolubility for the generated Cosmos throughout the whole of time.

Moreover, the above declaration of indissolubility would similarly not have been required if Ananke were merely mechanical motion in the world, since, as a part of the Cosmos,[137] it would then be subject to the same conditions as the Cosmos and would not be viewed as a threat to its continued existence. The possibility that Ananke is some sort of brute disorganizing force existing within the generated world (not without) must also be rejected. It would then have to be accounted for in its entirety as such a force, which Plato does not do. To the contrary, he goes to great lengths to disavow this by depicting Ananke's nature solely in metaphysical terms, whilst depicting the body of the world in physical terms and as being perfect and subject neither to age nor disease.[138] Although Plato suggests in the *Timaeus* that there are vestiges of disorder remaining in the Cosmos,[139] these would remain only as traces of the original chaos, since it was never completely subdued and accordingly, point only to the works of Ananke and not to Ananke itself. Plato's depiction, moreover, of the fixed stars as being divine and immortal,[140] is consistent only with ultimately equating their immortality with the concept of the indissolubility of the Cosmos throughout all time. The fixed stars are divine because they were made by the Demiurge. They are immortal because they are divine and being immortal, they are eternal. However, they are eternal only *a parte post* because, like all things generated, they, too, had a beginning. Thus, if they are to remain immortal, it must be the case that they can never be dissolved throughout all of time. They are immortal. Therefore the Cosmos in which they exist is indissoluble throughout all of time.[141] Finally, it is not sufficient to argue that the Demiurge designed the Body of the World, the Heavenly Bodies, and the World Soul to be immortal in order

137. *Ti.* 33a6-b1.

138. Ibid.

139. See *Ti.* 48a2-5; 53b5-7; 69b2-5.

140. *Ti.* 40b4-6.

141. The fact that the fixed stars by introducing measure form part of the mechanism which creates the notion of 'time' and are themselves immortal is another argument for the indissolubility of the universe throughout all of time, since as long as the stars exist (and they are eternal), so, too, will both the time which they create and the universe which houses them.

to imitate the Perfect Living Being,[142] the Form after which they are modeled. If this were the case, their immortality would point directly to that Form. However, the Perfect Living Being is not the only truly eternal object (i.e. is eternal both *a parte ante* and *a parte post*), but so are Being, Space (the 'Nurse' of Becoming), and Becoming,[143] as well as the primary causes, Ananke and the Demiurge.

1.3.14 The World Soul

In completing his picture of the formation of the body of the Heaven, Plato proceeds to place the World Soul securely within the Cosmos as the source of its "unceasing and intelligent life lasting thoughout all time":[144]

> (   ) ψυχὴν δὲ εἰς τὸ μέσον αὐτοῦ θεὶς διὰ παντός τε ἔτεινεν καὶ ἔτι ἔξωθεν τὸ σῶμα αὐτῇ περιεκάλυψεν, καὶ κύκλῳ δὴ κύκλον στρεφόμενον οὐρανὸν ἕνα μόνον ἔρημον κατέστησεν, δι' ἀρετὴν δὲ αὐτὸν αὑτῷ δυνάμενον συγγίγνεσθαι καὶ οὐδενὸς ἑτέρου προσδεόμενον, γνώριμον δὲ καὶ φίλον ἱκανῶς αὐτὸν αὑτῷ. διὰ πάντα δὴ ταῦτα εὐδαίμονα θεὸν αὐτὸν ἐγεννήσατο. (*Ti.* 34b3-9)

> And in the center He put Soul, extending it throughout the whole of it and further wrapped its body round about with Soul on the outside.[145] And so he established a single universe, a Circle revolving in a circle, solitary, but because of its excellence, able to keep its own company and requiring no other besides, sufficient to be its own acquaintance and friend. On account of all of this, that which He brought into being was a blessed God.[146]

Besides depicting here the placement and relation of the World Soul to the Cosmos (i.e. as the source of intelligent life), Plato makes a distinction between the eternal God, i.e. the Demiurge, and the Cosmos which has just been brought

---

142. See *Ti.* 37d3-4. "But inasmuch as the nature of the Living Creature was eternal, it was impossible to bestow this attribute in its entirety on what was generated" (Bury, 75; Lee, 51). From this passage it is clear that it was not thought possible to make the universe eternal simply by copying its Form. This further argues, therefore, for the position that Plato had another agenda, a *metaphysical* one (stated above) for attaching eternality to the universe beyond the cosmological position which was limited to simply bestowing a "moving image" of it.

143. *Ti.* 52d2-4.

144. Bury, 73.

145. Also, therefore, the Demiurge enclosed it with Soul.

146. When translating this passage, I have adapted the same use of capitals as suggested by Bury in his translation (65) so as to articulate the divine underlying Plato's metaphysics here.

into existence, calling the latter a blessed god because of its singular wholeness, circularity of motion, overall excellence, and self-sufficiency, stressing that it is for all of these reasons that it, as a whole, is a blessed god. Prior to this, at *Ti.* 30b6, he notes that the Cosmos has come into existence in very truth as a living creature endowed with soul and intelligence due to the providence of God. Again, a little further on at 30d1-31a1, he states that it was God's purpose to use as his model the highest and most completely perfect of intelligible things, "and so he created a single visible living being, containing within itself all living beings of the same natural order."[147] What is important to consider in each of these passages is that it is the whole which has become a God, having been imparted with overall perfection, e.g. it is round, contains the sum total of the four elements out of which everything is made, is self-sufficient, revolves in uniform circular motion on the same spot,[148] and related to the latter and perhaps most importantly, is contained by and contains and thus is completely diffused by the presence of ψυχή in the form of the World Soul (as well as human 'souls'). It is this diffusion of ψυχή or 'soul,' moreover, wherein dwells its intelligence, the measure by which Plato judges something to be superior (which in this case is the highest and the best there is of created things)[149] and through which it shares in the divine.[150]

Although Plato ascribes divinity to the Cosmos, declaring it a 'blessed god,'[151] he does not assign divinity similarly to each human being as a whole, i.e. he does not say Man or men are blessed gods, even though Man also partakes in soul and intelligence. One reason for this is that human beings are given over to be the handiwork of the celestial gods precisely so that they remain mortal,[152] thereby helping to complete the organized universe according to the Forms that Reason (nous) perceives exist in the eternal Living Being, which is its model (*Ti.* 39e6-9). Thus, the Demiurge declines to make man himself[153] precisely so that man as a whole is not divine, but remains, except for his soul (and only part of that), mortal (41c2) in keeping with the model envisioned. A second reason why it can be argued that for Plato man cannot be a god is that while man as an individual partakes of soul, his soul is not pure, described as being only to the second and third decree pure (41d4-7). A god for Plato necessarily includes the notion of perfection (34a8-b9) and not just that it partakes of soul. Thus, man remains a mortal being forming a part of the Cosmos, and whilst the latter is

147. Lee, 43.

148. This motion was deemed to be the motion most properly belonging to intelligence and reason. See *Ti.* 34a1.

149. See *Ti.* 30b1-3.

150. *Ti.* 92c4-9.

151. See *Ti.* 34a8-b9.

152. See the Demiurge's speech at *Ti.* 41a7-d3.

153. I have used the masculine pronoun here in keeping with Plato's metaphor in depicting the Demiurge as 'Father of this Universe' (*Ti.* 28c2-3).

wholly divine and wholly good given the factors noted above, only the intelligent part of man is divine (having been made by the Demiurge) and to the degree that his soul, in which nous or intelligence dwells, remains ordered, is it good.[154]

Soul, as a "divine source of unending and rational life for all time" originally belonged properly to the Cosmos as a whole (36d8-e5), extended by way of the World Soul. In a passage which soon follows, Plato depicts the mixing of the human soul as being made by the Demiurge from leftovers of what he had previously used for making the World Soul, only this time less pure (41d4-7) and which soul material is subsequently divided up into individual souls. He has already explained when describing the original mixing of the World Soul that his narrative was bound to reflect man's, and thus his own, contingent and accidental state (34c2-4), and equally so when describing the mixing of the human soul. What becomes evident from this narrative is that mankind as a whole is not a god, unlike the Cosmos, which, because it was made by the Demiurge in its entirety and not just because of the presence of the World Soul, is, as a whole, declared to be a blessed god. Pointedly, Plato's physical Cosmos resembles *as nearly as possible* a model world, a Form wherein there is no imperfection and thus no need or requirement to explain any aberrations in terms of Ananke. This is not the case, however, with regard to the generated universe or the copy itself, which was modeled after this perfect Form.[155] Here there would be a need to explain imperfection, since everything in the Cosmos that was not generated or ordered directly by the Demiurge lacked perfection. Thus, whereas the Cosmos as a whole was called divine or a god due to its provenance and perfection, its parts were not, specifically not man or any of the three kinds of living creatures, winged, water, or land. Indeed, some of the content of the universe was generated *after* the Cosmos was made, namely, the three kinds of living creatures. Moreover, other content appears to have evolved *alongside* generation, e.g. natural objects made up of the elements.

1.3.15 Soul or ψυχή

With regard specifically to soul or ψυχή, although Plato does not explicitly associate any particular αἰτία in the *Timaeus* with ψυχή, not having yet established this connection, he does state that intelligence is impossible without soul.[156] Moreover, he will describe two souls in *Laws X*[157] that bear a remarkable

154. See *Ti.* 43e8-44a5; also see *Ti.* 42b2-d2.

155. *Ti.* 31a2-b3.

156. *Ti.* 30b3. With regard to soul in the *Timaeus*, Plato is primarily concerned with the generation of the World Soul (*Ti.* 34b10ff).

157. There is no evidence to suggest that the *Laws*, specifically *Laws X*, preceded the *Timaeus*. On the contrary, the accepted view given the evidence is that Plato's *Laws* are "the latest of all Plato's writings" (see A. E. Taylor, *The Laws of Plato*, lxiii ff.), with the

similarity to, and I will later argue are an unmistakable development of, the figures of the Demiurge and Ananke from the *Timaeus.*[158] If, as Archer-Hind has done, by interpreting the second and third causes, namely, Ananke (Necessity)[159] and the accessory causes,[160] as having no genuine metaphysical status and

---

possible exception of the short dialogue *Epinomis*, meant as an appendix or supplement to the major work. Supporting this view of a later dating for the *Laws* is the fact that Book IV (711a-b) of the *Laws* contains historical details indicating that it must have been written after 360 B.C. when Plato returned from Syracuse for the last time. Moreover, it was traditionally maintained from later antiquity that the received text was never finally revised, being passed into circulation by scholars after his death (ibid., xiii). As Cornford has written, "The *Timaeus* belongs to the latest group of Plato's works: *Sophist* and *Statesman, Timaeus* and *Critias, Philebus, Laws*," with the *Laws* being "the only dialogue that is certainly later than the *Timaeus* and *Critias*" (see Cornford, *PC*, 1). I, too, assume a late dating of these works. Cornford hypothesizes that Plato, being too old (close to 80) and having too much material to finish properly his proposed trilogy of the *Timaeus, Critias*, and *Hermocrates*, leaves unfinished the *Critias* in mid-sentence and begins anew with "The *Laws*," a subject much closer to his heart, whereupon he "continues the story" (*PC*, 7–8). For a different dating of both the *Timaeus* and the *Laws* see G. E. L. Owen, "The Place of the *Timaeus* in Plato's Dialogues" in *Studies in Plato's Metaphysics*, R. E. Allen, ed., 313–38. In this work Owen argues that both of these works belong to Plato's middle dialogues. Cherniss, however, in an article from the same volume, "The Relation of the *Timaeus* to Plato's Later Dialogues" (339–78) concludes upon examination of the case presented by Owen that this view is mistaken in that evidence is lacking to prove the points which Owen wishes to make and, with respect to the *Timaeus*, its philosophical doctrine, no matter what its true chronology is, is not at variance with or in any essential point modified by those dialogues considered to be late, e.g. the *Sophist*, the *Politicus*, and the *Philibus*.

158. Mason's conclusion regarding the Demiurge in his dissertation *Reason and Necessity in Plato's Timaeus* (66–67) that the Craftsman in the *Timaeus* "represents" the reason responsible for this ultimate ordering [of the Cosmos] is wrong in as much as the craftsman does not, I suggest, *represent* anything, but is in fact the name given by Plato (within a necessarily mythic setting) to the primary ordering cause in his cosmology, of which 'reason' or 'nous' is a divine attribute, but which attribute is never spoken of by Plato or understood by him as the *only* divine attribute. Mason's further assumption that the Demiurge is a 'soul' ("though himself a soul") is given no support in the *Timaeus*, nor is his admission of the formal Aristotelian principle of separating an object (e.g. 'soul') into its genus (the Demiurge) and its subclasses (the 'World Soul' and 'human souls'). In the *Timaeus* the idea of 'soul' is strictly limited to the 'World Soul' and to 'human souls,' both of which are described as generated. Although *Laws X* offers a more developed understanding of 'soul,' so that included in this notion is now the idea of all non-physical causal powers from the primary causes down to the psychic powers of man, as I argue later (see Part II, §2.3.2.1.6) this is not a substantial change in Plato's understanding from the *Timaeus*, but is primarily a change of aspect.

159. Archer-Hind, Cornford, and Taylor all refer to Ananke as 'Necessity.'

160. See *Ti* .46d4. Only with regard to the accessory causes, but not Ananke (Necessity), does Plato state that they are not true causes, although he affirms that they are mistaken for such. He describes the accessory causes as being completely incapable of reason or intelligence, arguing that of themselves they are without soul since soul is

thereby understanding their respective roles as representing physical aspects of nature, i.e. mechanical motion or subjects of that motion, and in so doing eliminating them as causes altogether, then, but only then, are there grounds for claiming that there is only "one cause" in the sense of a genuine metaphysical cause.[161] However, I would argue that Archer-Hind has not successfully done this, since he must ignore the causal scheme Plato has set up in his cosmology to explain the genesis of the Cosmos, a genesis, moreover, which is explicitly given a backward-looking teleology, a fourth or final cause which Plato announces as its first, namely, that God was good and he wished everything as far as possible to be like himself.[162] Archer-Hind's analysis suggests an interpretation of the *Timaeus* which is highly revisionist, resulting in a strongly reductionist view. He writes:

---

invisible, yet they (fire, water, earth, and air) are all visible bodies. What Plato is arguing against here is the notion of some sort of innate rationality or 'nature' belonging to material things themselves apart from the ordering presence of 'soul' and the divine. 'Soul' for Plato is not simply order, but rather, order is the result of the action of 'soul.'

161. Borrowing Aristotelian terms for distinguishing amongst the different kinds of causes, a description of the Demiurge as being either the material cause or the formal cause of the Cosmos would not appear to be arguable since the material of the universe is merely organized by the Demiurge and the Forms or Ideas for Plato are themselves eternal, forming the pattern *after* which the form or essence of a generated object is determined, a function which does not involve the Demiurge in any demonstrable way. Although it can be maintained that the Demiurge is an efficient cause, it is clearly not sufficient as a cause on its own, coercing Ananke, at least in the myth, as a participant, albeit, as an arguably 'hostile' player. Further, one could also view the Demiurge as the final cause, if, as the Plato argues, the purpose of the Demiurge in generating the Cosmos was, in being good, he wanted to create a Cosmos which would reflect this goodness. Finally, marking the Demiurge as the first cause, however, is a tendentious move, favoring a notion of creation from nothing, rather than what Plato depicts, that is, generation of the Cosmos by the Demiurge from the pre-existing 'stuff' already present in the unformed universe, its chaotic content thereafter organized after the pattern of the eternal Ideas or Forms, which, like the Demiurge, are outwith the universe or Cosmos itself. Therefore, this last choice would have to be carefully qualified to include these co-factors in the consequent generation. The idea that the Demiurge is the Supreme Cause, αὐτὸ ἀγαθὸν, as affirmed in *The Republic*, is, I would argue, a better rendering for this metaphysical cause, as this title can easily include the idea of efficient causality without precluding another causal factor, i.e. Ananke, as well as the ideas of 'final end' and 'superiority.' Moreover, the supernality of the Demiurge, while assumed, is left essentially unchallenged by Plato, although not by Archer-Hind or Cornford. This quality of supreme excellence, however, is ably captured in the title 'Supreme Cause.'

162. See *Ti.* 29d7ff: "Let us state then the reason (αἰτίαν) for generation and for this Universe which the Framer constructed. He was good, and in which good no envy whatsoever concerning anything is possible. Being therefore free from this he wished all things to be as much as possible like himself."

> Consistently with all his previous teaching Plato here [*Ti.* 29e1] makes the αὐτὸ ἀγαθὸν the source and cause of all existence; this in the allegory is symbolized by a benevolent creator bringing order out of a preexisting chaos. Of course Plato's words are not to be interpreted with crude literalness. The cause of the existence of visible nature is the supreme law by virtue of which the one absolute intelligence differentiates itself into the plurality of material objects: that is the reason why the world of matter exists at all: then, since intelligence must needs work on a fixed plan and with the best end in view, the universe thus evolved was made as perfect as anything material can be.[163]

In designating the Demiurge as a symbol in an "allegory" Archer-Hind removes from Plato's account the key player in the metaphysical explanation given in the *Timaeus* for the Cosmos, a player, moreover, who at no time is explained away as a mere symbol in an allegory by Plato, himself. By doing this the denouement that Archer-Hind leaves behind as Plato's cosmological explanation is essentially a mechanistic account of the world. With the preternatural explanation suggested by a real and existent Demiurge removed, what is left is material in motion, governed by laws that exist in nature amongst the plurality of the material objects. However, the account for which Archer-Hind argues is more along Kantian lines where causality itself is a law of nature giving rise necessarily in time to an occurrence. Hence, he refers to "supreme law" and "one absolute intelligence," with intelligence requiring a fixed plan and evolution, rather than arguing for the Cartesian sense of mechanism where physical phenomena are explicable purely by mechanical laws. Cornford will ultimately agree with Archer-Hind, adopting, however, the starker Cartesian sense of mechanism where the Cosmos is governed entirely by mechanical laws, but not without first obscuring his account by divinizing Reason.[164]

---

163. A-H, *TP*, 91, comment re line 12.

164. "[T]here is no doubt he [the god] stands for divine Reason for ends that are good [Cornford, *PC*, 38]," and further, "We may ask how this divine Reason in the world is related to that divine Reason symbolised by the Demiurge [ibid., 39]." Although Cornford does not answer this question, and consequently holds back from equating what he calls 'divine Reason' symbolized by the Demiurge with divine Reason in the world, choosing to confine his attention to the world and the 'reason' it contains, nevertheless, by denying the Demiurge metaphysical reality, he simply transfers the problem to what he now calls 'divine Reason,' but once again chooses to ignore this, concentrating his attention on 'reason' with a small 'r.' Cornford has made a metaphysical distinction, nonetheless, and the problem remains as to the nature of this reality.

1.3.16 Plato's teleology

Whereas it is possible to see Plato's backward-looking teleology as not a teleology at all, but rather, as grounds for a mechanistic view with the future being explained in terms of the past and the past seen only as story-telling involving myth and metaphor,[165] such an interpretation, I suggest, is an unnatural and biased reading of Plato's cosmology, who stresses from the beginning of the *Timaeus* that it is a *likely* account limited necessarily by language when discussing such a topic. If Plato was dealing only with the physical world and was not genuinely trying to communicate his belief in the role that the metaphysical or divine had to play in its genesis, then he could easily have done so and would not have needed to invent such a story, a story, moreover, which, if taken only to depict the physical world, is extremely obscure. *Laws X* bears out his utter rejection of anyone positing a purely materialistic or mechanistic view of the world irrespective of how this would be communicated.[166] It can hardly be argued that he would find this more acceptable if it were reformulated as myth. Further, he makes no attempt in *Laws X* to present his position as a refutation of an earlier position, e.g. of a supposed mechanistic or materialistic view from the *Timaeus*. This strongly suggests that Plato's understanding of the metaphysical and the divine in *Laws X*, although developed and re-focused, remains substantially unchanged from the *Timaeus* and that neither the Demiurge nor Ananke are purely descriptive of the world as we know it.

It could also be argued, however, that Plato's teleology, relying heavily on religious assumptions articulated within his metaphysical description, rather than merely providing a picturesque vehicle for a mechanistic account of natural causality, was, in fact, arrived at purely out of Plato's ignorance, that is, in lieu of a more scientific approach and understanding of the world. This type of ignorance, so the argument goes, would lead naturally to mistaken identifications, as in the case of natural objects being seen as gods. Although Plato no doubt rejected many of the earlier anthropomorphisms and deifications, preferring to offer his acknowledgment of them without giving his assent to their validity,[167] he could

---

165. See Cornford, *PC*, 27 and 31–32.

166. See *Lg.* 886d2 and 887c5 ff.

167. See *The Apology* (*Collected Dialogues*, ed. by Hamilton & Cairns), 27c4-d1. Although purported to be Socrates's defense written down by Plato, it is likely that Plato was influenced by the views of his mentor and that this dialogue captures some of his own views as well. Accordingly, just as Socrates had a very strong belief in and attachment to a singular 'God' (over others), so did Plato (the 'Demiurge'). With respect to *Ap.* 26d3-4 ("he says that the sun is a stone and the moon a mass of earth"), however, although Socrates appears to deflect this position as not being his own, Plato willingly adheres more closely to it in the *Timaeus*, but with qualification (see *Ti.* 38e3-39e2). In this passage Plato describes the sun, the moon, and the planets as being made by the Demiurge, each with a physical body, bound by ties of 'soul.' For Plato's understanding of

nonetheless be seen as falling into other traps. Thus, in Plato's case it could be argued that he might have understood various natural processes as Demiurgic causal activity, when in fact they were not. Thus, a certain amount of hypothesizing or speculation would take place but wrongly, as for instance in earlier times when a bang of thunder might have been mistakenly understood as the angry voice of one of the gods. Although Plato rejects outright natural objects as being gods, his error could occur, for example, or so the argument goes, when what is really only natural ordering or causation is mistakenly seen by him as the work of a cosmic Mind or Demiurge, and so forth. I would counter, however, that Plato's theological basis in the *Timaeus* is *ipso facto* not a case of ignorance following upon and followed by mistaken identifications. It cannot be argued that the Demiurge is the mistaken personification for the principle of causation if Plato's claim is in fact inclusive of the claim his opposition is making, namely, that the Demiurge, along with Ananke, are the two primary causes and thus, the very principles of causation themselves, the former described as the 'best of causes' and the latter, 'operating without reason, producing effects casual and random.' Thus, the claim that Plato is purportedly denying is in fact accepted by him. The Demiurge and Ananke are both primary causes for Plato, and are articulated as such in the *Timaeus,* nothing more and nothing less. The main opposition to Plato's understanding, I would suggest, lies in his proposed ontology for these causes, which is rejected on two fronts, namely, first, for the position that they are both eternal and had existence before the generation of the Cosmos and second, that there is an axiology or moral assignation attached to these causes, that of perfect goodness to the Demiurge and its opposite, evil, to Ananke.

The first aspect of Plato's ontology which is rejected is that the Demiurge and Ananke exist as primary powers or forces of causation apart from all considerations of time and space connected with the generated Cosmos. In other words, what is rejected is that they do not exist simply as principles to be discerned amongst ordered objects, but as *powers* separate from those principles found to be operative within the Cosmos. The second aspect of Plato's ontology, equally important to him as the first, but which is also rejected by those claiming an error of "anthropomorphism," is Plato's assessment of perfect goodness given to the Cosmos and hence, the Demiurge, and contrarily, his assessment of evil given to Ananke as the ultimate source of disorder prior to and remaining in the Cosmos and the human soul. As stated above, however, the Demiurge and Ananke are not merely principles of causation in an already generated Cosmos,

---

how soul moves the heavenly bodies see *Lg.* 898e8-899a4 and 899b3-10. What is certain is that for Plato the elements making up the physical bodies were not themselves gods (see *Ti.* 46c7-d7). Although Plato appears to be disputing the atomists in this passage, the essential position which he is rejecting is the belief that any of these elements have power of their own, either natural or divine.

but are the powers to which these principles can be traced for their origin. This is not merely a religious assumption on Plato's part, for although his beliefs may have given him some impetus for making the inquiry, he nonetheless arrives at his conclusions and hence, understanding, logically and deductively.

Plato's reasoning can be outlined as follows, beginning with his understanding of the ontology of the primary causal powers: First, what is ultimately the cause of something cannot be in any way a part of it, otherwise it would not be its cause; second, everything within the Cosmos has a cause; and third, nothing can be or is the cause of itself. Thus, Plato argues that none of the physical causes[168] within the Cosmos can be ultimate causes, but merely participate as accessory causes, since they timeously govern (other events) and are themselves governed and as such, cannot be true or ultimate causes.[169] The remaining source of true, although, not ultimate, causation in the world after its generation Plato allots to 'soul,' which soul, while designed solely for this purpose by the Demiurge, exists prior to the generation of the Cosmos.[170] As with the primary causes in the *Timaeus*, 'souls,' too, in *Laws X*, are designated morally as being capable of good or its opposite, evil.[171] Just as with the primary causes, however, Plato's understanding of 'soul' is not a case of mistaken identity for some sort of natural phenomenon, since there is nothing equivalent to 'soul' for which it might have been substituted at the natural level. 'Soul' in effect is *sui generis* and is a preternatural kind, just as the primary causes, existing always at a metaphysical level and never as a natural kind subject to physics. Although 'soul' permeates the Cosmos through and through, the mark it leaves is rationality or intelligent ends within the universe, not to be confused with the physical motion that is so impacted and directed. Plato does not claim to understand *how* 'soul' achieves this, only that it does and was designed to do so.[172] According to Plato's ontology what is apprehensible by intelligence or reasoning does not change, and hence, exists necessarily apart from the world of change and what is perceptible by the senses.[173] The vehicle of this rational (or irrational) art, however, can and does co-exist with the world by the fact of the rationality and hence, order (or disorder) discernibly present within the Cosmos and is understood by Plato as 'soul,' whilst the source of rationality remains for him the Forms (and irrationality, lack thereof) and the means of their discernment, the human mind.

---

168. Subject to time, these would be events occurring after the generation of the Heavens.

169. *Ti.* 46c7-e6.

170. See Part II, where at §2.3.5 it is argued that 'soul' is more properly understood as having been begotten by the Demiurge rather than generated, the depiction of its generation forming part of the storytelling aspect of Plato's cosmology.

171. See *Lg.* 896e4-897d1.

172. See *Lg.* 898e8-899b10. Also see Part II, §2.3.2.1.1–5 on Plato's understanding of the 'soul' in the *Timaeus*.

173. See *Ti.* 27c1-29d3.

Secondly, with regard to Plato's moral understanding, 'order' (being the opposite of 'disorder') in no way participates in 'disorder.' Moreover, 'order,' being life-generating and supportive of the life which is generated (i.e. the generated Cosmos is 'ordered' and by this order has and remains in existence) is judged by Plato to be completely good with respect to the Cosmos. Hence, the 'goodness' that is known of and identifiable with the Demiurge lies in the fact of the Demiurge being both source[174] and cause of this 'ordering.' Thus, to the extent that the Cosmos is ordered, it is perfectly (completely) good and therefore its cause is also deemed to be completely good. Conversely for Plato, to the extent that the Cosmos remains or returns to disorder, this disorder is completely bad or evil, and so, too, is its cause.

1.3.17 A theological basis

The theological basis, therefore, which Plato gives to his cosmology is not a case of mistaken identity, since Plato's universe is neither speculative nor hypothetical, nor, as I have argued earlier, is it metaphorical, any of which would have to hold for a non-metaphysical interpretation of the Demiurge and Ananke to be tenable. Moreover, Plato's acknowledgment of the various nature gods[175] and gods within the Pantheon were those commonly accepted by Athenian society. Nonetheless, even these were open to investigation by Plato,[176] although reverently, and some, at the very least, to revision.[177] Thus, whereas Plato's belief in or at least acknowledgment of such gods is subject to the above criticism of ignorance and/or mistaken identity, the status of some of the gods was subject long before to his own criticism and for the same sort of reasons, which would have resulted in the revision, or perhaps demotion, of at least some of them as gods.[178]

However, the same is not true for either the Demiurge or Ananke, the reason appearing to lie in their ontology. Whereas the nature gods play no role in the *Timaeus* at all, the Demiurge and Ananke are the central metaphysical figures in

---

174. In Part II, §2.3.2.1.4, I articulate the relationship of the Forms to the Demiurge as their source.

175. See *Ti.* 40d6-41a2. See also 48d4-e1. It is uncertain which, if any, of the popular 'gods,' although acknowledged, Plato actually believed in. Plato states clearly in the *Timaeus* that fire, water, earth, and air are all visible bodies governed by soul (*Ti.* 46d5-7), which idea again appears in *Laws X*, where the governance of heaven and earth and their whole circuit is described as being conducted by soul (*Lg.* 897b7-c2).

176. See *Lg.* 888a1-2. See also Plato's *Apology* 18c1-3, where he portrays Socrates as inquiring into such matters.

177. See *Lg.* 886 b10-e2.

178. See Socrates's defense in Plato's *Apology* at 26c1-9; also 27c4-d1, where Plato, through Socrates, suggests that his belief in the Gods was related to his belief in supernatural activity, which principle, no doubt, would delimit the gods in which he believed.

Plato's cosmology. Moreover, and important to both sides of the argument—i.e. the claim of only metaphorical and symbolic importance for these figures versus a claim for their genuine metaphysical reality—acknowledgment of the Demiurge and Ananke arises for the first and only time in the *Timaeus*. There is no earlier precedent set, nor is there a later one, strongly suggesting that the existence of the Demiurge and Ananke are both specific to and necessary for the generation of the Cosmos. Thus, Plato is not trying to adapt old cosmogonical figures to a new cosmology. Leaving the old cosmogonies aside, he begins anew. The question which must be asked, however, is whether Plato deductively arrived at what he considered must be the case regarding the world and its beginnings, given the Cosmos as he then saw it and believed it to be, or whether, he began with a certain set of core beliefs and ran a version of the world past it, developing a cosmology in its wake.

An argument for positing that Plato began with at least some unshakeable assumptions rests on that fact that not everything comes together in his cosmology. In other words, not all aspects of his cosmology make sense when an attempt is made to bring its separate strands together into a coherent whole, and concerning which whole, in spite of obvious problems, Plato does not attempt to rectify or to adjust in order to make it work better. These assumptions are primarily religious and/or metaphysical in nature. Some of the more important ones are (1) the equivocation between Nous and the Demiurge: In spite of acknowledging Nous as playing a powerful role in the generation of the universe, the Demiurge is still figured separately as the primary cause; (2) Ananke is described as being "persuaded" by the Demiurge in generating the Cosmos, and yet it is also suggested that it is Ananke's evil 'will' which would move the Cosmos into disarray should the Demiurge not be on constant guard; (3) Being, Space (the receptacle and functionally, the 'Nurse' of Becoming and change), and the elements of Becoming are all described as existing before the Cosmos was generated and yet, whereas Space and the primary elements are entirely used up in its generation,[179] nothing is said further of what becomes of or wherein exist the Forms; (4) the 'Nurse of Becoming' is described as having a puzzling sort of intelligibility[180] and self-movement, quite distinct from the Demiurge, Ananke, Nous, or Soul,[181] and (5) similarly, whereas the elements comprising the 'Becoming' which is taking place in this receptacle are described as being able to gather enough momentum of their own accord (and also in conjunction with the movement of the receptacle) to form the basic building blocks of the Cosmos and as also possessing traces of their own nature (indicating cer-

179. There is no space or elements, which, after the generation of the Cosmos, do not form a part of it.

180. See *Ti.* 51a4-b2.

181. *Ti.* 53a2-5.

tain organization or evidence of intelligence[182] apart from and prior to any movement or intelligence originating from the primary metaphysical sources), these same elements are also described as being marked into shapes by God by means of forms and numbers.[183]

These few instances are representative of some of the inconsistencies or difficulties found within Plato's cosmology. In response to the above, I suggest that these examples are less problematic than they would otherwise be if their presence in the *Timaeus* can be seen as contingent upon Plato's belief in the quite separate metaphysical existence of both the Demiurge and Ananke, rather than as a poor attempt by Plato at hypothesis or speculation in describing a coherent set of physical principles with which to describe the Cosmos. To this end I offer the following possible explanations: First, the Demiurge as a figure could have been removed if all he represents is Reason[184] or Intelligence in the universe, as Nous is already adequately described to play this role. The notion of 'Maker,' on the other hand, is not necessarily contained within the notion of 'Nous,' being an entirely separate issue, although indeed requiring supreme intelligence as Plato has noted. The bottom line for Plato is that the Cosmos had a beginning and what has a beginning must have a cause, and this cause or maker he names variously the Demiurge, the Maker, and the Father.[185] The metaphysical assumptions which Plato makes here, therefore, are that the world has a beginning and this beginning has a cause, which cause is a powerful, eternal, good, and intelligent cause, existing outside the world it generates. While requiring supreme intelligence to carry out its task, moreover, this is too narrow an understanding for Plato's Cosmic Maker, the Demiurge, to be equated with this aspect alone.

Second, a plausible explanation for why Plato describes Ananke as having been both persuaded to a good end by the Demiurge, but also at the same time forever ready to usurp that end has already been explicated above, namely, 'persuasion' is metaphorical language for the fact that the order initiated by the Demiurge did in fact prevail and not the disorder as originally maintained by and now consequently re-sought by Ananke.[186] It should be noted, however, that

---

182. *Ti.* 53b2: ἴχνη μὲν ἔχοντα αὐτῶν ἄττα, [. . .].

183. *Ti.* 53b4-5.

184. Cornford, *PC*, 197: "It becomes more than ever difficult to resist the inference that the Demiurge is to be identified with the Reason in the World Soul."

185. *Ti.* 28c2-3. It is noteworthy that Plato writes ποιητὴν καὶ πατέρα in this passage, further qualifying the notion of a 'maker' with that of a 'father' or 'originator,' suggesting a stronger claim than merely that of a 'builder.'

186. For G. R. Morrow, "to say that the cosmos comes about by persuasion means that it results from the working of the powers inherent in the materials of which it consists, each of them bringing into being the effects natural to itself, and none of them being under any constraint by a power outside nature" ("Necessity and Persuasion in Plato's *Timaeus*" in *Studies in Plato's Metaphysics*, R. E. Allen, ed, 431). However, Plato does not state that the cosmos came about through the workings of the powers inherent in the

while metaphorical language is being used here, it is only used to depict what for Plato was impossible to show in any other way, in this case being descriptive of non-physical powers prior to the world being fully ordered, rather than what has been suggested by commentators, namely, colorful language for describing physical powers inherent in the world following its generation. The latter is not in keeping with Plato's ability to describe fully the most mundane physical matters of the world and the causes associated with them. The metaphysical assumption Plato is making here is that preceding the state of order there was disorder, which disorder had a cause and which disordering cause is variously named by Plato as the 'wandering' or 'errant cause' and 'Ananke.'

Third, whereas Being, Space, and Becoming are all described by Plato, along with the Demiurge and Ananke, as each existing before the Cosmos was generated,[187] with especially the notions of 'Being' and 'the Nurse of Becoming' (akin to the modern notion of 'space') quite easily interpreted as primitive attempts by Plato at empirical speculation re cosmic origins, this simply cannot be the case, as it must ignore Plato's given rationale for their articulation, which must be considered in the first instance.

1.3.18 Building a reasonable Cosmos

What is more likely is that Plato has in mind an argument from which he proceeds to build both the Cosmos and the metaphysics supporting it.[188] It is an

---

materials of which it consists. To the contrary, this is precisely what Plato rails against in *Laws X*, namely, the view that there are any powers whatsoever belonging to the materials which comprise the universe and is why, in both the *Timaeus* and in *Laws X*, he goes into such great detail explaining the differences between the secondary and primary causes. If, as Morrow suggests, Plato consciously borrowed the language of Democritus in choosing the term 'ananke' (424), which seems reasonable enough, Plato's intention in so doing could hardly have been to emulate Democritus at one level by excluding all "cosmic" purpose from the physical world (one step up from Democritus who excluded all purpose), while correcting him at another by including purpose at a purely physical level—the whole of the *Timaeus* bespeaks of 'ends' at a cosmic level brought about by a Demiurge outwith the universe it generated, thwarted to some extent by another cosmic power, Ananke. This is effectively, however, what Morrow's position defends in seeking to delimit Plato's understanding of the natural order to "causal sequence" (423). Thus, Plato's cosmology is effectively rewritten by Morrow with no need for myth, its causal powers and story of generation rendered meaningless, and the telling of a "likely" tale made unnecessary. A more or less exact account would have sufficed. This physicalist view, however, is precisely what Plato flatly and repeatedly denies and with everything at his disposal, be it myth, argument, or rhetoric. A more likely reason for Plato borrowing the term 'ananke' from Democritus, should this indeed have been the case, when naming the disorganizing causal power within his cosmology, would be to attach to this power the complete lack of 'nous' already associated with this well-known term.

187. *Ti.* 52d2-4.

188. *Ti.* 48e2-49a4.

argument, moreover, which proceeds in stages, requiring more players or metaphysical support as it progresses in order to be fully comprehensible and thereby arrive successfully, by Plato's standards, at its conclusion, each player having its own rationale and becoming a premise in an overall argument where the conclusion is the organized universe. So completely foreign to positivist thought is the idea that one can actually know anything prior to or about the universe by proceeding in this fashion, that there is the immediate move to relegate such writings to metaphorical language for physical or mental phenomena, but for Plato the physical reality of the generated Cosmos is a secondary reality, reflected not only in the ordering of the *Timaeus* itself, where it is presented last, but also explicitly described as such by Plato himself.[189] Moreover, after establishing his first premise, namely, that there is a primary cause, which for the sake of argument Plato names as the 'Demiurge,' it is upon consideration of his second premise, namely, that there is a perfect pattern or system of Forms[190] upon which the world and its contents have been fashioned, that his argument begins to move forward, with everything which follows becoming consequent upon this idea. The logic which moves Plato to make the assertion that there is an intelligible and unchanging model upon which the Cosmos is patterned is that the objects of opinion and irrational sensation, in other words, everything which comes to be and ceases to be, cannot possibly be the original models used for objects like themselves, as the results would not have been good since they would be based upon unstable and unreliable patterns, whereas for Plato the generated world as it stands now is completely good. The inference, therefore, which is being made is that the Cosmos which would have resulted had the Demiurge used the present world as his original model, albeit a world observed as being wholly good by Plato, would have been a universe which not only came to be and ceased to be, but one which was now full of error and imperfection, the further implication being that a natural regression would occur. Thus, the logic of Plato's argument so far, takes the following form:

A[191] (i) The world changes;

A (ii) Everything which changes must have a cause;

Premise 1: Therefore, there is a cause named Maker, Father, and Demiurge.

A (iii) A maker must have a pattern;

A (iv) A perfect pattern used by a perfect maker yields a perfect creation;

---

189. *Ti.* 27d5-29d3.
190. Also called 'ideas.'
191. A = Assumption.

A (v) The world is perfect;

Premise 2: The world must have been made upon a perfect pattern, which perfect pattern he collectively calls 'Being,' the 'Forms,' or 'Ideas,' suggesting that each object has an individual form after which it has been fashioned.

A (vi) There must be a place in which such objects are then put or located;

Premise 3: This place Plato names 'the Nurse of Becoming,' akin to the modern notion of space.

A (vii) All things located in 'Space' are perceived by the senses;

A (viii) Things perceived by the senses constantly change and have no stability;

Premise 4: All things located in Space (or the Nurse of Becoming) Plato calls 'Becoming,' further suggesting that to be truly accurate, due to the instability of such objects, it would be better to call them only by their qualities and not as 'this' or 'that' particular object.

What is noteworthy is that Being, Space, and Becoming, while described as existing before the Cosmos was generated and each logically required by Plato's argument (see above), are not metaphorically or anthropomorphically drawn in any way whatsoever and in this respect emerge quite distinct from the prior respective presentations given for both the Demiurge and Ananke, the other two metaphysical objects described by Plato as having existence before the Cosmos was generated. This suggests that strictly speaking a para-physical or supra-physical relationship to the world for these remaining three objects is perhaps a better interpretation than a metaphysical one,[192] which latter description I am

192. It is tempting to interpret 'Space' and 'Becoming' as descriptive of the early physical cosmos and hence, having simply a *physical* relationship, albeit a more basic one, to the world that was later generated. However, this is to deny Plato certain distinctions which he specifically sets out in his cosmology, as he will deny all self-movement and intelligence, apart from soul contained therein, to the generated universe, yet he clearly grants to both Space and Becoming a degree of self-movement and intelligibility, however nominal. I would like to suggest, therefore, that 'Becoming,' as it existed *per se* before the generation of the Cosmos and prior to the Demiurge organizing it into shapes by means of forms and numbers, like 'Space,' was entirely used up in the process of generation. Whereupon what remained were the organized universe or Cosmos and its objects, which objects Plato then devotes the remainder of the *Timaeus* to describe, and to describe, not as qualities, but as concrete objects.

arguing is properly attributable to the Demiurge and Ananke.[193] Strengthening this claim is that whereas Being, Space, and Becoming are requisite components of the Cosmos in Plato's scheme for its generation, the Demiurge and Ananke are not components of it at all, but are causes which exist completely separate from the object which they have just generated and which have no ongoing physical role in the Cosmos afterwards.

1.3.19 The Nurse of Becoming or Space

Returning again to the assumptions behind Plato's understanding of the generation of the Cosmos, the Forms are particularly problematic, for although they are logically required by him, their actual ontology is extremely difficult and which difficulty, moreover, Plato does not concern himself with in the *Timaeus*. However, owing to the importance of the Forms in Plato's overall metaphysics, they will be discussed later. With regard to 'the Nurse of Becoming' as laying claim "in a most puzzling way" to a certain amount of intelligibility in the process of becoming,[194] this appears at one level to be a fairly straightforward natural account of just how objects eventually come to be formed into objects out of their primary elements. It is arguable, moreover, that there is an underlying assumption here, at least of initial stages, of evolution. Further, while Plato's description of the 'Nurse of Becoming' is cautious when ascribing some intelligibility to this apparently innate, but necessary object in his metaphysical scheme, the caution is not unreasonable, given the high regard which Plato has already granted 'intelligibility,' combined with its previous assignation as a singular attribute of Demiurgic activity.[195] This particular passage, therefore, is one in

---

193. I am using the term 'metaphysical' not in the modern sense of the most basic building block(s) of the universe, which together with the less basic, form a kind of continuum of understanding and content for the universe as a whole, but rather, to denote that entity or those entities, which become apparent and distinguishable in the *Timaeus*, namely, of being(s), that are entirely separate from the physical world and in no way form a continuum with it. The suggestion of the term 'para-physical' or 'supra-physical' for describing the ontology of Space and Becoming (at least, but perhaps at most, in their early stages) is to take into account both their physical nature and their respective descriptions given to them by Plato as possessing a degree of self-movement and intelligibility. The Forms, however, appear to be uniquely placed in their ontology and will be discussed later.

194. *Ti.* 51a1-b2. See also *Ti.* 52a8-b2.

195. For C. Ritter (*The Essence of Plato's Philosophy*, 209) 'space' becomes the cause of the irrational or what constitutes 'necessity' in Plato. Thus, he writes, "It [the *Timaeus*] is in search of a general principle of explanation for these inconsistencies [which arise in the structure of the world of concrete objects] and it finds it in the resistance of matter to the purpose and the efforts of this formative rational power. Space appears as the essence of matter. Thus space becomes the cause of the irrational in the individual appearances."

which a theory of natural law, whether of the rationalist or positivist sort, might be thought to apply to Plato's writings. However, I would argue that this is extremely unlikely given the complete lack of any development on Plato's part of this idea beyond a mere suggestion of evolution and Plato's reticence in separating 'intelligibility' from its divine source and his condemnation of those who do.[196] What is more likely to be the case is that Plato, in granting prior existence to an object, in this instance the Nurse of Becoming or Space, he naturally assumes the object to carry some degree of divinity, and thus, intelligibility. This, moreover, is entirely in keeping with his belief in other gods and the apparent intelligence he grants to both Ananke[197] and the Forms,[198] and supremely so to the intelligence he grants the Demiurge. This in turn argues not for a natural law theory, but rather, for a supernatural, in the sense of preternatural, or divine law theory in which the Cosmos itself is a god (having been fashioned out of all things divine), which proposal is entirely in keeping both with the above-noted logic and pointedly, to Plato's claim that this was in fact so.[199] The evolution or emergence of the object which takes place, therefore, is entirely a consequence of divine ordering, not natural ordering, since behind all ordering is intelligence and behind all intelligence is a divine being or god.

The fact that Plato has the Demiurge further ordering the elements into various geometrical configurations, while suggesting at the same time that 'Becoming' emerged or evolved naturally out of Space due to certain motions and inclinations involving these elements—Space, along with the process occurring within it, described by Plato as the 'Nurse of Becoming'—is, once again, in keeping with Plato's inherent logic and style in the *Timaeus*, namely, of restructuring an explanation as he requires greater detail to complete his description of the Cosmos.[200] First, it should be noted that whereas the Demiurge gives greater organization to the elements of which Becoming is comprised, via geometrical designs, these elements are described by Plato as already obtaining, to some degree, their own natures and their own movement, and hence, as I have argued above, a share in divinity. Second, therefore, and consequent to this first point, Plato cannot be described as being either a strict physicalist or creationist, since in the former you remove the Demiurge and in the latter, you remove the pre-existence, self-movement, and intelligibility given by Plato to both Space and Becoming in their early stages before the generation of the Cosmos.

---

196. See *Lg.* 889b1-890a9.

197. I will later attribute only 'will' to Ananke, but not intelligence as we know it and will argue that the Forms are a reflective element of Nous and are not separate from it.

198. See *Ti.* 48e4-6.

199. *Ti.* 34b8-9.

200. *Ti.* 48e2-4.

1.3.20 Ananke and the natural world

Returning now to my criticism of the commentators, when Archer-Hind suggests that Plato did not have in mind the idea of a blind force existing in nature, referring here to ἀνάγκη, this is certainly true, but not, I would argue as Archer-Hind concludes, namely, because there is no such cause, but rather, because this force, as represented by ἀνάγκη, is neither blind (or if blind, one might ask in what sense) nor is it a force existing in nature. Rather, according to the *Timaeus*, this cause exists, along with the figure of the Demiurge, as well as Being, Space, and Becoming, *before* the generation of the Cosmos. As such, Ananke exists *before* nature and the advent of any laws within nature. As I have argued earlier,[201] the text suggests through its imagery that it is the force created by the clash between the Demiurge and Ananke, which brought about the Cosmos and not the primal causes themselves which brought it into existence.

I would like to suggest, moreover, that there is an unspoken equivocation by both Archer-Hind and Cornford on the meaning of 'intelligibility,' as well as the words 'reason' and 'law' when they are referred to as being *in* nature. These terms are not defined or properly clarified, at times being substituted one for the other with little or no explanation. Is the meaning which is to be derived or assumed when these terms are used as a reference to natural law theory—i.e. a rationality inherent in both the perceiver and the perceived—surely a proposal which Plato would abhor given his insistence on the instability of matter preempting any kind of true knowledge? Or rather, is the reference that is really being made by these terms to the causal laws themselves in the natural order, but again, a position that both the *Timaeus* (see *Ti.* 46d4-7) and *Laws X* (*Lg.* 891b8-d4) deny, and the latter vehemently so? Whereas Archer-Hind chooses to understand the Demiurge as intelligence in the Cosmos with a capital 'I' and Cornford as reason with a capital 'R,' neither explain what they actually mean by these epithets. Although such titles suggest that these figures represent the intelligibility or reason inherent in nature, with the high regard given to them represented by capitals, it seems intuitively out of place that causal laws in nature would warrant this capitalization (moreover, Cornford explicitly refrains from identifying "divine Reason" in the world with divine Reason "symbolised by the Demiurge").[202] Yet it seems equally implausible for either Archer-Hind or Cornford to understand these in terms of natural law theory where intelligibility is itself a genuine and sufficient cause, and thus, is not accounted for in the causal laws of sensible realities, where such realities themselves have their own genuine causality. If, however, what is meant by Archer-Hind and Cornford is that 'Intelligibility' and 'Reason' are somehow separate from the sensible realities and equally separate from the causal links between these realities, then what we are left with is a theory of mind as equally difficult to understand as Plato's

201. See my discussion of *Ti.* 48a1, beginning at §1.3.1 in Part I.
202. Cornford, *PC*, 39.

left with is a theory of mind as equally difficult to understand as Plato's account, but minus the possibility of a metaphysics by which to interpret it, since this much they have both denied. However, I would suggest that in fact what we are left with by this equivocation is an understanding of 'Intelligence' or 'Reason' that is as metaphysical as the supposed mythic material which each have respectively rejected, unaccounted for by Plato's physics and simply left without a context.

In *Laws X* Plato acknowledges at least two ψυχαί and in the passage where he makes this claim he also makes clear reference to a negative or evil soul, or as I have translated it, the 'Non-beneficent Soul' (what is contrary to the 'Beneficent Soul'). He never mentions a third or 'accessory' ψυχή, thereby paralleling the three causes of the *Timaeus*. This omission would provide some support towards Archer-Hind's claim that there is only one cause, being suggestive of the further inference that the accessory cause(s) as described in the *Timaeus* is not a true cause, representing as it does features of the natural order after the Cosmos has been generated and Plato clearly states as much.[203] With this much I would agree, but not with Archer-Hind's larger claim that there is only one cause equated with the one soul. There are, in Plato's words "not fewer than two, one Beneficent, the other capable of contrary effect"[204] and thus, no less than two causes. If one wants to speak of the force created between the two causes, i.e. the "clash" spoken of earlier, as the efficient cause of the Cosmos, then one can speak of one cause, but only in this manner.

Moreover, if Plato's principal concern in introducing the figures of the Demiurge and Ananke in the *Timaeus* was to show the provenance of the forces existing prior to and responsible for the generation the Cosmos, the specific purpose of the dialogue being to show the origin of the cosmic system down to the origin of man, then there would be no need to speak of them further in the dialogue once the Cosmos has been generated since their role for his present purposes would have ended and there would be very little he could say either empirically or on rational grounds about these two causes. In fact, this is exactly what he does do: they are removed from the scene. Thus, the stated purpose of the dialogue[205] is sufficient to account for their subsequent disappearance from it. However, although no longer the focus of the remaining dialogue, the respective ontologies of the Demiurge and Ananke can neither be done away with nor

203. See *Ti.* 46c7-e6.

204. *Lg.* 896e4. Taylor's concern in his footnote to this passage in his commentary on the *Laws* (289, footnote 1) is based on another consideration, namely, that modern interpreters have used this passage in *Laws X* to claim that there are two souls *of the world* [italics my own] or two World Souls, which he rejects. I would agree with Taylor in this regard as there is nothing to indicate that either of the souls described here are 'of the world' in the sense of being a feature of the world, rather, they are presented by Plato, I suggest, as universal causes of what is good or evil in the world, *Lg.* 896d5.

205. *Ti.* 27a2 ff.

revoked, since their being, whatever that may entail, is now thoroughly entrenched and irrefutably so in Plato's account, the *Timaeus* having established early on that it was through their 'striving' that the physical Cosmos was created. Thus, whatever the Demiurge and Ananke may or may not be, each had being before the Cosmos was generated and hence, are not physical or mental in the sense of being a descriptive part of the physical universe, both being prior with respect to it. To argue from the *Timaeus* that Ananke is physical force, whereas the Demiurge is Reason or Intelligence, is myth-making beyond what Plato's account can support.[206] What remains, however, is a possible metaphysical account of these pre-physical entities, which the *Timaeus* does not go into, but which will surface in Plato's discussion of "at least two souls" in *Laws X* and which I will detail in Part II.

Continuing with his explanation of ἀνάγκη at *Ti.* 46e Archer-Hind writes:[207]

> What does he mean by this? It is idle to treat the physical forces of nature as causes, since in themselves they have no intelligence or purpose. They are indeed designed and set in motion by Intelligence for the best ends; but the conditions of their action may be such that sometimes their immediate results are not good, and they have no power in themselves to avoid such results; they must operate inevitably according to the law of their natures.

There are at least four issues here that must be addressed. First, the *Timaeus* itself argues specifically against the position that Ananke is a physical force of nature in the sense of 'in nature.' This view is also denied strenuously by the Plato in *Laws X.* Second, one must ask what Archer-Hind meant by "the conditions of their action may be such that sometimes their immediate results are not good." Third, Archer-Hind, unlike Cornford, equivocates on the meaning of νοῦς and while translating νοῦς as 'reason,' explains it in terms of 'intelligence.'[208] If he means 'Intelligence' in some special sense, having referred to it with a capital 'I,' then he has failed to provide a metaphysical or ontological basis for its supposed increased status while diminishing that of ἀνάγκη. And finally, Archer-Hind's understanding of νοῦς to mean 'Intelligence,' while translating it as 'reason,' is further complicated by his description of νοῦς shortly thereafter as a power which "matter *qua* matter" possesses or 'has.' To wit, in discussing νοῦς at *Ti.* 47e-48a, he states:[209]

---

206. Taylor's understanding of myth is better suited to Plato's account than Cornford's, which latter demands a complete reinterpretation of the text. See Taylor, *PT,* 59; see also Cornford, *PC,* 174–76.

207. A-H, *TP*, 162.

208. A-H, *TP*, 167.

209. Ibid., 166.

> For this reason we must not suppose that there is in matter as such any resisting power which thwarts the efforts of νοῦς: this is an absolute misconception. Matter *qua* matter, being soul-less, is entirely without any sort of power of its own: whatever power it has is of ψυχή.

Taking these issues one at a time, we will look at the first, namely, the claim that ἀνάγκη is physical force in nature, which force does not have intelligence or purpose, and therefore, as such, cannot be a cause. The problem with this argument is that Plato explicitly asserts the conclusion which Archer-Hind denies. At *Ti.* 48a6-7 (τὸ τῆς πλανωμένης εἶδος αἰτίας) Plato, in referring to ἀνάγκη, states that it is a cause, albeit, an errant or indeterminate cause. Moreover, as a cause, it is not described in this passage as being without reason or intelligence, rather, it is described as being "persuaded" by intelligence (*Ti.* 48a5), which obscure passage is indicative at the very least of some sort of relationship of ἀνάγκη with respect to the reason or intelligence in order for it to be persuaded (i.e. subdued) at all, which I argue is the possession of a 'will' (or its like) and the ultimate overpowering of its will by the stronger will of the Demiurge. None of the accessory causes are described as being so "persuaded," since, having no power of their own, they lack both the need and means to be persuaded, putting ἀνάγκη in a class of its own as a cause. In other words, Ananke, while retaining the character of a force, is not, I suggest, a blind one, since ἀνάγκη, along with νοῦς, in being depicted as bringing about the best end for the most part of the things which came into existence (which I argue for Ananke was only by way of submission of its will),[210] are at the same time attributed by implication, at least to some degree, of having similar faculties to that of 'intentionality' and 'will,' neither of which can exist without mind or soul, wherein reason and intelligence lie.[211] Plato, on the other hand, does speak of blind forces, i.e. "those that operate without reason and produce effects which are casual and random," but the reference here is not to ἀνάγκη—this is a misreading of the text—but rather, to the accessory causes and specifically in this case, to the visible bodies fire, water, earth, and air, which he has adamantly just declared "are completely incapable of reason or intelligence."[212]

Nonetheless, ἀνάγκη clearly remains a problem for Plato who is unwilling to ascribe to it ψυχή and hence, his aforesaid seemingly obscure reference to 'persuasion by intelligence.'[213] At the same time, however, he is equally unwilling to let νοῦς be the sole cause of the Cosmos and hence, its partnership with

---

210. *Ti.* 48a3.

211. The idea of a 'disordering causal force' being closely associated with the movement of 'soul' is a problem, which Plato will deal with more specifically in *Laws X.*

212. *Ti.* 46c7-e6. For a translation see Lee, 64–65.

213. This reference is not obscure if Ananke (Necessity) is understood as one of the primary causes.

ἀνάγκη. This again points to the suggestion that ἀνάγκη, as the source of brute force in the Universe, nonetheless shares at least some of the characteristics of the divine in its own right. Conversely, it can be asked whether νοῦς is the source of intelligent force, i.e. that which moves by persuasion. Plato, however, is consistent in his description of ψυχή, or of that which has ψυχή, as being the only source of motion. Still, what is actually meant by this needs clarifying. Are the ideas of 'force' and 'motion' separate?[214] Moreover, is it one thing to be the force or the motion (if indeed these are separate ideas) and quite another to be the source of that force or motion? These questions need to be asked and if possible, sorted out. Thus, specifically, if ἀνάγκη is the cause, or one of the causes, of physical force, is it or can it be the force itself? How does this relate to motion if ψυχή is the only source of motion? Importantly, where do the accessory causes fit in to Plato's scheme and how? With respect to the latter question, the only so-called causes which Plato describes as being specifically without intelligence, and therefore not genuine causes at all, are the accessory causes. It is in relationship to the accessory causes, moreover, and their description, beginning at *Ti.*46c7, where we get a clear description of what for Plato constitutes a genuine cause. See the next chapter for an in-depth discussion of the accessory causes and explication of this passage.

Finally, reverting back to our discussion of ἀνάγκη as a force, at no place in the *Timaeus* is ἀνάγκη ever described as being a force within nature. Rather, it is specifically rendered as a force, along with νοῦς, which *preceded* the generation of the physical Cosmos. To interpret the *Timaeus* to accord with one's own understanding is one thing, which both Cornford and Archer-Hind have clearly done, reflecting a fairly modern scientific and non-religious account and bias, but to claim that this is what Plato meant or held to be true is quite another matter and has to be rejected. To be sure, the *Timaeus*'s prime concern is about the generated world, but implicit in this inquiry is how it came to be—the generation itself—which, although difficult to depict, Plato sets about to show. In fact, it is so difficult that Plato has to explain that it is only a likely account and not an exact one. There is little, if anything, in the text to support an understanding that omits this inherent difficulty. On the contrary, there is much to support a metaphysical account of Ananke. Modern metaphysics often restricts itself to talking about facts of the world, with the language of metaphysics having been appropriated to this end. Plato, however, did not work within these restrictions or within any other metaphysical scheme, but established his own, based upon and imbued with a deep, religious sense and an explicit belief in the divine and

214. Cornford makes no such distinction, agreeing with Grote when referring to Necessity (Ananke): "It is Force, Movement, or Change, with the negative attribute of not being regular," Cornford, *PC*, 172. The capitals, however, are noteworthy, but which Cornford neither explains nor dismisses. However, by being highlighted in this manner these words attain the status of a proper name and require an explanation.

the gods. Plato's metaphysics speaks to a universe not yet ordered and to a world not yet generated, and therefore, preempts in its consideration all facts concerning the world and thus, necessarily, cannot be about the world. Nor is Plato's metaphysics simply a continuation of what has been handed down to him, having rejected the causal claims of the atomists on the one hand and what he considered to be false cosmogonies on the other. Plato's metaphysics, particularly when dealing with the ontology of the Demiurge and Ananke, as well as the generation of the world which is dependent upon these, must be viewed on its own terms and then be rejected or accepted but not transformed into what is more metaphysically acceptable, which latter I argue both Archer-Hind and Cornford have done.[215] This is history of philosophy misappropriated.

---

215. Luc Brisson, like Cornford, understands 'necessity' as existing wholly at the corporal level. It is a term, he explains, employed by Plato to capture what resisted reason or was acted upon by the action of reason in the world and was never an autonomous reality. Understood in this way, it is a relative term comprised of three 'stages,' where only the last one, the assistant or secondary cause, is real or as Brisson puts it, represented for Plato 'pure necessity.' The first and second stages are imaginary early depictions of 'necessity' in a world drawn as generated for the sake of understanding, representing both the imaginary resultant incidences of movement created at the level of corporal nature by a world soul and later, the semi-tamed movement which formed traces of the four elements taking shape in 'space,' "la cause adjuvante [additive]." These first two stages, he argues, were introduced by Plato to gain a better understanding of the relationship between 'necessity' and 'reason.' The third stage, the only "real" one according to Brisson's understanding, takes place in the fully formed world and is where 'necessity' can no longer be clearly separated from reason, behaving as a secondary cause in a chain of reactions under the guidance of reason, or in Plato's imaginary terms, under the action of "l'âme du monde, dominée par le cercle du même" upon the mass of corporeal elements. See Luc Brisson, *Le Même et l'Autre dans la Structure Ontologique du Timée de Platon* (Paris: Éditions Klincksieck, 1974), 524–25. This is also Lutoslawski's understanding of "blind Necessity" (*The Origin and Growth of Plato's Logic*, 485). The essential difference between 'necessity' and 'reason' at this final stage appears to lie in the absence of reason and design from physical causation when considered alone and apart from the aims of reason. The problem with Brisson's understanding of 'reason' and 'necessity' in the *Timaeus*, as with Cornford's (and as regards 'necessity' only, also Lutoslawski's), is that he offers no explanation for the presence of these causes in the world, other than that they are. His explanation becomes a description of these causes as they would already be understood by the modern mind. I would challenge, however, that Plato attempted, at the very least, to offer insight into the provenance of these causes, both as to their ontology and their metaphysics, and with respect to the world, their power and their influence, and to mitigate this attempt is to offer less than what Plato intended.

Similar to Brisson, Mason's understanding of 'necessity' within Plato's overall theory is that it "refers to the nature/powers of material things, and that when Reason 'persuades' necessity this means that it takes control of, and exploits, the nature of material things. When governed by reason these things can be a factor in producing orderly effects" (A. S. Mason, e-mail transmission, 22 November 2002, University of Edinburgh, where he is currently teaching). With respect to the *meaning* of 'necessity' itself, Mason has enlarged his position from that of his original doctoral thesis (*Reason and Necessity*

1.3.21 Plato's metaphysics

The problem, however, which one soon faces in explicating Plato's metaphysics on its own terms is that it is not entirely rational. There is, as earlier discussed, a belief system underpinning it.[216] Is this, therefore, where philosophy must leave off and accept this basis or belief system as Plato's *a priori* beginning or does one try to give a rational account of the belief system itself, or failing this, is the belief system simply removed and either retold as myth (Cornford) or eradicated (Archer-Hind) more or less unceremoniously in order to better accommodate modern beliefs? As I have argued above, none of these approaches do justice to Plato's account of the Cosmos in the *Timaeus*. Plato begins with a core set of assumptions that are either religious, metaphysical, or arguably, supra-physical in nature and therefore are specifically *not* physical. Plato's assumptions in turn proceed to become premises in a wholly rational account or argument designed to explain the universe, which rationality is entirely dependent upon key metaphysical players around which these assumptions are built, namely, the Demiurge, Ananke (Necessity), the Forms (Being), the Copies (Becoming), and their receptacle or Space (the Nurse of Becoming).[217]

For Plato the physical world has no rationality of its own. Its rational movement or the design towards ends found within the Cosmos relies solely on a generated World Soul. Thus, it is this additional divine source of rational life, that of the World Soul, which Plato articulates and understands to be the source of the generated world's ordered movement,[218] and not physical or mechanical force. To speak, therefore, of religious or metaphysical assumptions *preempting* a fully rational account becomes nonsensical, since the key metaphysical players themselves in these assumptions are the sole participants in 'nous,' that is, in 'rationality' or some aspect of it, i.e. either 'rationality' or 'will,' or both. For example, whereas Ananke has a 'will to disorder,' the Demiurge has a 'will to order,' participating not only in the 'willing' aspect of nous, but also in its rational compo-

---

in Plato's Timaeus, 1990). Thus, he agrees now not only with the causal interpretation of Morrow, i.e. 'necessity' is called what it is because "one event follows necessarily upon another," but additionally with the factual interpretation of Crombie, i.e. 'necessity' is called what it is because "it is a 'given' or 'brute fact,' an inevitable feature of the world which reason could not eliminate." The main problem with Mason's understanding of 'necessity' is that any position which understands material things to contain powers of their own is directly opposed by Plato in the *Timaeus* and vehemently so in *Laws X* (see *Ti.* 46c7-e6 and *Lg.* 889b1-e1) and in fact, this is the *very* position which Plato opposes.

216. See §1.3.17 of Part I, where beginning at "the question which must be asked," I specifically address the issue of Plato's core beliefs and religious assumptions.

217. One could arguably include 'the gods' or heavenly bodies, but these are generated and like the World Soul and 'soul' generally form part of the generated Cosmos, which is not the source of its own rationality.

218. *Ti.* 36d8-37c5.

nent. The World Soul, on the other hand, participates solely in the rational component of nous, itself subject to the will of the Demiurge. Although there is nothing particularly rational in positing a 'Demiurge' or any other of the metaphysical players, it is not unreasonable to do this and in so doing Plato is able to account for the various factors in the Cosmos. He strongly asserts that the Cosmos was generated from a chaotic universe, but due to the immensity of his task in describing such generation, storytelling or the drawing up of 'a likely account' was found necessary.

To speak, therefore, of Plato's cosmology as preventing a fully rational account or preceding such an account, or in seeking to find rationality within Plato's belief system to then reinterpret his story as a mythic account of entirely physical phenomena, or yet again, to remove it as unessential, must all be summarily rejected.[219] To the contrary, Plato's pre-cosmic assumptions are where his philosophy begins and from whence the generated Cosmos proceeds.

---

219. Gregory Vlastos correctly argues that the *Timaeus* is unique among Plato's myths and that it is a mistake to put it on a level with those contained in the *Gorgias*, *Phaedo*, *Republic X*, the *Phaedrus*, and the *Politicus*. He writes: "The *Timaeus* uses none of the devices by which all of these disavow the scientific seriousness of major features of their account. . . . The sober, systematic, prosaic tone of his discourse [meaning that of Timaeus, the Astronomer, who in Socrates's estimation has reached the highest summit of all philosophy (*Ti.* 20a)] contrasts sharply with Critias' earlier reminiscences. This all but irrelevant introduction sets the fanciful myth over against the scientific myth. . . . A *mythos* is a tale. Not all tales are fictions." See G. Vlastos, "The Disorderly Motion in the *Timaeus*" in *Studies in Plato's Metaphysics*, R. E. Allen, ed., 380–82. I would go further than Vlastos, however, and state that for Plato those tales which are not fictions include for him not only a telling of the physical world, but also the metaphysical and to conclude otherwise risks delimiting and distorting that vision from what Plato intended.

Chapter 2

2 *TI.* 46C7-E6

2.1 Original Greek

Ταῦτ' οὖν πάντα ἔστιν τῶν συναιτίων οἷς θεὸς ὑπηρετοῦσιν χρῆται τὴν τοῦ ἀρίστου κατὰ τὸ δυνατὸν ἰδέαν ἀποτελῶν· δοξάζεται δὲ ὑπὸ τῶν πλείστων οὐ συναίτια ἀλλὰ αἴτια εἶναι τῶν πάντων, ψύχοντα καὶ θερμαίνοντα πηγνύντα τε καὶ διαχέοντα καὶ ὅσα τοιαῦτα ἀπεργαζόμενα. λόγον δὲ οὐδένα οὐδὲ νοῦν εἰς οὐδὲν δυνατὰ ἔχειν ἐστίν. τῶν γὰρ ὄντων ᾧ νοῦν μόνῳ κτᾶσθαι προσήκει, λεκτέον ψυχήν—τοῦτο δὲ ἀόρατον, πῦρ δὲ καὶ ὕδωρ καὶ γῆ καὶ ἀὴρ σώματα πάντα ὁρατὰ γέγονεν—τὸν δὲ νοῦ καὶ ἐπιστήμης ἐραστὴν **ἀνάγκη** τὰς τῆς ἔμφρονος φύσεως αἰτίας πρώτας μεταδιώκειν, ὅσαι δὲ ὑπ' ἄλλων μὲν κινουμένων, ἕτερα δὲ ἐξ **ἀνάγκης** κινούντων γίγνονται, δευτέρας. ποιητέον δὴ κατὰ ταῦτα καὶ ἡμῖν· λεκτέα μὲν ἀμφότερα τὰ τῶν αἰτιῶν γένη, χωρὶς δὲ ὅσαι μετὰ νοῦ καλῶν καὶ ἀγαθῶν δημιουργοὶ καὶ ὅσαι μονωθεῖσαι φρονήσεως τὸ τυχὸν ἄτακτον ἑκάστοτε ἐξεργάζονται.

2.2 Translation

Accordingly, these are all assistant causes which serve God, who uses them in carrying out, as far as it is possible, the idea of the best. But the majority believes them not to be assistant causes, but rather, to be causes of all things, producing[220] by cooling and heating, solidifying and dissolving, and all such processes. Yet none are even capable of having either an opinion or a thought. For, of existing things, that alone which has acquired Mind, one must claim to be soul—but this is invisible, whereas fire and water, earth and air, all come into being as visible bodies—and therefore, the lover of thought and knowledge has to follow closely as primary the causes[221] of a rational nature,[222] whereas for those which are moved by others or on the other hand, which in turn move others, as secondary. We also then must treat these accordingly and discuss both kinds of causes,[223] separating those which, with Mind, are makers of the beautiful and good, from those which, being devoid of thought, bring about random disorder each time they produce.

2.3 Commentary and Notes

2.3.1 Discussion of συναίτιον / the 'accessory cause'

The etymology of τῶν συναιτίων at *Ti.* 46c7 when taken at face value is clear, συν + αἰτίων. However, to clarify its meaning, Plato presents an argument for why αἴτια is prefaced by συν. Only Mind is capable of thought and opinion, which latter are requisite in bringing about a best end, and hence, is properly called a 'cause.' The only generated object that contains Mind, however, is soul and soul is invisible. The συναίτια, on the other hand, take into account physical phenomena often mistakenly assumed to be causes themselves. Plato's argument is that like the soul, the accessory causes are part of the ordered universe, but unlike soul, they are visible, e.g. fire, water, earth, and air, along with their respective processes (cooling, heating, solidifying, and dissolving, etc.). Thus, by the very fact that they are visible, the opposite cannot also be

220. I.e. 'producing effects.' *LSL* refers to another passage in the *Timaeus*, namely, 29a (cited as "28e al." in the *LSL*), where the meaning of ἀπεργάομαι is translated as 'to cause' or 'produce.' See also Bury's translation of the above.

221. These primary causes Plato is about to separate into two kinds.

222. In other words, the invisible ones are to be understood as primary and the visible, as secondary.

223. Plato is referring here to the primary causes, which he immediately divides up into two kinds, as he does not recognize the accessory causes (those without soul) as being true causes, let alone constituting a 'kind,' but rather, they are understood as affects or effects, always working in tandem with the Demiurge who produces the beautiful and the good (see *Ti.* 46c7-9 above); see also my discussion which follows.

true, namely, that the accessory causes are identifiable with invisible soul containing Mind. If, therefore, they are to be called causes at all, it must be by virtue of the fact that they are composite, in other words, are in some way[224] joined with invisible soul containing Mind, which alone is capable of bringing about an end. Thus, following Plato's line of thought, there must be something with which these αἴτια are joined—hence, the compound συναίτια—and which is also both invisible and capable of thought. In other words, the αἴτια or cause must be joined with soul (containing Mind) in order for the things subject to these accessory causes can be shaped in the "best possible way" as a consequence of this contact. This rules out grounds for understanding συναίτιον simply as a 'causal factor,' which understanding omits or at least buries Plato's strong denial of active powers of agency for any of the accessory causes. I would propose, therefore, that συναίτιον as a compound entity is best understood in either of the two following ways, first, as a cause, but only by virtue of being joined to invisible soul when it is so acting and as a consequence is described by Plato as a secondary cause or alternately, not as a cause at all, but when joined with or moved by a true or primary cause (συν-αἴτιον) is able to assist in bringing about a desired end. This passage supports both readings, which in turn converge to argue that only soul is capable of being a true cause within the generated world.

This argument aside, what requires to be established is exactly what Plato wants to include under the heading of the συναίτια. In the passage preceding the one under discussion, at *Ti.* 45b2-46c6, he has just discussed eyesight and how it works, but he does not wish to claim that either the eyes or eyesight are accessory causes. Rather, he goes on to explain that the 'accessory cause' gives the eye its power or eyesight. Beginning at *Ti.* 46c7 he writes Ταῦτ' οὖν πάντα ἔστιν τῶν συναιτίων, clarifying shortly thereafter that the accessory causes are not causes of a rational nature, but are ones which operate through bodies whose motion is derived from others or is passed on to others. Consequently, the accessory cause described in this passage can be understood in one of three ways, first, as including both the body which acts or is acted upon and the processes which help to carry out the act, or alternatively, simply the body itself, excluding anything external to it, or contrarily, not the body at all, seen only as part of the *modus operandi* for the cause so acting, but rather, the process or processes themselves which as a whole carry out the act. While a more abstract notion, I would suggest that the latter meaning approximates most closely the claim being made here. At *Ti.* 46e6-7 Plato refers to the auxiliary causes as contributing to the power or ability that the eyes possess:

224. The 'how' Plato chooses to leave within the context of myth telling, and hence, unknown or at least not fully unexplained. See *Ti.* 34b10-37c5.

> Τὰ μὲν οὖν τῶν ὀμμάτων συμμεταίτια πρὸς τὸ σχεῖν τὴν δύναμιν ἣν νῦν εἴληχεν εἰρήσθω·

Earlier at *Ti.* 45b2 when Plato begins to describe how sight takes place in the eyes, the eyes are depicted as the organs through which the many processes come together to produce sight. In so describing the eyes and eyesight, it soon becomes apparent that what Plato has in mind as comprising the accessory cause is not, for instance, the eyes themselves or any object made up of the basic elements of fire, water, earth, or air, which latter merely serve to qualify or make up the bodies through which the various accessory causes must act,[225] but rather, under the umbrella of 'accessory cause' reference is being made here to the many interrelated processes or movements which must take place in order for visible objects to have their particular power or δύναμιν (46e7), e.g. the eyes, their sight, the ears, their hearing, etc. The distinction which I am making, i.e. of process separate from the body, although not specifically drawn out by Plato, is almost certainly being made here given his insistence that the elements themselves are bereft of all causal powers. The text remains clear in this regard, presenting nothing problematic or in opposition to this claim. By the process of elimination, moreover, one can also arrive at this conclusion, as there is nothing left, given Plato's account, for an accessory cause to be. Thus, 'cause,' whether of a primary (e.g. generation of the Universe) or secondary (e.g. cooling) nature, is always in relation to movement. Finally, I would propose that Plato's main purpose in explaining the accessory cause in this passage is to clarify its intermediary, i.e. secondary, and not primary nature. What is also made clear from this passage, and most importantly, is that Ananke is not an accessory cause, nor can it possibly be one.

2.3.2 Ananke and the accessory cause

This last point is a crucial one, but requires some clarification, as it can easily become obscured both in the English translation and in the Greek, obfuscated not only by the brevity of the passage in question and by what precedes and follows, but also by the difficulty of the passage itself. At *Ti.* 46e2 Plato writes:

> ποιητέον δὴ κατὰ ταῦτα καὶ ἡμῖν· λεκτέα μὲν ἀμφότερα τὰ τῶν αἰτιῶν γένη, χωρὶς δὲ ὅσαι μετὰ νοῦ καλῶν καὶ ἀγαθῶν δημιουργοὶ καὶ ὅσαι μονωθεῖσαι φρονήσεως τὸ τυχὸν ἄτακτον ἑκάστοτε ἐξεργάζονται.

First, with respect to ταῦτα, this word may simply form part of a variation of the single adverbial expression 'in this way' (referring back to the actions of a

225. *Ti.* 46d6-7.

'lover of thought and knowledge' [46d7-8]), κατὰ ταῦτα, usually written in the singular, i.e. κατὰ τουτό. However, should this not be the case and Plato has chosen the plural for a specific reason and with a specific referent in mind, then the meaning of this sentence becomes equivocal, both with regard to its principal referent and its content, and thus becomes problematic. If, on the other hand, the adverbial expression stands, we are still left with a problem of meaning with regard to the passage that immediately follows. Both Bury and Taylor treat κατὰ ταῦτα as the adverbial expression 'in this way.' Cornford similarly translates this passage as "on this principle." Archer-Hind, on the other hand, translates ταῦτα as a demonstrative singular pronoun 'this then," introducing what follows by a colon, with Jowett following likewise.[226] In the prose that then follows, moreover, Plato does not help matters by leaving within the text a certain duplicity. When introducing *Ti.* 46c7, he begins with Ταῦτ', which referent, while clearly the συναίτια, is nonetheless left somewhat ambiguous. It can be understood as either referring back to the several processes just expounded upon in the prior passage describing how sight takes place in the eye or alternatively, to the accessory causes soon to be listed. These latter are specified within a more general discussion which follows on such causes, i.e. "cooling and heating, solidifying and dissolving, and producing all such effects," and wherein Plato defines very precisely what he means by these secondary or assistant causes, comparing them with the primary causes. It is also possible, however, that Plato meant the reference to apply both ways, the first being specific and exemplar of what is then generally discussed. As the above-mentioned translations by the various authors of this text bear out, there is no need to link the ταῦτα which next follows at line *Ti.* 46e3 with what has just preceded; moreover, it would be awkward to do so, as both the accessory or secondary causes and the primary causes, each in the plural, have now already both been discussed in the preceding passage at *Ti.* 46c7-e2. What I believe is important to consider, therefore, is that if this is the case, namely, that the ταῦτα at *Ti.* 46e3 introduces a new section or idea, then the two causes referred to in the line which now follows (*Ti.* 46e3), described here as 'the two kinds of causes,' are precisely that and do not refer back in any way to the accessory causes, which for Plato were not true causes. I would further argue, moreover, that this is not just a possibility, but is in fact what Plato intended, where at *Ti.* 47e he will soon introduce these causes as νοῦς and ἀνάγκη and attempt to explain their role in the generation of the world. Thus, the inherent sectioning of the text by Plato also supports this reading.

The reasons for concluding that neither of the two causes now mentioned refers back to the accessory causes are as follows. Firstly, the accessory causes are never described as μετὰ νοῦ, whereas the first mentioned cause at *Ti.* 46e4 is. This is only descriptive of 'soul': τῶν γὰρ ὄντων ᾧ νοῦν μόνῳ κτᾶσθαι

226. Bury, 107; Lee, 64; Taylor, *PT*, 293–94; Cornford, *PC*, 157; A-H, *TP*, 163; Jowett, *The Dialogues of Plato* (hereafter abbreviated *DP*), 733.

προσήκει, λεκτέον ψυχήν. Moreover, Plato specifically states that the accessory causes are *incapable* of possessing reason and thought *for any purpose*. This is strongly stated and unequivocally so:

λόγον δὲ οὐδένα οὐδὲ νοῦν εἰς οὐδὲν δυνατὰ ἔχειν ἐστίν.

Secondly, and what I believe to be the strongest argument in rejecting either cause at *Ti.* 46e3 as referring to the accessory causes, is that the accessory causes produce nothing of themselves, yet Plato refers at *Ti.* 46e4-6 to causes which produce, and in the second instance, to causes which also produce, but randomly and without order, but nonetheless produce. The verb used in this second instance is from the root ἐξεργάζομαι and can mean both to bring to completion and to destroy, and while a fitting verb for a disordering primal cause, is not appropriate for a secondary or assistant cause, which, being moved or in turn, moving others, would be quite incapable by itself of either bringing something to completion or destroying it, except by proximity to that event. The assistant[227] causes are always located, if we take Plato's description as evidence for their manner of existence, as part of the process of something produced and which latter are explicitly declared not to be random, e.g. eyesight, hearing, speech, music, all of which are cited as examples of those things which are produced with the help of the accessory causes. Moreover, and importantly, these are esteemed to be *extremely good*, 'sight' described as having been invented and given by God, θεὸν ἡμῖν ἀνευρεῖν δωρήσασθαί τε ὄψιν (47b6), to bring about the greatest benefit for man, φιλόσοφος, and with high regard also granted to the other accessory or assistant causes and for similar reasons. Thus, when at *Ti.* 46e3 Plato refers to the ὅσαι μονωθεῖσαι φρονήσεως as being devoid of thought and tending to random disorder each time they produce, he cannot be referring to the auxiliary causes, which of themselves produce nothing. Moreover, although he has affirmed that the auxiliary causes are incapable of any plan or intelligence for any purpose (*Ti.* 46d4) and as not possessing ψυχή, they are nonetheless always, by their very provenance, attached to and contribute to the power of objects of generation (*Ti.* 46e6). By definition, therefore, they are always and remain 'assistant' causes, helping to bring about the innumerable ends of which the Cosmos is comprised. Descriptive references regarding the accessory causes in *Laws X*, moreover, invariably stress their proper function and thus, their secondary but necessary roles in bringing about various good ends. This argues strongly against the passage at *Ti.* 46e3 as in any way referring to these causes, leaving Ananke, understood as a primary disordering cause, as the sole possible referent.

---

227. The terms 'assistant,' 'auxiliary,' and 'secondary' cause all refer to the accessory cause.

A further point worth considering is that to be described as being "destitute of reason" (from μονόω = to forsake or leave alone) is not the same thing as being described as *incapable* of reason (οὐδεν δυνατὰ). These are separate issues. The former is a much weaker notion and can simply imply a derogatory judgment towards the object in question without actually intending to qualify it in a more literal way or to its full extent.[228] In support of this reading is the fact that in *Laws X* Plato will treat the Non-beneficent Soul, the counterpart to Ananke in that work, similarly.[229] It should be noted, moreover, that the idea of being 'destitute of reason,' rather than being 'incapable' of it, is in keeping with passages elsewhere in the *Timaeus* where Plato appears to attribute to Ananke the possession of a 'will,' i.e. the ability to be persuaded by Mind (implying an independent 'will'), as well as having a weaker will than that of the Demiurge (in order to be persuaded).[230] When Plato later develops and makes explicit an argument in *Laws X* in support of the idea of a negative power, arguing that while all actions have soul in them and contain much virtue, likewise, some actions contain much vice, and since soul is the cause of good and evil, there must be at least two souls,[231] this closely parallels Plato's earlier distinction made above in the *Timaeus* of two kinds of primary causes, those which are causes of things fair and good and those which are causes of things which are not. This rules out categorically, I suggest, any direct identification whatsoever of these causes with the accessory causes, which latter, according to Plato, have absolutely no power independent of other causal power(s), thus, cause nothing of themselves, always are associated with and contribute to bringing about an end (which end implies 'nous,' organization and hence, 'goodness') and contrary to any notion connected remotely to that of disorder or 'evil,' can include amongst their numbers "the greatest benefit for mankind."

2.3.3 Ananke and "persuasion"

Furthermore, Plato's odd ascription to Ananke, namely, after having already been set out as a cause completely contrary to good order and reason, of then being capable of something which is normally thought of as requiring reason-

228. I argue that Ananke is an independent force for disorder, possessing a 'will' (i.e. a will for disorder) and hence, participates in none of the organizational powers of Nous as understood to be associated with the Demiurge. I do not argue with regard to the Demiurge or Ananke, however, that the possession of 'nous' (meaning specifically the noetic powers of 'reason' which bring about order) or an abuse or lack of it, comprises the total identity of either of these powers, but rather, only *that* part to which man's intellect is privy.

229. See England, *Laws X*, 904a6ff; Taylor, 297–99. See Part II for an analysis of the relevant passages of *Laws X* and their impact on and relation to ἀνάγκη.

230. See *Ti.* 47e3ff and 41a7-b6.

231. See *Laws X*, 896d5-897d1, Taylor, 288–90; see also *Lg. X*, 904a6ff; Taylor, 297.

ableness, i.e. "persuasion," does not present as a singular instance in the *Timaeus*. At *Ti.* 51a7 Plato describes the Receptacle of Becoming:

> ἀλλ' ἀνόρατον εἶδός τι καὶ ἄμορφον, πανδεχές, μεταλαμβάνον δὲ ἀπορώτατά πῃ τοῦ νοητοῦ καὶ δυσαλωτότατον αὐτὸ λέγοντες οὐ ψευσόμεθα.

It is spoken of as participating in, "in a most puzzling way," 'intelligibility' and also as being a concept that is 'hard to grasp.' Here there is no pejorative sense or hesitation, but nonetheless there is an admission of great difficulty in understanding it, but beyond this admission, Plato states nothing more. Thus, exactly how Ananke can be persuaded and how the Receptacle of Becoming is possessed of intelligibility are left unaddressed. These are left unattended, I would argue, not because a metaphor has been thought more apt to describe a physical aspect of the ordered universe—Plato proves himself fully prepared to describe in great detail all such aspects and does not need a metaphor for this—or, as in the case for the Receptacle of Becoming, because Plato lacks a proper understanding of how the senses or perception works—the Receptacle or Nurse of Becoming (Space), according to Plato, is apprehended *without* the senses[232]—but rather, he refrains from giving any further explanation, having followed the same pattern established earlier at a similarly difficult place[233] and refraining for essentially the same reason, because of the great difficulty involved in describing and fully comprehending these metaphysical realities. Ananke, as a disordering primal power, existed for Plato before the physical world was generated and so too, the Receptacle of Becoming, its existence explained as logically required for the world of things to have been generated and thus, as eternal and indestructible as Ananke, and as difficult to explain.

There is a further passage in the *Timaeus* where once again Plato indicates a semblance of order and hence, reasonableness, amidst the disordered and unreasoned forces and pre-existing stuff of the universe before generation began, albeit, only vestiges of reasonableness. This appears in his description of the primitive chaos, where beginning at *Ti.* 52d2 Plato explains how the elements or 'qualities' of fire, water, earth, and air, bear some traces of their proper nature, ἴχνη μὲν ἔχοντα αὑτῶν ἄττα,[234] even before God has marked them out in shapes by means of forms and numbers, while at the same time being ἀλόγως καὶ ἀμέτρωῆς.[235] The significance of this passage, I suggest, lies in Plato's repeated willingness, although drawn somewhat obscurely, in allowing, to at least some degree, order, intelligibility, and Mind to be present in the universe as

232. *Ti.* 51e6-52a4.
233. *Ti.* 28c5.
234. *Ti.* 53b2.
235. *Ti.* 53a8.

it stood prior to its formal ordering by God (the Demiurge) and hence, the deployment of fully rational causes. It is here, moreover, where the God is again described as constructing *as far as he could* everything to be as fair and as good as possible.[236]

2.3.4 Τὸ . . . δυνατὸν / 'as far as he could'

2.3.4.1 Discussion of phrase

The phrase τὸ . . . δυνατὸν, 'as far as he could,'[237] or ones with similar meaning or intention are scattered throughout the *Timaeus*. Significantly, this passage at *Ti.* 53b5-7 is an echo of the one found at *Ti.* 46c7-d1, both of which appear to curtail or in some way define the power of the Demiurge, possibly in some way by the opposing powers of Ananke or its effect ascribed thereto. However, another reading of either passage could take the reference which is being made there as that of the logical limitation of a copy to its Form,[238] rather than being suggestive of any abatement of the Demiurgic powers. A further reading, moreover, is possible, namely, the continued expression of the original desire by the wholly good Demiurge to share his goodness as *fully* as possible (i.e. to its *full* extent) with the Cosmos he was shaping, the idea here being that of the expanse of his goodness to include objects of generation and not to any lack (imperfection) on the part of the generator or the object he generates. In other words, the Demiurge's goodness is meted out to match and enable the continuing expanse of the universe as it is being molded. Minimally, Plato's understanding of God's goodness is his desire and ability to achieve form out of chaos. Thus, this reading preempts the idea of the universe and the Demiurge being one and the same thing or a pantheistic reading, while at the same time imbuing the generated Cosmos with the perfection of its creator. The God who generates is greater than any generated work. In other words, it is the Demiurge who shares his goodness, thereby generating and enabling the object. Although there is a notion of limitation here, it is not due to a lack of excellence, but rather, the idea suggested is that to create or generate something is to form or *de*limit it, thereby giving it boundaries, which in turn *enables* it to be this or that particular object and secondary to this, to be generated is to be subject to (be less powerful than) the generator. Therefore, the limitation that is understood by this reading is one of boundaries or the limitation of an object (in this case the Cosmos being formed by the Demiurge) by its form in order for it to *become* a particular object, and as such does not refer to any sort of lack or deficiency. In contrast to the other two

236. *Ti.* 53b5.

237. In the analysis which follows, for the sake of comparison, I will translate this passage '*as far as possible*' rather than '*as far as he could.*'

238. See the quotation from Plato's argument at *Ti.* 52c 5-d1, which follows.

readings, this understanding of limitation includes the idea of *perfection* as an outcome of the limiting action, without which, it (the Cosmos) would not be and thus, not to any notion of imperfection. The Cosmos as a finite object is perfect with respect to itself, while being limited with respect to the chaos out of which it is formed and the powers with which it has been bestowed (i.e. it did not generate itself, nor can it destroy itself, both of which minimally separate the Cosmos from the Demiurge). This sense of limitation, therefore, is not negative and is not due to any imperfection on the part of either the Cosmos itself (in the sense of lacking in goodness with respect to itself or lacking in form with respect to its Form) or its maker, but rather, is a positive result of the Generator's generosity in having wanted to share his goodness and perfection, as Plato explains at the beginning of the dialogue. Thus, a reading of the phrase "as far as he could" need not have any intended reference to either the Forms (a copy understood as being less perfect than its Form) or Ananke (the Demiurge did the best job he could given his materials) and thus, to lack of any sort, formal or otherwise. Accordingly, both the Demiurge and the Cosmos are perfect, the Generated no less than the Generator.[239]

There are problems with isolating any one of these views, however, when seeking an understanding of these passages and they will be reviewed shortly. What each of these readings ultimately suggests is that this phrase (or similar ones), in whatever way it is read, contributes to and is indicative of what overall for Plato is a realist perspective in his account of the generation of the world in the *Timaeus* and is also indicative of his main concern, that is, to give a 'likely' account of how the world began in relation to how it now stands. It is, therefore, not, I would argue, a directly metaphysical or speculative comment about the nature of God, although there are metaphysical ramifications, nor is its primary purpose to acknowledge the opposing influence of Ananke or disorder during the ordering of the Cosmos, although again, there are metaphysical consequences concerning Ananke. The question of whether reference is being made to the relationship, perhaps made askance, of a copy to its Form is also possible, but cannot be answered fully until the question of the reality of the Forms is addressed, which will follow later.[240] The context of the passage suggests, however, that the principal focus of this comment is directed at describing the state of the world as it then stood or indeed, when observing the world, as Plato saw it, since beyond his initial assessment of the Demiurge as being wholly good,[241] and this largely, if not entirely dependent upon the fact that the world which

---

239. This reading is congruent with Plato's closing words of the *Timaeus* (*Ti.* 92c4-9) where it is stated: "We can now claim that our account of the Universe is complete. . . . It has thus become a visible god, supreme in greatness and excellence, beauty and perfection, a single, uniquely created heaven" (Lee, 124).

240. See Part II, Chapter 2, §2.3.2.1.4.

241. *Ti.* 29e1-2.

Plato is about to describe in detail has already been judged by him from the beginning of the narrative, and repeatedly so to its very end, to be the fairest of all things generated,[242] he has already dismissed the possibility of ever truly knowing or being able to fully explain the nature of God.[243] Plato's reasons for coming to both of these conclusions, however, namely, that there is a Demiurge and secondly, the great difficulty in discussing the Demiurge's nature, are at once rationally and empirically based, derived in part from his empirical observations and in part from his metaphysical and theological assumptions concerning these observations and the arguments derived therefrom. Thus, while such assumptions and arguments do provide a philosophical framework and physical reference for his supervening, although barely discussed, metaphysics,[244] they do not account for the remaining disorder either in the world or in the generated soul, both of which are noted in the *Timaeus*. Plato is mindful of this disorder, but he must look elsewhere for an explanation. At no time, however, does he seem concerned with the apparent contradiction of his perfect god—the generated Cosmos, thoroughly infused with its perfect soul—in containing such disorder. My second comment, therefore, that to show the opposing influence of Ananke or disorder during the ordering of the Cosmos is not the intention behind Plato's comment here and thus, is not the focus of his attention in delimiting the Demiurge's power, needs now to be addressed.

There are several passages in the *Timaeus* containing references to disorder, to a lack of perfection, or to disequilibrium in the generated Cosmos. These references, moreover, depending upon where they appear in the dialogue, refer to one of three perspectives: First, to impending imperfection (or disorder) which will be left in the Cosmos due to extant powers or prior factors; second, to untoward conditions involving its generation and which will necessarily remain as an effect in it afterwards; or alternatively, third, to imperfection in the form of disequilibrium or disorder found extant in the Cosmos following its ordering. This third kind of disorder is not understood by Plato as necessarily following upon the other two kinds. Rather, its depiction forms part of the descriptive involving the process of ordering, either with respect to the Cosmos in the way in which (showing *how*) the elements moved from disarray to order within the Nurse of Becoming[245] and also later (in the Cosmic process) in how they disassemble and reassemble in the constant process of becoming[246] or alternatively,

242. See *Ti.* 29a5, 34a8-b9, 55d4-6 (θεόν is missing from the Greek in the Bury translation), 68e1-6, and 92c4-9 (the last paragraph of the narrative).

243. *Ti.* 28c3-5.

244. *Ti.* 34b10-c4. This passage is exemplar in showing Plato's awareness of the metaphysical scuttling required to bring what is contingent and accidental (his topic at hand) to the surface and in focus.

245. See *Ti.* 52d2-53b6.

246. See *Ti.* 49a7-d4.

with respect to the Soul as it acclimatizes to its bodily state[247] or is excessive with regard to its body.[248] Thus, this kind of disorder is always a result of process and is always moving towards perfection or *vice versa.* Moreover, not only are there three kinds of disorder distinguishable within the *Timaeus*, these passages fall roughly into three categories with regard to their meaning (or can be made to do so, i.e. upon rewording or rephrasing): First, in the sense of the superlative, e.g. *as far as possible* meaning 'the very best'; second, *as far as possible* used in reference to imitation of the Forms; and third, *as far as possible*, referring specifically to the limiting influence of imperfection in the form of disorder or disequilibrium. Finally, these references are made either directly or obliquely, and while some are quite broad or general in their scope and are included for the sake of completeness with regard to all references made, others are very central to the discussion and provide crucial insight for the conclusions that follow.

2.3.4.2 List of relevant passages in the *Timaeus*

These can be listed as follows:

29e2-3	Being therefore free from this [envy] he wished all things to be *as much as possible* [μάλιστα] like himself;
30a2-3	God, therefore, wishing . . . *as far as possible* nothing imperfect;
30c5-7	resembles *as nearly as possible* [above *all things*: τούτῳ πάντων];
32d1-33a1	that it *might* be, *so far as possible* [μάλιστα], a Living Creature, perfect and whole;
37c6 -d2	he rejoiced and planned to make it still more like its pattern; and as this pattern is an eternal Living Being, he set out to make the universe resemble it in this way too, *as far as was possible*;
38b7-c1	and it was made after the pattern of eternal nature, to the end that it might *be as like thereto as possible*;
39d8-e2	so that this world should in its imitation of the eternal nature resemble *as closely as possible* [ὡς ὁμοιότατον] the perfect intelligible Living Creature;

247. See *Ti.* 42e5-44d2.
248. See *Ti.* 89d2-90d7.

40a3-4	bright and beautiful to look at *as possible* [using superlative of the adjectives];
40b3-4	that each might be as perfect *as possible* [μάλιστα];
41b2-6	since you have been created, you are *not entirely immortal and indissoluble*, etc.;
42e2-4	and guide the mortal creatures to *the best of their powers*, except in so far as it should be *a cause of evil* to itself;
43e8-44b1	they made *wrong judgments* of sameness or difference, and lapsed into falsehood *and folly* . . . And because of this the soul . . . is as much *without reason today* as it was in the beginning;
46c7-d1	god uses in shaping things in *the best way possible* (the passage from which this discussion partly arose and whose meaning will be discussed separately);
48a2-3	Intelligence controlled necessity by persuading it *for the most part* to bring about the best result;
53b5-7	We must thus assume as a principle in all we say that God brought them to a state of *the greatest possible* perfection, in which they were not before (also one of the originating passages for this discussion and will be discussed separately);
56c5-7	and brought them in every way to the most exact perfection [δι' ἀκριβείας] *permitted by the willing consent of* Ananke;
58a1	The cause of disequilibrium is inequality, whose origin we have already described;
69b2-5	to the *degree and extent that they were capable* [ὅπῃ δυνατόν] of proportion and measurement;
71d5-7	For our makers remembered that their father had ordered them to make mortal creatures *as perfect as possible* [ὡς ἄριστον εἰς δύναμιν] (see also a separate discussion of this passage);
82a1-7	The body is composed of four elements—earth, fire, air, and water; and *disorders* and *diseases* are caused by an *unnatural excess or* deficiency of any of them, by their shifting from their proper place to

another, by any part of the body taking in an unsuitable variety either of fire or another element (for there are several varieties of them), and *by similar disturbances*;

83a2 and *no* longer maintain the *natural orderly sequence*, etc. (reference is being made here to the onset of disease and its description);

86d7-e3 no one wishes to be bad, but a bad man is bad because of *some flaw* in his physical make-up and failure in his education, neither of which he either likes or chooses;

87d1-3 For health and *sickness*, virtue and *vice*, the proportion or disproportion between soul and body is far the most important factor;

88d6-89a1 but if he imitates what we have called the nurse and foster-mother of the universe . . . by such motions he can *reduce to order and system* the qualities and constituents that wander through the body according to their affinities, in the same way that we have described in speaking of the universe; and so he will not leave foe ranged by foe to produce *conflict and disease* in the body, but friend by friend to produce health.

With reference to the first division, that of perspective, i.e. before, during, or after the generation of the Cosmos, the passages can be divided as follows:

Before the generation of the Cosmos: *Ti.* 29e2-3, 30a2-3, 30c5-7, 32d1-33a1, 39d8-e2, 40a3-4, 40b3-4;

During the generation of the Cosmos: *Ti.* 37c6-d2, 38b7-c1, 41b2-6, 42e2-4, 43e8-44b1, 46c7-d1, 48a2-3, 53b5-7, 56c5-7, 69b2-5; and finally,

After the generation of the Cosmos: *Ti.* 58a1, 71d5-7, 82a1-7, 83a2, 86d7-e3, 87d1-3, 88d6-89a1.

2.3.4.3 Division by perspective

These passages fall fairly evenly into the above three perspectives, possibly arguing for a conscious ordering on Plato's part of the three perspectives. More important to our argument here, however, is the next triadic division proposed, namely, that of division according to meaning, which can be outlined and constituted as follows:

(1) First, in the sense of the superlative '*as far as possible*'

With regard to the superlative, the passage in question may or may not have any reference to disorder or to a lack of perfection, depending on its context. The passages that fall quite clearly within this category are as follows: *Ti.* 29e2-3, 30a2-3, 32d1-33a1, 40b3-4, 82a1-7, 83a2, 86d7-e3, 87d1-3, and 88d6-89a1. Passages 46c7-d1 and 53b5-7 can also be included in this category in respect to one of their composite meanings. See my discussion in the fourth division entitled Composite Meanings.

(2) Second, in the sense of imitating its model or Form '*as far as possible*'

There is a natural limitation as to how far an object can be like its model or like any other object, for which Plato provides an argument at *Ti.* 52c 5-d1. Here he writes:

> whereas to the aid of the really existent there comes the accurately true argument, that so long as one thing is one thing, and another something different, neither of the two will ever come to exist in the other so that the same thing becomes simultaneously both one and two. (Bury, 125)

This argument affirms the position that two differently identified objects cannot be one and the same thing, made logically impossible by the fact of their difference(s). Responding further to this idea, Plato immediately reaffirms that Being, Space, and Becoming existed, three distinct things, before the Heavens came into existence. He does not describe either Space or Becoming as being in any way inferior to Being in this passage, as it is their separateness which he wishes to express here and not their relative value. Thus, in the passages which fall within the scope of this division, where imitation of an object to its model is being described, i.e. of 'becoming' to 'being,' emphasis is being given, I would suggest, not to imperfection or disorder inherent in one object as compared to the other, but rather, to the logical relationship which exists between the Form and copy in being distinguished separately, the Form in this case being prior and eternal and incapable of change, while the copy, being in time, subject to the finitude of change. This argument suggests that the copy is understood by Plato as being inferior only when compared to its Form and not in respect to itself, since it is only a copy as a result of its different ontology and not from any flaw in its ontology. Thus, although the Forms are understood by Plato as being perfect with respect to themselves, this argument also allows the copy the same perfection with respect to itself, since the argument is essentially an ontological one. The meaning, therefore, which 'as far as possible' suggests in this category is perfection of the object in question to the fullness of its capacity as a distinct object and separate identity apart from its Form and not to any inherent lack when compared to its Form. That this reference, moreover, need not be understood as referring to a lack of perfection of the copy in relation to itself is drawn out by Plato's reflection on the copies themselves at *Ti.* 50c4-6:

> But the forms which enter into and go out of her [the Nurse of Becoming] are the likenesses of eternal realities modeled after their patterns in a wonderful and mysterious manner.[249]

There is nothing "wonderful" about copies, which are essentially flawed. This understanding of the phrase 'as far as possible,' moreover, provides a basis for the closing words of the *Timaeus* when the Cosmos is described as "supreme in greatness and excellence, beauty and perfection, a single, uniquely created heaven."[250]

As regards to Plato's overall valuing of the ontology of the Form and its copy (the value of each with respect to the whole, which includes both Form and copy), Plato no doubt considered the Forms to be superior objects, since they are without change, to any copies made of them, which copies constantly come into being and disappear. However, my suggestion in setting up this division is not to oppose this, but to acknowledge the subtle comparison which is being made when passages are so qualified, which is not to the perfection of a copy as it measures up to its Form, but rather, to the perfection of any object in respect to itself, in other words, as it is realized to its fullest extent or 'as far as possible.' The essential difference in this division, therefore, when compared to the others is one of emphasis, the emphasis lying with the perfection of an object distinguishable from its Form and not with its lack of perfection in comparison with that Form.

The listed passages that fall within this category are: *Ti.* 30c5-7, 37c6-d2, 38b7-c1, 39d8-e2, 40a3-4. Passages 46c7-d1 and 53b5-7 can also be included in this category (see the fourth division entitled *Composite Meanings*). Plato has already offered at *Ti.* 51e6-52d4 a prior argument for why the Forms, in addition to the objective world, must exist in the first place, which in summary can be outlined as follows: First, what comes into being and thus is changeable, must have a cause. Second, if the world is beautiful (and it is the fairest of all generated things) and the maker good (he is the best of causes), then he had to have his eye on a pattern that is eternal and unchanging, apprehended by reason and understanding alone. And third, it therefore follows that the world is a likeness of something else.

(3) Third, in the sense of imperfection, disorder, or disequilibrium '*as far as possible*'

---

249. Jowett, *DP*, 737.
250. *Ti.* 92c7-9; Lee, 124. See also Jowett, *DP*, 780.

In this third main category either a direct or indirect reference is being made to a distinct lack of perfection and thus, to some degree of disorder or disequilibrium in the subject. Thus, reference is not being made specifically to either a copy to its Form or to the possibilities of perfection. Those passages which fall within this division are: *Ti.* 41b2-6, 42e2-4, 43e8-44b1, 48a2-3, 56c5-7, 58a1, and 69b2-5. Both 46c7-d1 and 53b5-7 can again be included here, with their inclusion explained under the next heading, 'Composite meanings.'

(4) Composite Meanings: *Ti.* 46c7d1 and 53b5-7 (Div. 1, 2 & 3) and 71d5-7 (Div. 1 & 3)

Out of a combination of the above three main categories, there arises also a fourth or hybrid grouping. It is within this fourth and final category, a composite of all of the above-noted meanings, where I propose that the passages initially queried, *Ti.* 46c7-d1 and 53b5-7, fall. As regards 53b5-7 and 71d5-7, while their contexts do suggest meanings which are composite, there is no direct reference to the Forms in the first instance and none at all in the second, thereby excluding 71d5-7 from division 2 of the suggested categories altogether. Thus, with regard to each of these three passages, there is present within all of them not only a reminder of the superlative goodness of the Demiurge and his desire to share it with his handiwork, but also a suggestion of curtailment in some way of this goodness or perfection and therefore, acknowledgment of disorder in some form. Whereas the presence of the first meaning is externally evident, although somewhat differently expressed each time, the third division of meaning, that of disorder, is not. Consequently, at *Ti.* 46c7-d1, it is the presence of other causes, i.e. of the accessory causes and of Ananke, in bringing about the Demiurge's desired perfection of the Cosmos, which makes absolute perfection impossible. The accessory causes involve movement, which in turn implies change, disavowing the possibility of absolute perfection, wherein there is, by definition, no change. Moreover, the effect of the presence of Ananke as a causal power of disorder is observed by Plato as never having been fully expunged. Ananke will soon be described as having been 'persuaded,' but only for 'the most part' by the Demiurge when ordering the universe.

With regard to the second two passages, *Ti.* 53b5-7 and 71d5-7, the unarticulated but, as I argue here, assumed existence of disorder remaining in the Demiurge's handiwork requires clarification. At *Ti.* 53b5-7 Plato has just described the primitive chaos "all without proportion and measure," after which he will then describe how God began 'reducing' this disorder to order, giving the elements, fire, water, earth, and air, a definite pattern of shape and number. It is at this juncture where he reminds the reader that as a principle it must be assumed that they were brought into the greatest possible perfection by the Demiurge. The reader has just been told, however, at *Ti.* 52d2-5, which is the opening of this section on the primitive chaos, that before the generation of the world there existed three distinct realities, Being, Space, and Becoming. Thus, although these two passages are not continuous, there has nevertheless been discreetly placed in the reader's mind a reference to the Forms or 'being,' albeit, somewhat removed, and is why I have included the meaning of division 2 as being also applicable to *Ti.* 53b5-7.

Specific mention is made in this passage of the fact that the elements of the Cosmos were not in this state earlier, i.e. an ordered state, not before God, the Demiurge, had touched them, the implication being that if there is imperfection remaining, any whatsoever, it is precisely because order has been brought out of prior disorder and no greater perfection is possible than what the Demiurge can give. As in *Ti.* 46c7-d1, the prior existence of disorder and change preempts the possibility of absolute perfection. The disorder referred to in the third passage, *Ti.* 71d5-7, is more specific, forming part of a larger discussion wherein Plato discusses man's lower or mortal parts of the soul (the emotions in the heart region and the appetites in the belly) and the various physical organs involved. The reader is left with an image of the gods doing the Demiurge's bidding by aiming for perfection when making man and who "did their best even with this base part of us." Reference here, of course, is being made to the liver and to the role it was assumed to have in divination. Again, the text is clear in articulating and intimating the idea that not only the Demiurge, but also the gods were working with imperfection and disorder, which would logically follow when considering a Cosmos in which the prior chaos, although subdued and shaped by the Demiurge, was not altogether eliminated.

A further word can be said concerning *Ti.* 71d5-7. This passage, like the other two composite ones, is similar in formulation: ὡς ἄριστον εἰς δύναμιν ποιεῖν, the main difference being that whereas earlier it was the Demiurge who is shown as bringing about an end which is as good as possible, here the memory of the Demiurge enjoining the previously-generated heavenly gods to bring about such an end when making mortal creatures[251] is brought to the reader's attention. What is of interest is that it is the Demiurge's *power* or *activity* (μιμούμενοι τὴν ἐμὴν δύναμιν περὶ τὴν ὑμετέραν γένεσιν) in making the gods themselves, which these created gods in turn are to take as their model, and not the Forms, when making mortal creatures. To take the Forms as their model would be logically impossible if the Forms are a vision privy only to the Demiurge precisely because of his perfection. Plato has been very explicit in stating that the heavenly gods are bonded together, albeit, directly by the Demiurge, and thus being themselves generated (δι' ἃ καὶ ἐπείπερ γεγένησθε) are not entirely immortal and indissoluble, and therefore, not being perfect, are capable neither of envisioning the Forms nor, in turn, of emulating them directly in their activity. This entire formulation, I would argue, is a logical move on Plato's part and involves very little, if anything, which is mythical. The Forms, it can be

251. See also *Ti.* 41c2-6.

concluded, are the vision logically required, according to Plato, of a perfect being, i.e. that of the Demiurge. The gods, I suggest, are required to take the Demiurge's power or activity as their model and not his vision of the Forms, being themselves incapable of such a vision. They will thus encounter all the vicissitudes that Ananke has previously impacted on the Demiurge's work. Consequently, at *Ti.* 71d7-e1, when reference is made to the gods rectifying that 'vile part of us' (τὸ φαῦλον ἡμῶν), this is not in any way a mythic presentation, but rather, is a logical consequence of Plato's ontology.

Finally, on returning again to *Ti.* 46c7-d1, this passage alone of the three passages listed above refers explicitly to 'the idea of the best,' τὴν τοῦ ἀρίστου κατὰ τὸ δυνατὸν ἰδέαν, and thus presents a direct reference by Plato to the Forms or the ideal state of perfection or being.[252] However, as mentioned above, immediately prior, this same passage also speaks in terms of the Demiurge's desire to make the Cosmos the very best, i.e. the superlative, albeit in terms of the Forms or the idea of the best, against which there is also a hint of curtailment preventing absolute perfection. It is in terms of this curtailment, moreover, expressed as disorder and change remaining within the Cosmos, from which I am arguing that *Ti.* 46c7-d speaks in terms of the superlative and not the ideal form of it, the latter expressing, rather, a standard, since the ideal world, given any remaining disorder or change, would be impossible. What is ultimately at issue here, however, is not the meaning of the phrase '*as far as possible*' (or similar formulations), be it a reference to a superlative aim, the Forms, or curtailment of the ideal, all of which possibly apply, or alternatively, to allusions of these sentiments however expressed, but rather, it is the purpose behind Plato's remark. I would propose that his chief aim in using the phrase '*as far as it is possible*' (or its equivalent) was to show the limiting effect which disorder and change had on the ideal world envisioned by the Demiurge, which because of his perfection could only be ideal and nothing less, although clearly, disorder or change as the limiting effect was not always the focus of individual passages. Plato's express purpose, however, was not one of emphasis of the ideal world, any more than it was a denial of a metaphysical position or affirmation of the physical, but was primarily one of emphasis of the real world, which for him included not only disorder and change and with that, the physical, but also a Demiurge[253] and the changeless ideal and with that, the timeless and seamless metaphysical. For Plato, the modern notion from the History of Science of the 'disenchantment of the world,' a point in time where myth, imagination, and spirit were cut off from what is now called 'scientific reality,' would have told only half the story.

---

252. See also *Ti.* 30c5-7, 37c6-d2, 38c1, 38c6-d2, 39d8-e2, and 40a4.
253. One amongst at least two gods, but possibly many.

2.3.4.4 The classical commentators

None of the classical commentators, Cornford, Taylor, or Archer-Hind, pick up on or comment upon this re-occurring theme of '*as far as possible*,' as it nonetheless appears and re-appears throughout the *Timaeus.* Thus, they have little or nothing to say in this regard, whether it concerns the less obvious passages noted above or the more important ones at *Ti.* 46c7d1, 53b5-7, and 71d5-7, either collectively or individually. This omission, however, is noteworthy and I would argue is primarily due to the label given of 'myth' to substantial parts of the *Timaeus* which in fact are not myth and secondly, to a much too early impacting and facile handling of Plato's early admission by Timaeus that his story is a likely tale, inserting an 'only' where there is none.

For Cornford, as discussed earlier, even the Demiurge is a mythic figure, following his purely positivist line of interpretation upon which his Commentary on the *Timaeus* is based. Thus, it is not surprising that there is silence with regard to any of the above-noted possibilities regarding the nature of the Demiurge, Ananke, and the Forms, since for Cornford these are not metaphysical realities, but are aspects of the physical or mental realm, redrawn as myth or abstractions because of their complexity. All references to likeness are thus interpreted as objects within the world as they compare to the ideal in the mind as objects of rational thought:

> We have seen that, although the creator god, as such, is a mythical figure, the relation of likeness to model none the less subsists between the visible world and the intelligible. . . . However we may interpret the divine Reason symbolised by the Demiurge, this model [the generic form of the Living Creature][254] is one among the objects of its thought. It is the ideal, whose perfection the visible universe, as a living being, is to reproduce in its own structure, so far as is permitted by the conditions of temporal existence in space. (*PC*, 40–41)

In response to the section at *Ti.* 32c-33b, Cornford comments upon certain 'powers' (δυνάμεις) in the sense of qualities or properties of bodies considered as having the 'power to act and be acted upon' (δύναμις τοῦ ποιεῖν καὶ πάσχειν), which discussion is not problematic (*PC*, 53 and 199). However, it is not δύναμις or the well-known use of the prepositional phrase κατὰ δύναμις which carries the weight of '*as much as possible*' here, but rather, μάλιστα, the superlative of μάλα or 'exceedingly,' reflecting the slightly different meanings, as discussed above, behind the same English translation. However, Cornford refrains from commenting on μάλιστα, just as he does not comment earlier at *Ti.* 30a2 when δύναμις is used with κατά or later at *Ti.* 37d2 when it is used with εἰς, veering away from considering δύναμις other

254. The square brackets and their content are my own.

than as he explains it above as an elementary force, even when its meaning has clearly changed when combined with such prepositions as κατά at *Ti.* 30a2, 38c1, and 42e2 or εἰς at 37d2 and 71d5-7, or again later with κατά at *Ti.* 46c7, but this time combined with its adjectival form, κατὰ τὸ δυνατόν.[255] *Ti.* 53b5-7 is an example of τὸ δυνατόν being used in combination with the superlatives, κάλλιστα ἄριστα, to indicate *the greatest possible* perfection. *Ti.* 71d5-7 follows similarly, combining the superlative ὡς ἄριστον with one of the prepositional phrases, εἰς δύναμιν. Cornford is silent regarding these passages.

Taylor comments on *Ti.* 56c5-7, stating that "there is a suggestion that the 'compliancy' of ἀνάγκη is not absolute," and further:

> 'When ἀνάγκη submits,' you get minute and exact conformity to a recognizable law—this is the point of the δι' ἀκριβείας—but when ἀνάγκη is not sufficiently compliant, you get only the imperfect approximation of this ideal. (Taylor, *PT*, 382)

By his comment, Taylor acknowledges a division 3 meaning for this passage, but he does not discuss any of the other passages listed in their relationship to the above-discussed theme of '*as far as possible*' or the various meanings that would attach.

Archer-Hind, on the other hand, would appear to place *Ti.* 30a2-3 in division 3, whereas I have placed it in division 1. At this section, he is commenting specifically on κατὰ δύναμιν where he concentrates on the notion of the inevitable presence of 'evil.' However, Archer-Hind remains a skeptic as to what this might be and places it firmly within the realm of matter:

> To make the material universe absolutely perfect would be impossible, since evil, whatever that might be, is more or less inherent in the very nature of matter and can never be abolished. (A-H, *TP*, 92)

Moreover, his reason for focusing upon 'evil' is unclear.

2.3.4.5 κατὰ δύναμιν and the notion of 'evil'

Up to this part of the *Timaeus*, which is still very much at its beginning, Plato has not yet introduced 'Ananke' and the word which is here translated by 'evil' is φλαῦρον, a form of φαῦλος. However, the meaning of this word much more closely approximates that of 'careless,' 'common' (in the sense of 'vulgar'), or 'poor.' Thus, there is no reason to think that 'evil' understood as some sort of, as yet, nameless negative force or effect forms the background thought of the text

---

255. From the deponent verb δύναμαι, 'to be able, capable, strong enough to do.'

here. Although Plato does mention that the universe was in discordant and disorderly motion when the Demiurge decided to order it, what is being focused upon and discussed is the God's goodness and complete lack of imperfection, and his desire to share that goodness as far as is possible in all that he orders. Thus, rather than contrasting that goodness with anything other (i.e. evil) than itself (i.e. goodness and an approximation to that goodness), what Plato appears to be doing is present a known (from men of understanding) principle according to which 'becoming' and the Cosmos came into being, the μάλιστ' ἄν τις ἀρχὴν. The spirit engendered here is one of generosity on the part of the Demiurge. If Archer-Hind were correct in his understanding of an assumed contrast of 'evil' with the goodness of the Demiurge, then the discordant and disorderly motion would have to be an example of that which is careless, common, or vulgar (the proper meaning of the word Archer-Hind has translated as 'evil'), which is meaningless both in this context and in relation to what Plato later states about the pre-existing chaos. Archer-Hind does not comment on any of the other passages.

2.3.5 The accessory causes

With regard once again to the accessory causes, however, Archer-Hind makes an interesting observation, which to some degree is at odds with his understanding above, where 'evil' in the *Timaeus* is understood by him to be the mitigating force or forces "inherent in the very nature of matter" (similar to what he describes as Democritus's understanding of a "blind unconscious force" [see below]) preventing the Demiurge from bringing about perfect goodness, but which position Archer-Hind now claims Plato set specifically out to argue against. He correctly notes with regard to the τῶν συναιτίων at *Ti.* 46c7 that Plato is guarding against being supposed to mean that the physical principles he has just laid down are the real causes, and are really the means *through which* the true cause works "viz., νοῦς operating ἐπι τὸ Βέλτιστον."[256]

His next observation is worth quoting in full. He suggests comparing the above with *Phaedo* 99b, and then states:

> The whole of this latter part of the chapter contains a polemic partly against Anaxagoras, partly against Demokritos. Anaxagoras did indeed postulate νοῦς as his prime force, but he used it simply as a mechanical agent, without attributing to it a conscious effort to produce the best result. Demokritos conceives a blind unconscious force, ἀνάγκη, to be the motive power of the universe. Thus whereas the

256. Superlative of ἀγαθός, 'best.'

> opposition between Demokritos and Plato is fundamental and essential, Plato's controversy with Anaxagoras is due rather to inconsequence or incompleteness on the part of the latter.
>
> (A-H, *TP*, 161, fn. 9)

It can be assumed that the comparison Archer-Hind is making here is with that section of the *Timaeus* which follows *Ti.* 46c7, dealing with νοῦς and ἀνάγκη, and not the *Phaedo* from 99c onwards, although the attention he draws towards Plato's disappointment and disagreement with Anaxagoras in this part of the *Phaedo* is apt. That the *Timaeus* contains an understanding of 'Nous' contrary to that advanced by Anaxagoras, or at least greatly developed and extended, is not disputed. Further, I would agree with Archer-Hind that this was due primarily to the incompleteness of the idea of 'Nous' as it had been developed by Anaxagoras. However, a few words should be said in this regard.

2.3.5.1 Anaxagoras

Firstly, it is questionable whether Anaxagoras, in fact, can be cited to be a 'physicalist' or 'physicist' in any strict sense, regardless of Plato's disappointment in him for not appearing to recognize the powers of 'Nous' beyond the strictly physical sense or that of mechanical agent or power, such metaphysical powers for Plato being the ability of Nous to bring about the greatest good and to be the cause of everything. Although, as the ancient historian Diogenes Laertius describes him, Anaxagoras indeed felt he had been born "for the study of the sun and the moon and heavens," he differs from other physical philosophers in a decisive way, namely, he believed definitively that 'Mind' was not only the first principle of movement, but also of order: "*All things were together. Then Mind came and arranged them.*" He became renown for this belief, in fact, so much so, that he was nicknamed 'Mind,' appearing in the *Silli* of such writers as Timon. Secondly, this was not a material mind, rather, Anaxagoras argued specifically for a non-material, non-physical Mind (from the first book of his *Physics*): "*Mind is infinite and self-controlling and it has been mixed with no thing, but is alone itself by itself.*" He argues that if anything had been mixed with it, "*in everything there is a present share of everything,*" this mixing would have prevented Mind from the absolute control it has (a view similar to that of Xenophanes), thus: "*For it is the finest of all things and the purest, and it possesses all knowledge about everything, and has the greatest strength. And mind controls all things, both great and small, which possess soul.*" (B12) However, Anaxagoras's Mind is not entirely transcendent and this perhaps is where he falls out with Plato: "*In everything there is a present share of everything except mind—and in some things mind, too, is present.*" (B11) Thus, although he could have been, and indeed was, accused of atheism or impiety, having regarded the heavenly bodies as objects of scientific inquiry, rather than in the traditional, religious sense as gods, Anaxagoras was certainly not a physicalist in the way it has been suggested Plato understood him to be from the *Phaedo*. Plato specifically states in the *Phaedo* at 98b7-c1 that Socrates's disappointment (assuming that he shares this view) in Anaxagoras lay primarily in Nous or Mind being assigned no causality *for the order* of the world, leaving such order to the elements (e.g. air, aether, and water) or to what Plato has now designated at *Ti.* 46c7 as the accessory causes and with it, omitting any ability by Nous to discover "how it was best for that thing to be."[257] However, Plato's assessment of Anaxagoras's position would appear to be somewhat biased in light of the latter's own writings where Mind is acknowledged to control all things that contain soul and who traditionally has been credited with being the first to introduce the teleological principle into the explanation of the natural world, with only Xenophanes possibly preempting Anaxagoras's right to this claim. As noted above, Socrates's primary disappointment in Anaxagoras was directed at his omission in acknowledging Mind as the *cause* of everything, which latter philosophical consideration, however, occupied neither Anaxagoras's main interest nor pursuit (nor perhaps ability, as Plato's dialogue implies), having already acknowledged his own scientific turn of mind, which would lead naturally, as it did, to discerning causation at the physical level, rather than at the metaphysical. While both Xenophanes and Anaxagoras serve as predecessors to Plato's more developed understanding of Mind, it is much too strong a claim to say that their views were counter to it.

2.3.5.2 Democritus and Plato – a comparison

However, although the same cannot be said concerning Democritus (and Leucippus),[258] at least not in any straightforward sense, care must be taken not to succumb to the broad generalization made here by Archer-Hind regarding an obvious "fundament and essential" opposition between Democritus and Plato.[259] There are enough points of similarity between these two philosophers to counter or at least, temper, any statements of this sort. Although Plato's religious and metaphysical bias strongly rejected strictly physicalist views, this is not to say that he did not absorb certain of these ideas into his own cosmology, while at the same time adapting them to his own metaphysical scheme. Further, although many commentators on Democritus or Plato tend to favor this starkly contrasted understanding of Democritus and Plato, he is never actually mentioned by Plato,

257. See Plato, *Phaedo*, 97c-99c.

258. I have adapted the traditional position of presenting the atomist philosophy through the writings of Democritus, but who in fact was a student of Leucippus, the actual father of the atomic theory, but about whom we know very little.

259. I have chosen the more modern English transliteration, 'Democritus,' rather than that of Archer-Hind, 'Demokritos.'

let alone in this light. Moreover, many of their views on the physical world are remarkably similar.[260]

Significantly, the lives of these two men closely overlap, with the younger Plato being born some thirty-two years later and dying only about twelve years after the atomist. Democritus, therefore, would have been a mature man and philosopher by the time Plato was born and his philosophy would have been widely known and his writings largely disseminated throughout the Greek philosophical community by the time Plato began his career as a philosopher. Thus, his ideas would have been thoroughly known to Plato, either directly or indirectly. The ideas that Plato appears to have appropriated into his own cosmology from Democritus's atomic theory and consequently used in the *Timaeus* are not insignificant.[261]

First, it cannot be denied that Plato has adopted very painstakingly a version of the atomistic[262] understanding of substance in forming the building blocks of his physical universe. There are too many similarities. The ideas of these two philosophers (Plato and Democritus) may differ, perhaps, as to the particulars in working out their individual theories, but the overall view that emerges is not discontiguous in respect to the other and can be outlined as follows:[263]

260. J. J. Cleary ("Plato's Teleological Atomism" in *Interpreting the Timaeus-Critias*, eds. T. Calvo and L. Brisson, 242) notes that both Empedocles (Fr.360KR) and Democritus (DK68A40) are credited with positing the vortex to explain the separation of like elements into different regions; Anaxagoras (DK12) also seemed to accept the same mechanism for separating things out. Cleary suggests, moreover, that the receptacle is Plato's alternative to the Atomists' void, "as a plenum in which change and movement can happen without the introduction of absolute non-being."

261. There are other similarities which have been noted in antiquity, e.g. regarding 'knowledge': "You cannot say that every impression is true, because of the reversal – as Democritus and Plato showed in their reply to Protagorus," (Sextus Empiricus, *Against the Mathematicians*, VII 389-390) in *EGP*, Barnes, 256–57; also with regard to Moral Philosophy: "Democritus and Plato both place happiness in the soul" (Stobaeus's Anthology, II vii 3l) in *EGP*, Barnes, 265. There are also many other areas of Democritus's philosophy reflected or shared in the writings of Plato.

262. All the following references in this section are from *Early Greek Philosophy* by Jonathan Barnes, Penguin Classics, 1987, ch. 20, 246–88 and will be hereafter abbreviated *EGP*.

263. This particular passage is Simplicius's quote of Aristotle, contained in his *Commentary on the Heavens*, 294.30-295.22.

EGP[264]	Democritus	*Ti.*	Plato
247	"Democritus thinks that the nature of eternal things consists in small substances,	53c4-8[265]	"In the first place, it is clear to everyone that fire, earth, water, and air are bodies, and all bodies are solids. All solids again are bounded by surfaces, and all rectilinear surfaces are composed of triangles."
	infinite in quantity,		From the *Timaeus* it is difficult to discern precisely what Plato's view is on this. However, there are indicators throughout this work favoring a view similar to Democritus in this regard, namely, infinitude in quantity. First, although Plato discerns only one Cosmos or universe, he gives a rational account for this and not a physical argument, following which he does not appear to delimit the extent of this universe (see *Ti.* 30c2-31b3). In other words, he does not posit one Cosmos[266] because there were only so many triangles to be used up by the Demiurge, but rather, one Cosmos is superior ontologically to a multitude of worlds, given that a wholly perfect Demiurge constructed the Cosmos and unity is argued to be superior over multiplicity ("nothing can be good which is modeled on something incomplete . . . and so he created a single visible living being"). Secondly, 'becoming' (comprising the whole of the physical universe and concerning which, the triangles form the basic building blocks or particles) mirrors 'being' (the Forms), which Forms are never limited in number by Plato. Thus, although the Cosmos appears to be finite with regard to the *kinds* of objects / of

264. None of the references with regard to Democritus's atomic theory are original fragments by him, but rather, are comments by others on his theory.

265. See Lee, 73; Bury, 127.

266. I use 'Cosmos' and 'world' here as substitutes for universe, as to speak of 'one universe' appears redundant and hence, slightly confusing, and yet the possibility of multiple universes or worlds was a concern which even Plato himself raises. He, of course, denies this possibility.

EGP	Democritus	*Ti.*	Plato
(247)	(infinite in quantity,)	(53c4-8)	of 'becoming' it contains (e.g. four living creatures) and even the number of a particular kind (e.g. mortal beings limited by the number of souls created or the effect of their evolution or transmigration), the basic particles themselves need not be limited and are in fact potentially unlimited outwith any particular object of being. Thirdly, Plato does not work through the ramifications or status of the original chaos from which the triangles in his theory originally came together, nor does he attempt to do so. His account suggests, however, that the original chaos, which itself contained some natural ordering not initiated by the Demiurge, remained even after the ordering of the Cosmos, with the result that it is not done away with in its entirety, nor are its limits spelt out or known. Exactly how this original chaos ultimately relates to the ordered universe is not discussed. There is nothing in the *Timaeus* to suggest that this original chaos, out of which the ordered universe emerged and was shaped, now forms a part of the universe itself or that the chaos was all used up (only the four elements were), although vestiges of it remain in the Cosmos, and understood in this way, constitutes an anachronism. To the contrary, Plato's account as it stands suggests that the ordered universe forms a part of the original chaos, separated by order and design. There is also a clear indication in the *Timaeus* (see *Ti.* 34a8-b9) that this perfectly round "blessed God" (Cosmos) "with the extremes equidistant in all directions from the center" has a discernible outer edge (*Ti.* 40a2-7) and "a smooth and unbroken surface" (*Ti.* 33b4-c), which outer edge would logically meet up with the original chaos. Again, precisely how is not worked out in the *Timaeus*. The fourth and strongest argument supporting / a

EGP	**Democritus**	*Ti.*	**Plato**
(247)	(infinite in quantity,)	(53c4-8)	a position favoring infinitude in quantity, however, is where Plato refers to the triangles as being infinite in variety (see *Ti.* 57d3-6), whereby he stresses that such variety must be taken into consideration when considering probable reasoning concerning Nature. 'Potentiality' does not enter as a factor here. An infinite variety would infer infinite quantity, but again, how this is to be worked out is left undiscussed.
	and for them he posits a place, distinct from them and infinite in extent.	52a8-b2[267]	"third, space, which is eternal and indestructible, which provides a position for everything that comes to be."
	He calls place by the names 'void', 'nothing' and 'infinite'		For Plato there is no 'void,' but it is clear from the reference to 'void' in Democritus that he is describing the space ('void' for Democritus, 'space' for Plato) where 'being' ('becoming' for Plato) would reside. Thus, when Plato states that there is no void, he is not stating that there is no place for 'becoming' ('being' for Democritus) to reside, but rather, there is no space without 'becoming' being contained therein and he explains why (*Ti.* 58a4-7). This is not at odds with what we are being told about Democritus's theory, where 'void' is simply the place for 'being' (Plato's 'becoming') to reside. For both philosophers this would appear to be a logical requirement, for which Plato offers the argument that "an image . . . needs to come into existence in something else if it is to claim some degree of reality, or else be nothing at all" and "so long as two things are different, neither will come to be in the other and so become at once both one and two" (*Ti.* 52c3-d1). Thus, based on this / reasoning

267. Lee, 71; Bury, 123.

EGP	**Democritus**	*Ti.*	**Plato**
(247)	(He calls place by the names 'void', 'nothing' and 'infinite')	(52a8-b2)	reasoning, Plato posits a separate entity called 'space,' which entity Democritus had named 'void.'
	and each of the substances he calls 'thing', 'solid' and 'being'.		It is in regard to the facing passage noting Democritus's naming of substances wherein lays Plato's main and significant divergence away from Democritus's physical theory and from those of all previous philosophers. This dissent resulted in Plato's introduction of his theory of Forms. Philosophically, such naming is at odds with Plato's understanding of the cyclical process that all substances constantly undergo, since it implies a stability that Plato denies at the physical level. This idea of 'instability' with regard to the physical Cosmos (i.e. the elements out of which it was comprised) was of such prime importance to Plato that it gave rise to a strong dictum by him against such naming, wherein he sets about to specifically address the related issues. First, Plato prefers to call such substances 'becoming,' rather than 'being' because of their constant physical transformation from one element into another and whose movement must have a cause, reserving 'being' for the postulated non-physical patterns after which each object of 'becoming' is shaped and the Demiurge (or those which the Demiurge directs) as their ultimate or efficient cause (*Ti.* 27d5-29d3).[268] Second, and for the same reason, Plato finds such objects so lacking in permanence and solidity that he suggests that it would be more accurate to refer to them as "having a quality," rather than thinking of them as solid objects or "being a thing" at / all

268. See Lee, 42–43; Bury, 49–53.

EGP	**Democritus**	*Ti.*	**Plato**
(247)	(and each of the substances he calls 'thing', 'solid' and 'being'.)	(52a8-b2)	at all (*Ti.* 49b2-e7).[269] He obviously, however, does not suggest that this is at the same time practical, referring himself to individual solid objects throughout the remainder of the *Timaeus* and not to their 'qualities' when discussing them, e.g. the various parts of the human body. Third, Plato suggests that we should not call something, not even when seeing them as qualities, 'this thing' or 'that thing,' as this suggests a permanent reality and stability, which substances and qualities just do not have. He suggests, rather, that when the demonstrative expressions 'this thing' or 'that thing' are used, they should only be used to denote *that in which* the process takes place and "in which these qualities appear for a time and vanish" (*Ti.* 49e7-50a4).[270] The essence of Plato's argument is that due to the fact that all objects are subject to constant change, no permanent designation would ever be truly accurate over time. I would argue, however, that the overall effect of the above considerations remains at a theoretical level for Plato. Stated in practical terms and for the remainder of his description in the *Timaeus* of the physical Cosmos, with the exception of his treatment on the soul and its role in causation, which again is dealt with at a highly theoretical level, Plato's view remains closely analogous to that of the Atomist position.

269. See Lee, 67–68; Bury, 113–15.
270. See Lee, 68; Bury, 115–17.

EGP	**Democritus**	*Ti.*	**Plato**
(247)	He thinks that the substances are so small that they escape our senses,	56b7-c3[271]	"We must, of course, think of the individual units of all four bodies as being far too small to be visible and only becoming visible when massed together in large numbers."
	and that they possess all sorts of forms and all sorts of shapes and differences in magnitude.	53c8-57d6[272]	Plato goes into great detail explaining how the differently shaped triangles bring about differences in shape and magnitude of the pure primary bodies or when amassed, the four elements.
	From them, as from elements, he was able to generate and compound visible and perceptible bodies.		As quoted above, for Plato all visible and perceptible bodies are composed of the four elements, which ultimately derive from specific triangles, and further differences, from variations in these triangles.
	The atoms struggle and are carried about in the void because of their dissimilarities and the other differences mentioned, and as they are carried about they collide and are bound together in a binding which makes them touch and be contiguous with one another but which does not genuinely produce any other single nature whatever from them;	56d1-6[273]	"When earth meets fire it will be dissolved by its sharpness, and, whether dissolution takes place in fire itself or in a mass of air or water, will drift about until its parts meet, fit together, and become earth again, for they can never be transformed into another figure."
	for it is utterly silly to think that two or more things could ever become one.	52c3-d1	"so long as two things are different, neither will come to be in the other and so become at once both one and two."

271. See Lee, 79; Bury, 137.
272. See Lee, 73–81; Bury, 125–41.
273. See Lee, 79–80; Bury, 137–39.

EGP	**Democritus**	*Ti.*	**Plato**
(247)	He explains how the substances remain together in terms of the ways in which the bodies entangle with and grasp hold of one another; for some of them are uneven, some hooked, some concave, some convex, and others have innumerable other differences.	56a6-b3[274]	In Plato's theory, the different figures are compounded of the various triangles, with some configurations more stable than others: "So to sum up, the figure which has the fewest faces must in the nature of things be the most mobile, as well as the sharpest and most penetrating, and finally, being composed of the smallest number of similar parts, the lightest. Our second figure will be second in all these respects, our third will be third.
	So he thinks that they hold on to one another and remain together up to the time when some stronger force reaches them from their environment and shakes them and scatters them apart.	57a5-6[275]	"but the process of dissolution continues so long as transformation produces a weaker mass to offer resistance to a stronger one."
247–248	He speaks of generation and of its contrary, dissolution, not only in connection with animals but also in connection with plants and worlds – and in general with all perceptible bodies. (Aristotle, *fragment* 208).[276]	56c8-57c6	Plato goes into considerable detail describing the physical process of generation and dissolution of the elements and their component particles, which, except for the soul, comprise the whole of the generated Universe.
248	Democritus too, when he says that a whirl of every kind of form was separated off from the whole (B 167) but does not say how and by what cause, seems to generate it spontaneously and by chance.[277]	52e5-53a7	"And its [that of the primitive chaos or 'Nurse' of Becoming] contents were in constant process of movement and separation, rather like the contents of a winnowing basket or similar implement for cleaning corn, in which the solid and heavy stuff is sifted out and settles / on

274. See Lee, 79; Bury; 137.
275. See Lee, 80; Bury, 139.
276. End of Simplicius's quote of Aristotle (contained in his *Commentary* on Aristotle's *On the Heavens*, 294.30-295.22).
277. Simplicius, *Commentary on the Physics*, 327.23-26.
278. Sextus Empiricus, *Against the Mathematicians*, VII 116-118.

EGP	**Democritus**	*Ti.*	**Plato**
249	For in the one case the whirling of the sieve separately arranges lentils with lentils, barley with barley . . . and in the other by the motion of the waves, oval pebbles are forced into the same place as oval pebbles . . . , as though the similarity in things contained some sort of force for collecting things together. [B 164] That is Democritus' view."[278]	(52e5-53a7)	on one side, the light and insubstantial on another: so the four basic constituents were shaken by the receptacle, which acted as a kind of shaking implement, and those most like each other pushed together most closely, with the result that they came to occupy different regions of space even before they were arranged into an ordered universe."[279]

Second, although the main difference between Democritus and Plato would appear to lie in the area of causation, it is mainly because Plato has developed a theory of causation, namely, that of the soul with regard to physical motion within the Cosmos and the Demiurge and Ananke with regard to primary causation, which Democritus failed to do.[280] With regard to causation, therefore, as with so much of Democritus's physical theory where he provided 'likely' detail, Plato has now filled in the gaps,[281] providing what he regarded as a 'likely' theory of causation. 'Likely,' I would argue, because Anaxagoras and Democritus had already set its precedent. Having adapted for his framework 'Mind' from Anaxagoras, where "*it possesses all knowledge about everything, and has the greatest strength . . . and controls all things, both great and small, which possess soul*" as the primary cause for good, which is synonymous with order and ordering, Plato now adds the "blind unconscious force, ἀνάγκη," from Democritus as the primary cause for evil, which is synonymous with disorder and disordering, and which becomes for Plato the motive power of the universe before its ordering by the Demiurge, and for all disorder thereafter. In neither case, however, does Plato account for mechanical motion in the generated Cosmos by either Nous (the Demiurge) or Ananke (Necessity), reserving the accessory causes (not true causes) for all but initial movements, which latter causal role he

279. See Lee, 72; Bury, 125.

280. "As for motion (whence and how existing things acquire it), they too, like the others, negligently omitted to inquire into it." (Aristotle, *Metaphysics*, 985b4-20) in *EGP*, Barnes, 248.

281. "Democritus too, when he says that a whirl of every kind of forms was separated off from the whole [B 167] but does not say how and by what cause, seems to generate it spontaneously and by chance" (Simplicius, *Commentary on the Physics*, 327.23-26) in *EGP*, Barnes, 248.

reserves specifically for the generated 'soul,' but which I will argue later[282] is a psychical source of movement and not a physical one.[283]

Taylor's[284] *Commentary* on the accessory causes, or as I have translated them, the 'assistant causes,' as they pertain to the *Timaeus* is insightful. Of note are his references, first, to the passage from the Phaedo (95b-99d),[285] which Archer-Hind also cites (but who leaves the passage unexplicated[286]) and second, to the early occurrences of the words συναίτιος, μεταίτιος, and αἴτιος in the extant literature.

2.3.5.3 The *Phaedo* 95b-99d

With reference to the first, Taylor points out that what is at stake here is the principle of agency and not merely the "formulation of mathematical laws of interconnexion between events."[287] Thus, Socrates's bones, muscles, and sinews are in no sense the real cause of his now being seated in the prison waiting to be executed, but rather, it is his notion of what is best for his soul, i.e. to face Athenian justice. Hence, when the purposive behavior of an agent, e.g. Socrates, aiming at a real or supposed 'good,' is distinguished from those things 'without which a cause would not be a cause,' the latter become distinguished as accessory causes. Thus, with regard to 'disorder' per se, either as a cause or an effect, specifically in its not requiring to attend to any other action or effect for its potential or real existence, it remains logically distinct from the notion of accessory cause.

---

282. See Part II, §2.3.2.1.3.

283. There is little doubt that Plato was influenced in varying degrees by the ideas of a great many thinkers. Of note, is the suggestion by Philo of Alexandria concerning the early influence of the poet, Hesiod:

> But some persons think that the father of the Platonic theory was the poet Hesiod, as they conceive that the world is spoken of by him as created and indestructible; as created, when he says, -
>
> "First did Chaos rule
> Then the broad-chested earth was brought to light,
> Foundation firm and lasting for whatever
> Exists among mankind;" (Hesiod, *Theogon*, 116)
>
> and as indestructible, because he has given no hint of its dissolution or destruction.

See Philo, *The Works of Philo*, trans. C. D. Yonge, Henrickson, Massachusetts, 1993, 708–9.

284. Taylor, *PT*, 290–94.

285. Ibid., 291.

286. A-H, *TP*, 161, fn. 9.

287. Taylor, *PT*, 291.

2.3.5.4 Early occurrences of the words συναίτιος, μεταίτιος, and αἴτιος

With regard to the early occurrences of the words συναίτιος, μεταίτιος, and αἴτιος, Taylor notes their juristic origin, where the important distinction being made is between the 'principal' agent and the 'accessories' or intermediate agents in acts which have legal or moral consequences and hence, is not between the primary agent and the consequent actions which follow thereon. In Aeschylus *Agam.* 1116 Casandra calls Clytaemestra συναιτία φόνον 'accessory' to the murder of her husband, the 'principal' being Aegisthus, as he afterwards boasts. In like manner, αἴτιον is that which is entitled to the legal credit or discredit attaching to an act, in other words, that to which an act can be 'blamed.'[288] Thus, the earliest uses of these words distinguish agency involving purposive behavior, but distinguish neither the ordering of the acts themselves that ensue therefrom nor the acts outwith such purposive behavior.

The idea of Ananke or ἀνάγκη as representing brute physical force in the cosmos is not raised by Taylor and neither is any mention made of ἀνάγκη in connection with the accessory causes. The closest Taylor comes to explaining Ananke in terms of mechanical motion is at *Ti.* 46e1-2 in reference to the prepositional phrase ἐξ ἀνάγκη, where he argues that the text is simply making a contrast between those things which cannot move or restrain themselves, so that they are ἐξ ἀνάγκη = 'without being able to help themselves,' which he then translates as 'mechanically' and those things which, in possessing a soul or ψυχή, are self-moving.[289] I would argue, however, that behind Taylor's explanation, which would appear correct in itself, is the further suggestion hinted at by his reference to νοῦς that the underlying contrast being made here is between νοῦς and its display of intelligent purpose and ἀνάγκη, not displaying such purpose and thus, with both causes appearing on an even, but opposing playing field and as primary.[290] I would argue, therefore, that the contrast being made here in using ἐξ ἀνάνγη is not between a true causal power, i.e. ψυχή or ψυχαί possessing νοῦς, and a secondary or non-causal power, i.e. mechanical motion, but rather, is between a motion displaying intelligent purpose and a motion which of itself does not. As Taylor has noted:

> The thought is a combination of the doctrine of the *Phaedo* that the 'good' is the real explanation of everything with the doctrine, first introduced in the *Phaedrus*, of the soul as the one and only 'self-moving' thing.[291]

---

288. Taylor, *PT*, 291. Also see footnote on same page.
289. "This is all that ἀνάγκη means here," Taylor, *PT*, 293.
290. See my earlier discussion regarding *Ti.* 46d7-e2.
291. Taylor, *PT*, 293.

It is only later, however, in *Laws X* where Plato will develop his theory of soul to explicitly include a non-beneficent causal power or soul. I would argue, therefore, that at this juncture care must be taken not to make the above comparison either more all encompassing or more discrete than that suggested, as a broader or more specific appeal is not supported by the text. Thus, the point of comparison or similarity which Plato makes, i.e. a lack of intelligent purpose between the individual secondary causes or mechanical motion(s) of generated objects and the disordering cause or ἀνάγκη, is where all such comparisons and similarities between these causes begin and end.

2.3.5.5 The mechanism of vision

In commenting on the accessory causes, Cornford begins with the passage from the *Timaeus* on the mechanism of vision at *Ti.*45b-46a, where he states:

> He [Plato] begins with the bodily mechanism of vision, for the sake of leading up to the contrast between these 'secondary causes' and the true reason or purpose, which is that man may learn number by seeing the heavenly bodies and so pass on through the sciences of number to all philosophy.[292]

Although Plato does give a clear statement as to why he thinks man was given the mechanism of vision by the Demiurge, he does not make the above statement regarding 'philosophy' in this passage, but in a subsequent passage beginning at *Ti.* 47a. Cornford has had to pass over important metaphysical detail given by Plato at *Ti.* 46c7-e6 to arrive at the conclusion that Plato did this "for the sake of leading up to the contrast between these 'secondary causes' and the true reason or purpose," with the result that the emphasis which is placed on the above-quoted contrast, and the contrast itself, are both contrived, but not by Plato.

There are two problems. First, the object of the purported "true reason or purpose" Cornford is suggesting here is unclear. For Plato to have included a discussion of the secondary causes for the sake of discussing vision, so as to further explain the origin of man's philosophic ability, is unlikely given that Plato's discussion of vision is exceedingly detailed and lengthy when taking into account the length of the *Timaeus* overall. This alone belies any move to claim that such a discussion was included 'for the sake of' some other purpose, which purpose, moreover, is barely touched upon before it drops from the discussion completely. To the contrary, Plato's inclusion of the secondary causes is a reasonable furtherance of his plan stated at the outset of the *Timaeus*, namely, to expound upon the origin of the cosmic system down to the story of man.[293] For

292. Cornford, *PC*, 152.
293. *Ti.* 27a2-6.

Cornford to have telescoped this general plan in an instance is quite rash and renders as discontiguous Plato's actual discussion which follows upon this section on the mechanics of vision, namely, the relative importance of the secondary causes in relation to the primary or true causes.[294] This brings into focus a second and ultimately more serious problem created by Cornford's analysis, namely, of removing Plato's metaphysics from the discussion at hand. Not only are the objects of the comparison wrong—clearly it is the primary and secondary causes which are being discussed at this juncture in the *Timaeus* and for their own sake—but the shift which Cornford suggests here, namely, towards a general emphasis on man's ability to do philosophy, is out of keeping with, and indeed, contrary to, Plato's stated purpose for writing the *Timaeus*, not only stated at its beginning as noted above, but also repeated in more specific terms later on:

> And if, for relaxation, one gives up discussing eternal things, it is reasonable and sensible to occupy one's leisure in a way that brings pleasure and no regrets, by considering likely accounts of the world of change. So let us now indulge ourselves and proceed with an account of the probabilities next in order.[295]

Thus, Cornford, by connecting and conflating the content of two separate passages, *Ti.* 45a6-46c6 above with 46e8-47b5,[296] and importantly, by omitting Plato's metaphysics wherein Plato does make an explicit contrast, moves the latter's attention away from the world of change (which is undoubtedly affirmed by him to be the object of his inquiry) to that of eternal Realities, wherein lie the objects, properly speaking, of philosophy. The contrast, however, which follows in the passage subsequent to *Ti.* 45a6-46c6, namely, 46c7-46e6, is not, as suggested by Cornford, between the means and the end, i.e. between the mechanism of vision and its purpose, but rather, as argued above, between the secondary or assistant causes and the Primary Cause(s). In other words, the contrast that Plato draws is between the secondary or assistant causes (those that operate through bodies) and the Primary Cause(s) or those of a rational nature. What is at issue here is not what Plato held to be more important, for there is no doubt as to the importance he placed on the purpose of vision or any other aspect of the physical world, his complaint of Anaxagoras[297] in the *Phaedo* confirming his desire to

294. *Ti.* 46c7-e6.

295. *Ti.* 59c5-d3. Lee, 84; also see Bury, 147 and 149.

296. Plato eschews the contrast suggested by Cornford when he writes: "As far as the eyes are concerned, we have said enough about the accessory causes" (*Ti.* 46e6-7).

297. See §2.3.5.1, Part I, on Anaxagoras. Plato's basic complaint in the *Phaedo* (97c-99c) is that Anaxagoras fails to recognize the power of Nous beyond the strictly physical or that of mechanical agent. It is important to Plato that the basis of order in the natural world be properly understood as resulting from the metaphysical power of Nous, which is an ordering that takes place first at the level of Mind. All physical power and its ordering, i.e. in the form of mechanical agents and processes, are not directly attributable to Nous,

properly understanding the physical world and its many processes. Rather, what is central to this discussion and to every discussion in this work is Plato's reason for writing the *Timaeus*, which is to offer a description and understanding of the physical world—the objects, apparatus, and phenomena which enable such purpose, and equally, the metaphysics behind the world, so as to gain an understanding of the source of its generation and support, and ultimately, the context of such purpose. It is significant that Plato's inclusion of his "disquisition on optics," under Cornford's analysis, seems to Cornford to be intrusive, which intrusiveness he softens by making reference to *The Republic* and Plato's interest in analogies between the bodily eye and the eye of the soul, and between the sunlight and the truth, when in fact it is Cornford's analysis itself which is the cause of any apparent intrusiveness.[298]

Moreover, the above-noted conflation and the errors which ensue from it soon lead, by the shear weight of the omission of Plato's full metaphysical infrastructure, to a more serious problem and hence, error, by Cornford, namely, his mistaking of the assistant causes referred to above (the various and sundry bodily mechanisms) and as discussed at *Ti.* 46c7-46e6 in the *Timaeus*, with the works of Necessity (Ananke) at *Ti.* 52d-53c.[299] Moreover, it is on the basis of this and similar errors, whereby Cornford is able to dismiss all notions of either a transcendent or immanent metaphysics for Plato, explaining away possible references to such meaning in terms of myth and metaphor,[300] and thereafter relating what would have been previously understood as metaphysical matters (e.g. the 'Demiurge,' 'Ananke,' the 'soul') almost exclusively, if not entirely, to the physical realm.

First, Cornford does note briefly the distinction drawn by Plato between the secondary and primary causes, referring to them as "subsidiary causes" and "rational purpose," respectively.[301] However, he only admits to this distinction at *Ti.* 47c4–e2, while completely ignoring any prior reference within the passage at *Ti.* 46c7-46e6. He does not elaborate further, introducing this subsequent reference only as a link to the second part of the *Timaeus* where a discussion on

---

but, rather, are a *consequence of* this metaphysical or noetic ordering and hence, are secondary to it. Importantly, without this metaphysical ordering, there would be no physical order. As regards their sphere of existence, however, these powers are completely separate. Thus, Anaxagoras's analysis is incomplete according to Plato, rather than wrong, the essential separation of the metaphysical world from the physical having not been articulated. Plato's focus remains at all times the proper understanding of the world and its processes, which however complex in their origin, must be properly understood, and is never a retreat into idealism.

298. Cornford, *PC*, 156.

299. See ibid., 157 and 209.

300. Such references, both explicit and inferred, to myth, poetry, and metaphor, appear throughout Cornford's *Commentary*, e.g. see *PC*, ix, 29–31, 40, 74, 173, 175–76, 209, and 361.

301. Cornford, *PC*, 159.

Necessity subsequently follows. Thus, first, by omitting any discussion of the Primary Causes referred to earlier, i.e. τὰς τῆς ἔμφρονος φύσεως αἰτίας πρώτας at *Ti.* 46d8 and τὰ τῶν αἰτιῶν γένη at *Ti.* 46e3, where, notably, such causes are referred to in the plural, and second, by further minimizing their importance by grouping them together and categorically describing them as "rational purpose" in the later reference at *Ti.* 47c4-e2, while third, offering no explanation for either their earlier dismissal within his discussion or the transformation of these Causes into 'rational purpose,' Cornford fails to properly exegete the Primary Causes in these passages as they stand and as they are referred to by Plato. The consequence of this is that Cornford's analysis in relation to these Primary Causes and what follows from his analysis, especially in what he says regarding Necessity following upon his discussion of the secondary causes and 'rational purpose,' while believable (if one fails to notice Cornford's move or simply accepts it) from the perspective of the terms that Cornford does choose (which are not Plato's terms) and interesting is not what Plato wrote, nor does it follow from what he wrote. This is clearly wrong, rendering Cornford's position untenable when considering the Primary Causes as they are actually addressed in the text.[302]

As mentioned above, the more serious problem for Cornford is when he ultimately identifies the assistant causes explained at *Ti.* 46c7-46e6 in the *Timaeus* with the works of Ananke described at *Ti.* 52d-53c,[303] blurring by apposition (substituting one for the other without adequate explanation) any logical connection or need for such a connection. He achieves this end in a slow and methodical way, moving inevitably but erroneously to the above conclusion. Thus, following the translation of Plato's discussion of Ananke at *Ti.* 47e-48e, Cornford repeats Plato's suggestion that a study of the irrational factors is now in order and follows similarly.[304] However, without explanation, Cornford's next move is to declare that these "irrational factors" (i.e. the 'works of necessity') are "at once connected with 'the nature of fire and air, water and earth'."[305] Against this move, however, besides not offering a rational basis for it, is Plato's explanation in the above passage that what he is concerned with here is the true origin of such elements as opposed to an overall silence regarding their generation, wherein they are usually posited as first principles, but which he openly scorns as being an unacceptable assumption to even those of little intelligence—"elements (as it were, letters) of the universe."[306] Thus, in introducing the Primary Causes, of which Ananke is one and Νοῦς or Mind is

302. See my full discussion beginning at Part I, §2.3.

303. See Cornford, *PC*, 157 and 209.

304. Ibid., 161.

305. Ibid. See also 199.

306. *Ti.* 48b8: στοιχεῖα was the regular Greek term for letters of the alphabet. Plato borrows this term in order to make a point. By using στοιχεῖα to describe the physical elements, he opposes the then current uninvestigated understanding of the universe, chid-

Primary Causes, of which Ananke is one and Νοῦς or Mind is the other, the very last thing Plato is proposing is to connect these causes with 'the nature' of any of the elements, and to the contrary, his purpose was to oppose, strongly, such a view.

2.3.5.6 The accessory causes and Ananke

Thus, Cornford's error lies in his ultimately identifying the assistant causes with Necessity (Ananke) and further, after abandoning the notion of 'Necessity' as a proper cause, identifying the assistant causes with the 'works' of Necessity. The root of Cornford's error lies in his over-emphasis of the modern adjectival understanding of an English translation of ἀνάγκη, namely, 'necessary,' but which usage in ancient Greek and specifically, consistently in the *Timaeus*, is rendered by the addition of the preposition of separation, ἐξ, i.e. ἐξ ἀνάγκης. As earlier argued, Plato explicitly uses the substantive ἀνάγκη in the *Timaeus* to depict the second of two primary causes, which causes, according to his theory, brought about the Cosmos. In doing so, however, Plato has borrowed a much earlier understanding of this compound word, which at that time would have had the meaning of 'force' or 'might.' Moreover, this idea of at least two primary causes is repeatedly and consistently maintained throughout the text. Cornford, on the other hand, has ultimately only considered a much later development of this word in his commentary on the *Timaeus*, with the result that he eventually arrives at a misidentification of the various movements of the elements with what he thinks is Plato's understanding of 'necessity.'[307]

2.3.5.7 An Empedoclean approach – 'necessity' as chance

At first Cornford does refer to 'Necessity' as a cause, but as he pursues his analysis and fails to garner a satisfying explanation from Plato, it becomes clear that Cornford wants to abandon this understanding in favor of a more tangible or specifically scientific interpretation. Thus, initially he examines Necessity and

---

ing that it would not appear as a reasonable representation even to someone of little intelligence, since for Plato such elements are neither first principles nor are they of so little importance as to be described after the manner of mere syllables. By playing into the richness of the language here, Plato turns possible ambiguity to his full advantage. While he next specifically eschews any attempt in this investigation to get at the true nature of first causes, due both to the difficulties involved and his current manner of proceeding, it becomes quite apparent as his treatise unfolds that a more in depth understanding of the elements was within its scope. Thus, Plato very clearly and from the outset, separates first principles or causes from the elements and their natures. See also Taylor's explanation in *PT*, 306–8.

307. This is not to say that Cornford did not choose this identification, perhaps knowing of the earlier etymology, but deciding to ignore it.

concludes that it is "the very antithesis of natural law," finding accord and support in Grote's view:

> 'This word (necessity)' he wrote [Grote], 'is now usually understood as denoting what is fixed, permanent, unalterable, knowable beforehand. In the Platonic *Timaeus* it means the very reverse: the indeterminate, the inconstant, the anomalous, that which can be neither understood nor predicted. It is Force, Movement, or Change, with the negative attribute of not being regular, or intelligible, or determined by any knowable antecedent or condition—*vis consili expers*' (*Plato*, iii, ch. 36).[308]

Just prior to this quote, Cornford has introduced Aristotle's discussion of the Empedoclean understanding of 'necessity' with its meaning of random chance, contrasting this with the idea of final causation in nature. Cornford does this in order to support his view that Plato's understanding of the term 'necessity' is of a lower kind of ordering in the form of the accessory causes or mechanical motion,[309] which in turn is a further development of this idea of random chance[310] or in other words, of order, but bereft of the higher ordering of soul. Aristotle, Cornford argues, clearly understood 'necessity' to mean random events, citing Aristotle's quote of the Empedoclean theory of causation (which both Aristotle and Plato opposed), where the opponent argues that natural phenomena, e.g. "rain falling," need not have an end or purpose, i.e. 'so that corn may grow,' but may fall ἐξ ἀνάγκης, "without purpose." However, by introducing this argument, Cornford once again moves the discussion in a non-Platonic direction, focusing not on the substantive ἀνάγκη, which for Plato was a primal disordering cause, but on the prepositional phrase ἐξ ἀνάγκης, which phrase, when Plato does use it, is undoubtedly used in its everyday sense of 'necessarily.' Aristotle's analysis here of hypothetical necessity stays clear of first causes, wherefore Plato had initially introduced the two primal causes, Nous and Ananke (Necessity). Equally, Plato's introduction of ἀνάγκη as a primary cause does not extend to his discussion of natural phenomena or ends within the Cosmos, which discussion occurs later on in his cosmology. With regard to these ends, moreover, Plato clearly acknowledges that they exist, both in the

---

308. Cornford, *PC*, 171–72.

309. Ibid., 172: "Here the lower type of causation, transmitting motion or change from one body to another, is, in the same breath, declared to proceed 'of necessity' and 'at random and without order'." As I have argued with regard to *Ti.* 46e, Cornford, by conflating the playing field in his interpretation of this passage, requires an equally conflated explanation of two of its key players, namely, Ananke and the necessary causes, which results in a rather absurd hybrid or "man-faced ox-progeny" in the form of random, but necessary order.

310. Arist. *Phys.* II 8. 198b17. See Cornford, *PC*, 165–66.

form of an overarching end (from which 'Ananke' by definition is excluded),[311] i.e. the good as exhibited in and extended by the Demiurge in the generation all things within an ordered universe[312] and also in the ends which are manifest in every generated object in the form of purposive design.[313] Plato's fairly extensive use of the phrase ἐξ ἀνάγκης in the latter half of the *Timaeus*, however, when he is discussing the physical phenomena within the already generated Cosmos, is never used with the meaning of random chance, but always and invariably so, with the meaning of exigency, i.e. that something necessarily follows. Thus, Cornford's introduction of Empedocles's understanding of the phrase ἐξ ἀνάγκης has no direct bearing whatsoever with respect to clarification of the Platonic text, although this most certainly was his intention. Plato does not use the prepositional phrase ἐξ ἀνάγκης when referring to the primary causal force of disorder or randomness, rather, he uses the term ἀνάγκη and when he does use the phrase ἐξ ἀνάγκης, he uses it entirely in the modern sense of 'necessity,' meaning 'necessarily so,' and never with the meaning of random chance or disorder.

Furthering this line of thought, Cornford's next move is to introduce from *Laws X* Plato's opposition to the early cosmogonies where it is thought that "Fire and water, earth and air . . . all exist by nature and chance, not by design," in other words, exist "by chance, of necessity" (κατὰ τύχην ἐξ ἀνάγκης)."[314] Cornford's purpose in introducing the passage is to highlight the existence of this earlier view, where 'necessity' is not only connected with chance, but now also with nature. Although in *Laws X* Plato does oppose the atheistic materialism of the early cosmologists, this is arguably a later work where his concern is not with the particulars of the cosmologies themselves, but

---

311. *Ti.* 29e1-3.

312. *Ti.* 31a8-b3.

313. There are many instances of purposive design in the *Timaeus*. The ordering of the heavens is perhaps the best known, which by their ordering (i.e. of the heavenly bodies), 'time,' generated as a consequence, becomes 'a moving image of eternity' (*Ti.* 37d5-7). Another equally important example, showing a refinement of and interdependence between ends, is that of eyesight, where, at *Ti.* 46e6-47c4, Plato describes our sight of the heavens as leading mankind to the derivation of numbers and from there, causing him to delve into the nature of the universe and ultimately, to discover philosophy, the "greatest gift of the gods." This same passage also shows a hierarchy between ends, where Plato eschews a discussion of lesser such gifts as being for those not interested in philosophy, claiming philosophy to be the greatest good of eyesight: λέγω δὴ τοῦτο ὀμμάτων μέγιστον ἀγαθόν· τἆλλα δὲ ὅσα ἐλάττω τί ἂν ὑμνοῖμεν, ὧν ὁ μὴ φιλόσοφος τυφλωθεὶς ὀδυρόμενος ἂν θρηνοῖ μάτην.

314. Cornford, *PC*, 167, referring to *Lg.* 888e-890b.

with the implications that such views would have upon belief in the gods.[315] The same objection would hold here, however, as it did for the Empedoclean distinction noted above, namely, that the prepositional phrase ἐξ ἀνάγκης, which occurs frequently in the Greek language and hence, in both *Laws X* and the *Timaeus*, is distinct in its usage from the substantive ἀνάγκη as used in the early stages of Plato's cosmology in the *Timaeus.* Nevertheless, a point which Cornford repeatedly tries to convey, namely, the connection of 'necessity' with spontaneity, coincidence, and chance as opposed to 'purpose' and its further identification with chance as opposed to 'design' in *Laws X*, is correct, but only insofar as there is no degree of ordering meant or indicated when Cornford uses these descriptive terms as synonyms for 'necessity.' Ananke, as has been argued, is the exact opposite of Nous as a causal power, Ananke being degenerative and thus disordering as distinct from Nous, which is generative and ordering.

How Cornford consequently errs is twofold. First, with respect to his understanding of 'chance'[316] from the Aristotelian passage noted above, although Cornford distinguishes this idea from any kind of *intentional* ordering (or disordering), he does not distinguish it from the ideas of 'ordering' and 'disordering' themselves—it can be either, as long as it is unintentional. However, inasmuch as he fails to do this, his analogy misses the mark. All ordering in the Cosmos is understood in the *Timaeus* as having been brought about by the desire or 'will'[317] of the Demiurge (although the carrying out of this activity is allocated in part to the gods) and therefore whenever order is present, so, too, is intention. All disorder (except re-generative disorder), on the other hand, is the work of Ananke (Necessity). Thus, most,[318] if not all instances of spontaneity, coincidence, and chance that also involve ordering within the generated Cosmos cate-

---

315. *Laws X* becomes relevant, but not with regard to the prepositional phrase ἐξ ἀνάγκης, rather, with regard to references to at least two souls, one beneficent and the other not.

316. Cornford appears to group the ideas of 'chance,' 'spontaneity,' and 'coincidence' together, so as to mean the opposite of any kind of intention, rather than a lack of order or disorder *per se*.

317. See Part II, §2.3.6.1 concerning 'will.'

318. At *Ti.* 69b6 Plato refers to things standing in no measurable relations "except by *chance*" (ὅσον μὴ τύχῃ) prior to the organization of the universe by the Demiurge. However, all order, intentional or not, is closely associated by Plato with 'nous,' being ultimately the source of all order (and as a result of which order, the Cosmos was generated and is sustained), whether wrought by chance or brought about through the noetic movements of the Demiurge and the gods, just as all disorder (re-generative disorder excluded) is formed by disassociation from nous, again, whether wrought by chance or brought about through the disordering power of Ananke. See *Ti.* 29d7-c1. Thus, although the reference to 'chance' above involves a pre-cosmic instance of ordering and accordingly, stands side by side with Ananke's disordered universe, this instance of order, I suggest, exists in spite of Ananke and not because of it. The notion of chance, therefore, refers exclusively to the absence of intention and not to the absence of nous.

gorically lie outside of Plato's understanding of 'necessity' (ananke) no matter how they come about.[319]

The second way in which Cornford errs is in relegating all subsequent disorder within the Cosmos to mere chance phenomena, having disregarded and dismissed Ananke altogether as an independent causal force before, during, or after generation. In rendering Ananke ('necessity') non-existent as an independent causal force, Cornford soon identifies 'necessity' with the accessory causes. This move, however, is contrary to the text and as shown, does not hold up well upon scrutiny. When Plato does refer to disorder as remaining in the Cosmos after its generation he acknowledges this as being due to the fact that the Demiurge did not entirely subdue Ananke as a causal force during its generation. Consequently, whatever remains in the Cosmos that is degenerative or unordered marks the residual active presence of Ananke, having never been fully subjugated by the Demiurge. This is not to say, however, that this marks Ananke's *only* existence. The picture of Plato's cosmology that Cornford presents is considerably less robust than that which the text presents, since as a causal force Cornford at best explains Ananke only in terms of its residual effects, whereas Plato is most explicit in presenting both Nous and Ananke in the first instance as primal causal forces. At worst, Ananke is ultimately identified and espoused by Cornford as falling within the concept of the 'accessory cause,' which is simply wrong.

2.3.5.8 Ananke understood as 'phenomena'

Cornford moves to his final understanding of Ananke or that of 'phenomena' (i.e. the accessory cause) in stages, namely: first, by abandoning the idea of Ananke as a primal causal force; second, by consequently concentrating exclusively on the usage of the prepositional phrase ἐξ ἀνάγκης; third, by next appealing to an understanding of ἀνάγκη in terms of that usage whereby Ananke is initially explained in terms of spontaneity and change, being both irregular and unintelligible;[320] and finally, fourth, by interpreting Ananke or 'necessity' in terms of phenomena exhibiting or bringing about change within the generated Cosmos.

Having already discussed the first two stages, I would like to address this third stage. While the text in the *Timaeus* presents Ananke as a primal mover or force, a cause operating without order, i.e. without 'nous' or intelligence, Cornford (in quoting Grote) alters the figure of Ananke substantially, changing it from an active agent outwith the universe to a passive, abstract noun within it,

---

319. As I have argued separately, the physical causes, that is the forces and processes within the Cosmos, lie outside the realm of the primal causation, which latter movements are entirely noetic or anoetic, depending on which primal cause is the source.

320. Cornford, *PC*, 172.

whereby Ananke instantly becomes the collective term for the impotent observed facts of irregularity and confusion within the generated world, i.e. "the indeterminate, the inconstant, the anomalous, that which can be neither understood nor predicted."[321] This leaves, however, the possibility that Ananke could still operate intelligently (only without our understanding) and reasonably (i.e. according to some scientific, but unknown law), and since not understood by us, not predictable. This possibility, I suggest, posed a problem for Cornford. Having already denied Ananke any sort of metaphysical status as a primary cause by not allowing it to have pre-existence as a force outwith the generated Cosmos, likely being too 'religious' or Christian in connotation for Cornford[322] and in further allocating all intelligence and understanding to Nous (or for Cornford reason in the universe personified as "Reason"), he consequently abandons this idea of Ananke as an abstract subject. What he consequently does is to then appropriate and further Grote's understanding of 'necessity' even more, where the subject is now switched and it is *man* who does not understand 'ananke' or is able to predict its actions. For Cornford this allows 'ananke' to become an object of inquiry and veers far enough away from the notion of a mysterious, forceful subject, an irrational position with which Cornford has already expressed discomfort and rejected as Christian conjuring, to become broachable within the comfortable context of natural phenomena. However, in doing this, Cornford winds up almost where he begins, i.e. with 'necessity' still possessing possible intelligence and remaining a force capable of acting independently, thus leaving in its wake this unintended and undesirable consequence, which, at very bottom,

---

321. Cornford, *PC*, 172.

322. Ibid. See also v, ix, 35–37 and 163–164. Cornford, from the very first page of the 'Preface' to his commentary, sets out that it is his intention to discuss "each problem of interpretation" which arises in his (or any) translation of the *Timaeus*. One such problem, however, which becomes a mainstay of Cornford's criticism, is what he describes as the "risk of falsifying the sense, especially by misleading reminiscences of the English Bible." What Cornford fails to discern, however, is that to remove such reminiscences simply because they are reminiscences is to risk equally falsifying the passages at hand. Thus, in his eagerness to cut out all similarities between Plato and Christian thought, he subsequently fails to allow room for any similarity and in so doing, distorts the text according to a new bias, a markedly non-Christian one. Thus, at page ix of the 'Preface' one finds Cornford decrying the postulated conclusion (of a hypothesized Christian no doubt) that "Plato (or Timaeus?) is at heart a monotheist and not far from being a Christian." This translates into an exaggerated response by Cornford as he writes his commentary, so that wherever potential similarities arise in the text he must cut them out with razor-sharp efficiency and insert an equally 'pagan' or alternatively, preferred modern scientific reading of the passage at hand. On balance, this is not good historiography, merely pointing to Cornford's impatience with Christianity specifically and to religious interpretation generally, although as I have pointed out, he will concede to a pagan reading, as long as there are no similarities to Christianity. This unfortunately is a weakness in Cornford's methodology in his commentary on the *Timaeus* and must be kept in mind if the reader is to distance oneself from this unabashedly studied bias.

is a position which Cornford would clearly deny, since 'necessity' for him is both irrational and powerless, and better explained by natural science, a position which he also sees Plato as sharing. The solution to this apparent conundrum, however, finds closure in Cornford's fourth and final move, which is to have 'necessity' redrawn as the 'works' of necessity, where the meaning of 'necessity' has now become an abstraction exacted from an analysis of the prepositional phrase ἐξ ἀνάγκης (not from the substantive ἀνάγκη as used in the *Timaeus*) and then to identify these 'works' with the συναίτια, which Plato has argued are only accessory and not true causes. Thus, the metaphysical has become the physical and 'necessity' has become the equivalent of mechanical motion.[323]

By introducing Aristotle's analysis of the Empedoclean understanding of ἐξ ἀνάγκης the idea of necessity for Cornford becomes irreversibly and intrinsically linked with the abstract notions of spontaneity, coincidence, and chance, not with the singular metaphysical or primal causal force first introduced by Plato in the *Timaeus* and contrasted with Nous at *Ti.* 47e5. Thus, following on from the line of thought introduced by Aristotle, Cornford describes " 'a necessary' result" as that which comes about by causes that cannot act otherwise than they do and are not directed by purpose.[324] In quoting Socrates at *Phlb.* 28d, Cornford cites Socrates, who asks: "Which are we to say, Protarchus—that everything, this 'whole' as we call it, is at the disposal of a force that works without plan, at random, and just as it may chance, or on the contrary . . . that it is an ordered system, guided by some admirable reason or intelligence?" At this point 'necessity' for Cornford is still linked to the idea of a causal force, but only just, for in quoting a similar passage immediately following from the *Sophist*, a second alternative to "admirable reason or intelligence" (i.e. 'divine craftsmanship') is then offered in place of the above-quoted "force that works without a plan," namely, a spontaneous cause within Nature ("Φύσις") that generates things without intelligence (*Sph.* 265c).[325] It is this latter alternative, "a spontaneous power of generation" from the *Sophist*,[326] supported by the passage from Aristotle (*Ph.* B, viii, 198b) and not the "force that works without plan," which Cornford will hereon in begin to identify as 'necessity' and contrast with design. Thus, by a literary slight-of-hand, Cornford has placed two alternative notions side by side, and through the literary device of apposition,[327] chosen the

323. Cornford, *PC*, 54; also see 164.

324. Ibid., 166.

325. Ibid., 167.

326. Ibid.

327. 'Apposition' is a literary device whereby a thing or person is more fully identified by a phrase, usually a noun phrase, being placed alongside it. The object so qualified is capable of being replaced by or identified with the descriptive with no further explanation expected or required (e.g. "that man, the coward"), as the truth is generally assumed. This is a powerful, although easily abused device, better suited to poetry or drama than

second. Thus, offering no reason for this choice, he simply appends it as the appropriate meaning for an altogether different work, namely, the *Timaeus*. Moreover, after adopting the meaning of 'necessity' for the *Timaeus* from the *Sophist*, Cornford then refines this borrowed meaning by opposing a "vaguely personified 'Nature',"[328] ascribed in the *Sophist* and the *Physics* to this 'spontaneous power of generation.' In this way, not satisfied with simply abandoning the meaning of Ananke from the *Timaeus* and appropriating a meaning from elsewhere, Cornford must duly edit this meaning as well, again excising all metaphysical content. However, there is little philosophical basis or argument, and none given by Cornford, for choosing this meaning either as an understanding of Ananke in the *Timaeus*.

Forming part of the progressive movement of his suggested 'fourth stage,' Cornford next identifies the above-noted 'spontaneous power of generation' with "their several powers (active properties, δυνάμεως)." Quoting Plato's summary of the atheistic materialism that is being opposed in *Laws X* (888e-890b), the inanimate elements, fire, water, earth, and air, are described as moving "by the *chance* of their several powers."[329] It is with these 'powers moving by chance' that Cornford now attempts to identify 'necessity.' There are, however, two problems with this. Firstly, as pointed out earlier and as again substantiated by his explanation of the phrase 'by chance, of necessity' (κατὰ τύχην ἐξ ἀνάγκης)[330] in the above passage, Cornford's understanding is based on an analysis of the common prepositional phrase ἐξ ἀνάγκης and not on the substantive ἀνάγκη, both of which are found in the *Timaeus* and are clearly distinguishable. Secondly, Cornford makes no attempt at clarifying what is meant later on in this passage by a further phrase describing these elements and the actions of their several powers moving by chance as "fit[ing] together with some sort of affinity—hot with cold . . . and in other mixtures that result,"[331] but simply glosses over it. Plato, however, was disturbed with this as an explanation, as he would certainly be now with Cornford's acceptance of it in like terms. The third problem, however, which stems from errors in both method and logic, leads quickly to a fatally flawed conclusion and culminates in Cornford's fourth and final move.

---

philosophy. Well-known examples are found in the *Psalms* and the *Song of Solomon* of the English Bible, but are also found in it elsewhere, e.g. "He raises up the poor from the dust; he lifts the needy from the ash heap" (*1 Sam.* 2.8). The poetical breadth and inherent richness of such passages are not permitted by the philosopher, who is expected to exact a pure meaning or alternatively, to substantiate the meaning of any second phrase so appropriated by careful analysis.

328. Cornford, *PC*, 167.

329. The italics are Cornford's.

330. Ibid.

331. Ibid.

First, by Cornford taking up the ideas associated with atheistic materialism from *Laws X*, clearly rejected by Plato, where the δυνάμεις or 'active powers' within the four elements bring about 'by chance' the existence of the heavens, the sun, moon, stars, and earth,[332] and then applying this whole interpretation (i.e. in its broadest sense, involving motions, and not simply borrowing the idea of 'chance') to Plato's understanding of Necessity in the *Timaeus*,[333] Cornford errs seriously in method if what is hoped for is an accurate vision and honest interpretation of how Plato understood 'Ananke' in the *Timaeus*. Plato rejected this move outright by rejecting the whole atheistic materialist agenda—'soul' is more than just some innate earthly power in *Laws X*. Whereas Ananke, as has been previously argued, is understood by Plato as a negative metaphysical force in the *Timaeus*, this idea is no longer relevant in *Laws X* and is consequently replaced by the notion of a non-benevolent soul(s)[334]—making the materialistic position engineered by Cornford blatantly false, although respectable as a contemporary explanation, in representing the metaphysical position persistently held by Plato regarding negative causal powers and causality generally. Whatever Plato may have thought in *Laws X* concerning ἐξ ἀνάγκης (he does not introduce the substantive ἀνάγκη outside of the *Timaeus* and he does not discuss this prepositional phrase), if the discussion is to have any bearing on the *Timaeus*, it must be subject in the first instance to what Plato thought about ἀνάγκη in the *Timaeus*, and not be made simply to replace it. This amounts to bad scholarship, not just poor method.

At this juncture, in drawing together the final pieces of his above-noted interpretation of 'necessity,' Cornford drops all references to Plato and concentrates solely on the "atheistic materialism" (which Plato has clearly rejected), taking Cornford through to the end of his discussion on 'Reason and Necessity,' moving all the while in concentrically smaller circles and ending with what can best be described as an approximation of a cross between the pantheism of Aristotle and an early animism, where nature is 'alive' and Reason is 'deified.'[335] It is worth quoting Cornford in full as he begins his closing argument concerning Necessity:

---

332. Cornford attempts to give credibility to this position earlier when he introduces Aristotle's summary of the Empedoclean objection to Plato's position. Although a slightly different argument concerning ends (the Humean-like causal idea of no ends as distinct from the idea of natural ends put forward here), it nonetheless supports the latter by also rejecting the metaphysical idea of 'the good' as an end and thus appears as a natural progression of thought, representing the earlier stage, moving from the pre-scientific to the scientific, and always with any metaphysical implication carefully rooted out.

333. Cornford, *PC*, 167–68.

334. See Part II on *Laws X*.

335. Cornford, *PC*, 168–77. Only at the very end does Cornford again refer to Plato, introducing yet another character in his place to provide interpretation, Thucydides.

> In this passage of the *Laws*, as in the *Physics*, we find necessity linked with chance, while law (νόμος) and order are connected with design. Chance and necessity, moreover, are associated with 'Nature', which is credited by the materialist with some spontaneous power of generation. This idea has survived from the earliest cosmologies, which had conceived the primary element or 'nature of things' as living. In consequence, the first physical philosophers had felt no difficulty about an original cause of motion. The divine and immortal substance of the world moved and gave birth to individual things, because it was alive. It was only later that this substance came to be reduced to the level of the bodily, which needs some external force to move it about. At this stage separate moving powers emerged: the Mind of Anaxagoras, the Love and Strife of Empedocles. These forces, however, remained part of Nature; they were not what we should call immaterial, but were extended in space. They retained that power of self-motion which had originally resided in the primary substance; but their motion was not directed by purpose towards any ideal of perfection in an ordered world. Even Anaxagoras' Mind, in spite of its name, had not been represented as working with conscious design for any good end, but only as giving the first impulse of mechanical motion to the revolution, or cosmic eddy, in which the world takes shape.
>
> In the last of these physical systems, the atomism of Leucippus and Democritus, the cause of motion seems to have entirely disappeared.[336]

There are several points from this passage that require addressing. First, a material "substance" as it is articulated above is foreign to Plato's thought in the *Timaeus*. In this work Plato was primarily concerned with investigating the being of things from the standpoint of their intelligibility and for him there was no other focus meriting investigation. He traces this intelligibility, first, from the object to the ultimate maker of things, the Demiurge, then to goodness, to Nous, to intelligibility itself, and finally to the individual Forms, the abiding universal and only constant of all things, thereby tracing intelligibility from what is fleeting to what is ideal. With regard to the latter, contrary to what Cornford states above about 'substance,' these Forms do not extend into space, such placement being reserved exclusively for the copies of the Forms, understood by Plato as the ever-changing and transient objects of becoming. Further, the copies themselves, although composed of the same elements noted by Cornford, are nonetheless constructed from an arrangement of basic triangles and have little in common with the above idea of 'substance.' Again, contrary to the above, Plato does not talk of these triangles as ever being 'alive' on their own account or self-moving. Rather, they were brought into proportion by the Demiurge by means of

336. Cornford, *PC*, 167–68.

forms and numbers,[337] with their movement maintained by a certain disequilibrium existing between them and also by the forces extant in the Cosmos.[338] Prior to this, the movement of the simple triangles and elemental 'qualities' were maintained by the general disequilibrium existing within the unordered universe. These triangles, therefore, were never immortal or divine, but were generated into more refined shapes and thus, had a beginning (as did the Cosmos) and were kept in motion by the interplay of their various constructions, assisted by the physical conditions in which they found themselves and the powers to which they were subject.

Second, Cornford, when speaking of "primary substance" as it "came to be reduced to the level of the bodily," would not be correct if he had in mind the idea of an underlying substance qualified by accidental properties, akin to Aristotle's idea of the individual, i.e. an individual 'this' consisting in the unity of its matter and form, since this idea is also foreign to Plato's account. Plato explicitly denies stability to objects of becoming when he writes:

> Whenever we see anything in process of change, for example fire, we should speak of it not as *being a thing* but as *having a quality*; water, again we should speak of it not as a *thing* but as *having a quality*. And in general we should never speak as if any of the things we suppose we can indicate by pointing and using the expressions 'this thing' or 'that thing' have any permanent reality: for they have no stability and elude the designation 'this' or 'that' or any other that expresses permanence. . . . We should only use the expression 'this

337. *Ti.* 53a7-b5.

338. *Ti.* 57d7-58a1. For Plato, motion can never take place in conditions of uniformity, motion and equilibrium always being disassociated. According to his theory of motion outlined in the *Timaeus*, before the generation of the Cosmos by the Demiurge the Nurse of Becoming, although characterized by the qualities of the various elements, had no homogeneity or balance of the forces within it, this disequilibrium taking the form of constant movement and separation of its contents. Plato's understanding of movement in the Cosmos after its generation by the Demiurge builds onto this same idea, but with various dissimilar groupings of order or homogeneity in the form of the various elements now added. Thus, the groupings of triangles forming the four elements, being dissimilar and assisted by the constant shaking of the Cosmos, change places with one another (except for earth), changing also in shape and size. The element with the fewest faces (fire) is the most mobile, lightest, sharpest, and the most penetrating, while that with the most faces (earth) can only wait to regroup with its own kind upon dissolution, being incapable of penetration or exchange with the other elements. When enough similar elemental shapes meet up, the separate elements are then carried by the various motions to their respective regions in the Cosmos (according to their size and weight). This separation is never permanent, however, due to their constant interaction with the other elemental shapes at their perimeter (hemmed in by the sphere of the heavens within which they form a continuum, there being no void) and the overall shaking motion of the Cosmos. The consequence is that there is a continuous, unending source of motion throughout the extent and breadth of the Cosmos that takes place.

> thing' or 'that thing' when speaking of that in which this process [of change] takes place and in which these qualities appear for a time and then vanish; we should never apply them to any quality, to hot or cold, for example, or any other contraries, or to any derivation of them.[339]

Thus, clearly, Cornford did not consider the resultant problems in suggesting that the idea of "divine and immortal substance" was Plato's understanding, particularly as this conflicts with the latter's explicit denial of material substance ever being "alive," but also with regard to his denial above of the existence of "individual things," having granted such existence only to the Nurse of Becoming, the voidless 'space' in which the process of becoming takes place.

Third, the very problem which Plato had with respect to Anaxagoras's understanding of Mind was that this 'separate moving power' remained in Nature. Thus, this aspect was precisely what Plato did not wish to embrace and yet this is what Cornford appears to want to recoup in his argument.[340] For Plato, the two primary moving powers or causes exist outside of the generated Cosmos, being logically prior. Cornford, however, as he continues with his historical scheme, picks up on a relatively new notion, namely, that of separate moving powers which are extended in space and are self-moving, and reminds the reader of their original locus, that is, of their residing in primary substance, which is ultimately a view he infers to share with Aristotle and on which his understanding of 'necessity' in the *Timaeus* ultimately comes to rest.[341]

Finally, Cornford expresses the complaint (*PC*, 168) that in the last of the physical systems, culminating with Leucippus and Democritus, the cause of motion seems to have entirely disappeared. Plato also complains (*Laws X*, 888e-890b), but again, we find Cornford siding with Aristotle, not with Plato, and consequently appropriating Aristotle's solution and understanding, and not Plato's. As already outlined and argued, Plato shared many ideas with the atomists, but like Aristotle[342] was dissatisfied with the idea of motion being left without a cause. This is where Plato radically departs from Aristotle's position (and ultimately from Cornford's), however. Aristotle was dissatisfied with the notion that the only principle left governing the motions of these particles or atoms by the Atomists was the unanalyzed axiom of the tendency of like coming together with like, and in this case, of atoms coming together in the vortices. As a bare, unexplained fact it was not acceptable as an explanation. Cornford, carrying on from Aristotle, states that this principle is still found "as an ultimate unexplained

339. *Ti.* 49d3-50a4; see Lee, 68.

340. See *EGP*, 235; Plato, *Phaedo* 97d6-98c1.

341. How exactly separate moving powers can be extended in space is problematic, but is beyond the scope of this book.

342. Aristotle, *Physics*, B, 4, 196a, 25.

assumption, at work in the chaos of the *Timaeus*."[343] This, however, is a gross over-simplification of Plato's position, for although Plato does give voice to the position of like coming together with like,[344] as does modern science[345] (but with its own explanations), he goes into considerable detail providing a possible explanation for this occurrence.[346]

2.3.5.9 Ananke, 'chance movement restrained'[347]

It is precisely at this point in Cornford's historical survey that he slips into the position of explaining 'necessity' (Ananke) as though it were chance movement which has been restrained by previous motion, or in Platonic terms, has become an 'accessory cause':

> From another point of view the result may be called necessary, in the sense that every motion takes place 'under constraint' (ὑπ' ἀνάγκης) of some previous motion: an atom receives a shock and blindly passes on. But the ancients had not discovered the laws of motion: to say that a movement happens 'by constraint' is not to say that it conforms to any law. Necessity, in fact, did not carry with it the associations of law and order, at any rate in the earlier phases of atomism.[348]

This final move, however, when considering the accessory causes as Plato presents them, apart from all earlier suggested errors in method by Cornford in relation to Ananke ('Necessity'/'necessity'), presents in itself an extravagant error of logic.[349] Still ignoring the underlying logic of his argument concerning the identity of Ananke with the accessory causes and wishing to strengthen his position further, Cornford introduces yet another player, Thucydides, who "believed in Fortune, defined as 'any non-natural agency which breaks in, as it were, from outside and diverts the current of events, without itself being a part of the series or an effect determined by an antecedent member of it'."[350]

---

343. Cornford, *PC*, 169.

344. *Ti.* 53a4-6.

345. It is interesting to note that Cornford appears to associate a 'scientific' conception of the world with that of a "strictly mechanical" one. See Cornford, *PC*, 170.

346. *Ti.* 55d6-58c4.

347. In Cornford's explanation of the Empedoclean idea of 'necessity,' necessity becomes associated with 'spontaneity,' 'coincidence,' and 'chance,' which three are closely linked in meaning and are all opposed to purpose. See Cornford, *PC*, 166.

348. Ibid., 169.

349. See Part I, §2.3.5.11.

350. Cornford, *PC*, 170.

2.3.5.9.1 Thucydides

Comparing this fifth-century historian with Plato, Cornford finds a certain analogy between the two: "Thucydides sees the field of human action divided between human foresight and chance; Plato sees the world of physical events divided between divine purpose and chance associated with necessity."[351] Although there are some grounds for comparison, the idea of 'chance' as presented by Cornford and as then developed and associated with the idea of 'necessity' understood from the perspective of the common prepositional phrase ἐξ ἀνάγκης, albeit in its larger historical framework, is a separate consideration from the idea of 'Ananke' as introduced in the *Timaeus*. In this work, Ananke, as a substantive is presented in the first place as the second of two Primary Causes existing as a disorder force outside of the generated Cosmos and therefore outside the realm of either 'the field of human action' or 'the world of physical events,' existing secondarily, however, as residual disorder or arguably, the Non-beneficent Soul, in both realms.[352] As regards function, 'Ananke' for Plato was the antithesis of Nous,[353] the ordering cause, with the idea of 'chance' adding a possible further descriptive to Ananke as long as this notion did not involve intentional order, only chance order and was not its sole meaning.

2.3.5.10 Ananke identified with the accessory cause

Cornford now drops his discussion of 'necessity' as associated with 'chance' and moves again to Plato's description of the secondary or accessory causes. As argued previously, however, Cornford conflates what Plato states in the *Timaeus* concerning the mechanical processes of eyesight (Plato having completed that discussion) with subsequent passages immediately following regarding both true causes and accessory causes. Thus, whereas Plato now clearly distinguishes between two *kinds* of causes, those that are causative in a primary metaphysical sense and those that have no power of their own, but only participate in a process, Cornford now mistakenly identifies the 'two kinds' of the "both kinds of cause"[354] as representing, firstly, the Demiurge, which stands for 'Nous' or Reason and, secondly, the accessory causes, leaving out the notion of 'Necessity'

351. Cornford, *PC*, 171.

352. See the relevant sections of this book regarding the ways in which Ananke permeates the generated world, either as a remnant of disorder remaining in the generated Cosmos or arguably, in its re-emergence into the world as the 'Non-beneficent Soul' (for this see Part II on *Laws X*).

353. Ananke, as the antithesis of the ordering power Nous, is, I suggest, a much larger concept than simply "the very antithesis of Natural Law" (Cornford, *PC*, 171), being the second of two causal forces which existed *before* the Cosmos or natural world was generated, wherein Natural Law was later seen to be at work.

354. See Cornford's translation, *PC*, 157.

altogether. As a consequence, however, and one which complicates matters still further, because Cornford does take seriously the discussion by Plato involving 'Ananke,' he develops an explanation (discussed at length above) whereby 'necessity' is now to be understood as representing in a general sense the 'accessory cause.'

Because of this confusion, Cornford's remaining discussion becomes quite inflated and to some extent even 'mysterious,' whereby he must end his explanation by trying to answer a large cortege of questions that his conclusion now raises.[355] In summing up this conclusion, Cornford, as before, now turns to Aristotle[356] and distinguishes two kinds of necessity, (1) the necessity which Aristotle calls 'hypothetical necessity' or the indispensable means to an end, e.g. food in relation to life (although it is not necessary that we should live) in contrast to 'absolute necessity,' and (2) the necessity residing in the properties themselves and governing their action (a quite different notion of 'necessity' from the notion of 'chance,' which Cornford has argued for above). However, rather suddenly and hence, oddly, Cornford attempts to explain this second notion of 'necessity' in terms of Plato's 'wandering cause,'[357] stating that although fire acts by constraint of its nature, e.g. it has the characteristic power of burning heat, "the fire is indifferent to my purpose and has none of its own." He then qualifies this necessity and further states that one can direct it to a "foreseen and purposed end" (e.g. to a griddle or chimney).[358] He then ties this second type of *extended* necessity (extended as it relates to the wandering cause, e.g. necessity as 'fire,' when subdued by the persuasion of Reason, becomes extended, e.g. a 'hot griddle') directly to Plato by referring to the *Phaedo* and Socrates's reference to 'that without which the cause would not be a cause.' Cornford then makes the further claim that Socrates's regret in the *Phaedo* that this distinction of causes should be applied to the explanation of the world as a whole, but which he himself was unable to attempt, is now taken up by Plato in the *Timaeus*. In making this final claim, Cornford ties the hypothetical necessity of the *Phaedo* and its explanation by Aristotle to his (Cornford's) explication of the prepositional phrase ἐξ ἀνάγκης from the *Timaeus* (and other sources), depending for its overall consistency on numerous errors of method along the way. He then sums up his conclusion, stating:

> And here in fact we find him [Plato] speaking of the Demiurge as making use of the lower kind of causes as auxiliaries (συναίτια) or

355. Cornford, *PC*, 173.

356. Aristotle, *Metaph.* Δ 5 ("where the various meanings of 'necessity' are distinguished," Cornford, *PC*, 174).

357. Cornford, *PC*, 174.

358. This example is my own, not Cornford's.

> subordinate instruments in his work of producing the best results possible (e.g. at 46c).[359]

Thus, although ingenious in some respects, i.e. in the sheer effort which Cornford has undertaken to arrive at his conclusion, the truism that it takes just as much genius to be wrong as to be right holds true for this extended argument, where Necessity in the *Timaeus* is identified by Cornford with the accessory cause. Too many errors in method have occurred and the underlying logic has been consistently ignored.

2.3.5.11 A fatally flawed conclusion

There remains, therefore, one more aspect to be discussed in Cornford's argument, concerning what I have already labeled as his 'fatally flawed conclusion,' and that is his error in logic resulting in an unsound argument. While it can be argued that Cornford's premises may indeed be false, e.g. his method badly flawed, it can still be claimed that his conclusion is still nonetheless true, based on other assumptions or perhaps on a broadly based inductive argument. However, Cornford's conclusion, equating 'necessity' with the secondary causes, when compared to the actual claims Plato made in the *Timaeus* regarding Ananke and the accessory causes (outlined above), has absolutely no basis in fact and therefore is false, creating an unsound argument which must be wholly rejected.

In order to put forth this position, however, a number of key facts must first be assumed. One, it must be admitted that Plato did in fact have something to say concerning the substantive ἀνάγκη ('Ananke'), either directly or indirectly, beyond its appearance and use within the common prepositional phrase ἐξ ἀνάγκης; two, that what he had to say concerning Ananke was of a metaphysical nature and not a physical nature; three, that his discussion of the accessory causes had nothing to do with the first two, namely, the substantive 'Ananke' or metaphysics generally; further, four, that Plato admitted to there being true causes and adjoins 'Nous' (identified with the Demiurge) and 'Ananke' as being the only true causes in the generation of the Cosmos, both infinite and capable of directing movement within the Cosmos; and finally, five, and conversely, that the accessory causes are decried by Plato as being in any way whatsoever true causes in themselves. Rather, they are assistant to the true generative causes, having no power of their own and existing finitely in time after the generation of the Cosmos. Admitting to any one of these facts, and I hold they all are true, makes it impossible to logically maintain Cornford's final position, namely, that 'necessity' and the accessory causes are one and the same thing. Admitting to all of these facts, however, as my argument does, presents

359. Cornford, *PC*, 175.

not only a valid argument concerning Ananke, but a sound one. It is, moreover, an argument in which Plato's claims are considered in the first instance and their textual integrity in the second, whilst historical interpretation and scientific reinterpretation are granted input only latterly,[360] given that the initial two first-order considerations provide the requisite congruity and clarity in themselves to put forth a sound argument and robust position.

With regard to the first two premises, namely, that Plato did speak to the subject of the substantive 'Ananke' distinctly and separately from its use within the common prepositional phrase ἐξ ἀνάγκης and that a distinctly metaphysical nature was accorded it, these have already been argued for extensively and consequently averred in Chapter 1 of Part I, beginning at §1.3.1, which chapter formally addresses that section of the *Timaeus* pertaining to *Ti.* 47e3 to 47d6. That Plato did speak to the general fact of Ananke, however, is not disputed by Cornford, at least in its mythic form, and is reflected in how Cornford sets up his commentary, heading his chapter on this subject as '*Reason and Necessity.*'[361] What is disputed are the terms in which he wishes to interpret 'necessity,' giving heavy weight in his interpretation to several non-textual sources, including later arguments by Aristotle, philosophical insight from the fifth-century historical writings of Thucydides, as well as earlier arguments by Anaxagoras, Empedocles, Leucippus, and Democritus, but settling for the most part on an extraordinary identity with the above-noted prepositional phrase, whilst ignoring the details of the text itself, where, paired with Nous, one is most apt to find the proper identity and meaning of ἀνάγκη.[362]

With regard to premise three above, namely, that Plato's discussion of the accessory causes had nothing to do with the substantive 'Ananke' or anything of a metaphysical nature, this at first glance would appear to be exactly what Cornford is maintaining. However, Cornford only achieves this, first, by denying, either by omission or directly, all substantive and pre-cosmic existence to 'necessity,' which, however, should such existence be admitted, would point to a metaphysical interpretation for 'necessity'; second, of further interpreting 'necessity' in terms of the prepositional phrase ἐξ ἀνάγκης; and third, then, but only then, by equating the accessory causes with all actions which happen 'of necessity.' Thus, by solely interpreting 'necessity' in descriptive terms as a modifier and by stressing the link which this modifier has, i.e. the necessary links of process, to movement and change, with further momentum given to this position by offering an historical understanding of chance in relation to change, Cornford effectively not only removes both Ananke and the metaphysical from his discussion of the accessory causes, but also effectively removes these from

360. In other words, the argument is placed within its historical setting as a fully formed argument.

361. Cornford, *PC*, 159–76.

362. See my discussion just prior, regarding Cornford's explication of 'necessity.'

the discussion altogether. On the contrary, however, the position which Plato articulates in the text and which I argue is to be taken at face value, is a position whereby Plato claims unequivocally that the Cosmos had a beginning, had a Maker, was generated, and was subject to two primal causes in its generation, one being 'Nous,' the orderly and ultimately more powerful cause, referred to as the 'Demiurge' and the other, 'Ananke,' the disorderly, and given the Cosmos, less powerful cause, referred to also as the 'Errant' or 'Wandering Cause.' Although Plato does not have all the scientific answers to his cosmology in place—indeed, there are huge anomalies that Plato simply does not address—to dismiss his explanation as myth and then to interpret its key players as elements within a non-generated cosmos, i.e. Reason and necessity as they exist *within* the cosmos, is a far more fantastical stance from a scholarly point of view given that Cornford's position is, at best, a modern version of the very view which Plato adamantly denies to be the case, not only in the *Timaeus*, but also in *Laws X*. Although one may be uncomfortable with the metaphysics that Plato's cosmology embraces and the dualism it appears to advocate, to remove this discomfort by simply denying its metaphysics, as Cornford has done, creates not only new problems, but also the making of a completely false and unsound argument to replace them.

The fourth premise which I argue must be assumed, namely, that Plato admits there are true causes and adjoins Nous and Ananke as being the only true causes in the generation of the Cosmos, being both infinite and capable of directing movement themselves, is rejected by Cornford. Plato's cosmic story is subsequently taken up by Cornford as myth, with the primal powers becoming objects requiring causal reinterpretation, which powers are quickly subsumed under a theory in which nature is depicted as being 'alive' and of having innate powers extant within it.[363] However, what rings true for the first three premises also rings true with the generative causes. They cannot simply be dismissed or subsumed under another guise without leaving that part of Plato's cosmology largely unexplained or blatantly wrongly explained. In positing the idea of innate powers within nature and of it being alive, Cornford more closely approximates that active aspect of nature that Plato ascribes to the Cosmos *after* its generation and which activity and causal powers he gives to the World Soul.[364] The World Soul, however, Plato clearly states was generated by the Demiurge for this very purpose, thus is subsequent to the presence of the primal generative causes, their existence being prior to anything they generated, i.e. the Cosmos. When Cornford further introduces the idea of the accessory causes as requiring 'of necessity' to act in a certain way, he moves what was originally presented by Plato as a primal disordering cause—Plato having even given it a name, 'Ananke' (as he did the ordering cause, Nous, naming it the 'Demiurge')—still

363. See Cornford, *PC*, 167–75.

364. See *Ti.* 37c6-d1.

one step lower in the ontological scheme of things, regarding it simply as a functioning of the mechanical aspect of the universe. To wit, this reinterpretation by Cornford of the primal causal powers, namely, of Nous viewed as akin to the activity of the World Soul,[365] in other words, representing its rationality and of Necessity regarded as a function of process, leaves them essentially unexplained and in an ontological quagmire, for this in no way touches upon their attendant ontology as primal causative powers when first introduced by Plato as 'Nous' and 'Ananke' and as having existence before the Cosmos and thus, being prior also to the World Soul, nor does it address the fact that the accessory causes as argued for by Plato are considered in entirely separate terms from those of the primal powers,[366] nor does it explain the attendant metaphysics of the latter in relation to the former, nor why Plato felt it necessary to introduce the idea of a World Soul, and a generated one at that, if both the world and its attendant rationality have always been in existence—he could easily have stated as much, but chose to state and rather eloquently, the opposite. Yet all of the above Cornford has had to ignore to arrive at his conclusion.

The fifth and final fact which must be assumed, namely, that the accessory causes are to be rejected as being in any way whatsoever true causes in themselves, but rather, are assistant to the true generative causes, having no power of their own and existing finitely in time after the generation of the Cosmos, can now be addressed and is the key premise which I argue determines Cornford's position as false. Textually there is no argument with the first part of this premise. Plato clearly states that the accessory causes are not true causes. What is being questioned, rather, is the identity by Cornford within this passage of the 'two primal causes'[367] as really being only one primal cause, wherein the second primary cause is consequently subsumed (and explicated at length by Cornford) under the auspices of secondary or assistant causes, and thus is named as secondary, rather than as second. As I have argued earlier, the text here gives rise to some confusion if not looked at carefully in terms of the content both preceding and following it due to the summary nature of the surrounding passages, but is both consistent with the *Timaeus* as a whole (having already introduced not one but two primary causes) and is understandable *in situ* when viewed in these terms. In conclusion, Ananke, in being presented by Plato as one of two primary causal powers in the generation of the universe, directing pre-Cosmic chaotic movement and which, like the Demiurge, is infinite, cannot at the same time be touted as the secondary or assistant cause, having no powers of its own and indigenous to the finite universe, and therefore, itself finite. This part of Plato's myth, even if accepted as pure story-telling, contradicts Cornford's conclusion regarding Necessity (Ananke), namely, that 'necessity' is a lower kind of cause

365. *Ti.* 36d8-e5.
366. *Ti.* 46c7-e6.
367. See *Ti.* 46e2-6.

or συναιτία that assists the Demiurge in "producing the best results possible,"[368] since even in the story, Ananke and the συναιτία are presented as entirely separate players, the former existing antecedent to the latter.

---

368. See Cornford, *PC*, 175–76. Here, again, in Cornford alluding to both the συναιτία (accessory causes) and the (so-called) 'persuasion' of Necessity by the Demiurge to produce the best results possible, the conflation of these text passages is transparent.

Chapter 3

3 *TI.* 29D7-30C1

3.1 Original Greek

Λέγωμεν δὴ δί ἥντινα[369] αἰτίαν γένεσιν καὶ τὸ πᾶν τόδε ὁ συνιστὰς συνέστησεν. ἀγαθὸς ἦν, ἀγαθῷ δὲ οὐδεὶς περὶ οὐδενὸς οὐδέποτε ἐγγίγνεται φθόνος· τούτου δ' ἐκτὸς ὢν πάντα ὅτι μάλιστα ἐβουλήθη γενέσθαι παραπλήσια ἑαυτῷ. ταύτην δὴ γενέσεως καὶ κόσμου μάλιστ' ἄν τις ἀρχὴν κυριωτάτην παρ' ἀνδρῶν φρονίμων ἀποδεχόμενος ὀρθότατα ἀποδέχοιτ' ἄν. βουληθεὶς γὰρ ὁ θεὸς ἀγαθὰ μὲν πάντα, φλαῦρον δὲ μηδὲν εἶναι κατὰ δύναμιν, οὕτω δὴ πᾶν ὅσον ἦν ὁρατὸν παραλαβὼν οὐχ ἡσυχίαν ἄγον ἀλλὰ κινούμενον πλημμελῶς καὶ ἀτάκτως, εἰς τάξιν αὐτὸ ἤγαγεν ἐκ τῆς ἀταξίας, ἡγησάμενος ἐκεῖνο τούτου πάντως ἄμεινον. θέμις δ' οὔτ' ἦν οὔτ' ἔστιν τῷ ἀρίστῳ δρᾶν ἄλλο πλὴν τὸ κάλλιστον· λογισάμενος οὖν ηὕρισκεν ἐκ τῶν κατὰ φύσιν ὁρατῶν οὐδὲν ἀνόητον τοῦ νοῦν ἔχοντος ὅλον ὅλου κάλλιον ἔσεσθαί ποτε ἔργον, νοῦν δ' αὖ χωρὶς ψυχῆς ἀδύνατον παραγενέσθαι τῳ. διὰ δὴ τὸν λογισμὸν τόνδε νοῦν μὲν ἐν ψυχῇ, ψυχὴν δ' ἐν σώματι συνιστὰς τὸ πᾶν συνετεκταίνετο, ὅπως ὅτι κάλλιστον

369. Written ἥν τινα in Bury, 54.

εἴη κατὰ φύσιν ἄριστόν τε ἔργον ἀπειργασμένος. οὕτως οὖν δὴ κατὰ λόγον τὸν εἰκότα δεῖ λέγειν τόνδε τὸν κόσμον ζῷον ἔμψυχον ἔννουν τε τῇ ἀληθείᾳ διὰ τὴν τοῦ θεοῦ γενέσθαι πρόνοιαν.

3.2 Translation

Let us state, then, the reason why he who brought about generation and this universe, constructed them. He was good, and in that which is good no envy whatsoever concerning anything comes to be. Being therefore free from this, he wished all things to be **as much as possible** like himself. Accepting this principle, then, from men of wisdom as being above all one of supreme authority, it can be accepted as absolutely true. For God had desired everything to be good, and nothing imperfect, **as far as it was possible**. Thus, then, **having taken in all that was visible**, which was not at rest, but moving about without harmony or order, he brought it into order out of disorder, believing that to be altogether better than this latter. But it neither was nor is right for Him who is best to do other than that which is most fair. Therefore, taking into account such was **naturally visible**, he found, taking each as a whole, that nothing without intelligence to ever be a more beautiful work than that having intelligence and furthermore, that it is impossible for intelligence to come to be in something without a soul. Consequently, because of this reflection, by placing intelligence in soul and soul in body, he was helping to construct the universe so that his completed work might be by nature both the finest and the best. Thus, then, according to what, upon reflection, **is likely,** one must affirm the cosmos to have come into being in very truth a living creature endowed with both soul and intelligence due to the providence of God.

3.3 Commentary and Notes

3.3.1 'As far as possible' = 'the very best'

There are three relevant issues in connection with the above passage (as shown in bold print), two of which have already been dealt with. The first issue deals with the idea of 'as much as possible' or 'as far as . . . possible' and the second, with the idea of something being 'likely.' The third issue, not yet discussed, is the idea of something being 'naturally visible.'

The first idea, that of 'as much as possible,' as mentioned, has been discussed at length earlier.[370] However, there are further reflections that can be made. What must be kept in mind with regard to this expression and to other

370. See §2.3.4 of Part I.

passages containing the same or similar phraseology is that, depending on the passage, these expressions take on various shades, configurations, and aspects of meaning and do not always refer to lack or limitation. With respect to this particular passage, I have previously argued that reference here is being made not to limitation, but rather, to the idea of the superlative, i.e. *as far as possible*, meaning 'the very best.' In fact, soon afterwards, in the same passage, Plato has the astronomer further affirm that it is in fact *impossible* for the God or Demiurge to achieve anything but the very highest. There are subsequent passages, moreover, which suggest that 'the very best' is what the Demiurge did in fact achieve.[371] Plato will later argue that because of certain factors which logically adhere to the concept of generation, as well as empirically to the generated object itself, e.g. movement and change (which includes the idea that what is bonded together can be potentially dissolved),[372] in addition to the threat suggested by the presence of a residual power lying latent in the universe ready to reduce the Cosmos's order back into primordial disorder, its powers appearing immanent, that relative to these factors the work of the Demiurge is challenged. However, Plato never comes out and states in the *Timaeus* that the work of the Demiurge was curtailed. Rather, his conclusion at *Ti.* 92c4-9 affirms the opposite:[373]

> Here at last let us say that our discourse concerning the universe has come to its end. For having received in full its complement of living creatures, mortal and immortal, this world has thus become a visible living creature embracing all that are visible and an image of the intelligible, a perceptible god, supreme in greatness and excellence, in beauty and perfection, this Heaven single in its kind and one.

It could be argued that this conclusion is mere hyperbole on Plato's part and that any imperfections and limitations in the Cosmos have been forgotten (or forgiven) by the writer. However, such a conclusion is virtually groundless and literally, very shaky. If the universe, as noted above, were "a perceptible god, supreme in greatness and excellence," then admitting to any imperfection whatsoever would be a contradiction in terms of this description. It is only *before* its generation or *after* its generation that Plato will allow imperfection or disorder to exist in the universe, not *during* it when only the Demiurge's hand is at work

---

371. E.g. see *Ti.* 30b4-6, 34a8-b9, 40a2-7, and 46c7-d1. Note that in this passage reference is made to shaping things "in the best way possible," a stronger notion than simply "as far as possible," *Ti.* 53b5-7 and 68e1-4. See also Plato's conclusion at *Ti.* 92c4-9. The idea continually suggests itself that given what the Demiurge set out to do, the best was achieved and in fact, nothing less was possible. This is contrary to Cornford's suggestion at *PC*, 159, where he states that: "the work of the most ungrudging benevolence cannot be perfect; it can only be 'as good as possible'."

372. See *Ti.* 41a8-b2.

373. Cornford, *PC*, 359; see also Lee, 124 and Bury, 253. Plato's description of the world from the very start of the dialogue at *Ti.* 29a5-6 as "the fairest of all things" argues for a concept of perfection that was maintained throughout the dialogue, from its beginning through to its conclusion.

and Ananke is left with no active role. It must be stated, however, that this does not therefore lead to the opposite conclusion, namely, that Ananke as a primal cause does not exist, standing in metaphor-like fashion, as Cornford holds, for some mode of physical change in an already existing cosmos. Indeed, as stated above, there is disorder both before[374] and after[375] the universe has been ordered. Rather, Ananke is the disordering[376] or second of two primal causes, while the first primal cause, the Demiurge, is given the sole role of being the primal ordering cause.[377] This ordering power, therefore, persisted successfully in bringing order out of chaos, with the result that the Cosmos became nascent and according to which order, by the very fact of its existence, chaos gave way. It is with this idea in mind, I suggest, namely, of the Demiurge being a stronger power than Ananke, that Ananke is described by Plato as having been "persuaded," fitting well with a 'likely tale' where human analogies abound, but more than this, I further suggest, nothing is meant.[378] Taking the idea of 'persuasion' literally creates such a gross anthropomorphism, involving as it does the two key metaphysical players in Plato's dialogue, that every consequent step thereafter in this work also becomes subject to the whims of supposed myth-making (under the umbrage of a 'likely-tale' but with the unfortunate meaning of a 'fairy tale'). Thus, if the natural logic present in this passage, which lies fairly near the surface, of one metaphysical power being stronger than the other prior to the generation of the Cosmos and thus prevailing, is glossed over, then so, too, is the consequent metaphysics. What ultimately results is that these powers then become this or that aspect of myth-making representative of primal *physical* powers in an already existent world, since the metaphysics enabling generation

374. See *Ti.* 30a2-6, 53a7-8, and 69b2-8.

375. See the earlier section regarding 'as far as possible' (§2.3.4); see also *Ti.* 81e6-92c3 (to the end of the *Timaeus*), where Plato deals with diseases of the body and soul or mind.

376. See *Ti.* 41a8-b6.

377. See my earlier argument with regard to Ananke being "persuaded" at §2.3.3.

378. See *Ti.* 47e5-48a5. If this is a human analogy describing the Demiurge's greater power over that of Ananke, then the question can be raised as to why Plato used a word meaning 'persuaded' and not, for instance, 'destroyed' or 'overwhelmed.' However, I would suggest that two ideas converge to make 'persuaded' the better choice. The first is that Plato's concept of the Demiurge is entirely positive and this power, being wholly good, is not a destroyer and hence, would not destroy. The second idea is that Ananke is not merely mythical or representative according to Plato, but is an equally real, but weaker (having no power for order) metaphysical power, which co-exists, and hence is co-eternal, with the Demiurge. The idea of 'persuasion' brings out clearly Plato's understanding of Ananke being a weaker power, while at the same time, and an idea which will persist strongly in *Laws X* (albeit via the idea of a Non-benevolent Soul), retaining the idea that Ananke is very much a real, if not latent, power outwith and to some degree within, the generated universe. Viewed in this light, therefore, the term 'overwhelm,' like 'destroyed,' undermines too greatly the actual power of Ananke for potential disorder, to which disorder *Laws X* later attests and of which Plato was fully aware even in the *Timaeus*.

would have been sufficiently eroded to allow this, but as I have suggested, erroneously.

According to Plato's cosmology, moreover, neither the Demiurge nor Ananke has a relevant role to play after the generation of the Cosmos. In fact, thereafter, when Plato begins to describe the objects of the generated universe, both of these primary powers quite literally disappear from the dialogue. This is significant for two reasons. First, it elevates and confirms their ontology as true causal powers, nothing more and nothing less. The Demiurge is not merely the deification of reason with a capital 'R,' nor is Ananke the personification of change or motion. Second, and perhaps the more important reason in terms of this dialogue, by the Demiurge and Ananke having exited from the stage as powers, all remaining powers and their effects in the Cosmos can now be identified, i.e. soul containing 'nous,' order, the secondary or subsidiary causes, movement, change, disorder, and disease, according to their own design and taxonomy. In other words, these latter can be now named and explained as part of the natural order, having already been either generated or effected after the generation of the Cosmos, allowing physics, rather than metaphysics, to come to the fore. Thus, with Plato no longer requiring references to causal powers of the 'eternal Reality' sort, the discussion is finally freed up so that he might indulge in the relaxing pursuit of what he describes as "likely accounts of the world of change."[379]

3.3.2 Two kinds of disorder – natural and unnatural

I would further argue that subsequent references in the *Timaeus* to disorder or instances of disorder within the Cosmos are one of two kinds, namely, either natural disorder or unnatural. Instances of *natural* disorder can be understood as regenerative, representing the decay and rebuilding activities which occur naturally within the Cosmos as it maintains itself, supported by the fact that the Cosmos has already been described by Plato as "designed to supply its own nourishment from its own decay and to comprise and cause all processes."[380] This is not contradicted by a slightly earlier reference in this same passage to the universe being ageless and free from disease (i.e. decay) as this refers to an earlier stage of generation where the Cosmos is described as a Living Creature, and although a 'single complete whole' consisting of parts that are wholes, it is not yet complete in terms of what it will contain as a completed universe, since it is still to be inhabited by the composite bodies.[381] The regenerative processes noted above involve these subsequent composite bodies, whereas the Living Creature, at least initially, appears to refer only to the larger structure which

379. *Ti.* 59c5-d3.
380. *Ti.* 33c7-d1; see Lee, 45; see also Bury, 63 and Cornford, *PC*, 55.
381. *Ti.* 33b2-4.

would later house and support these bodies without itself decaying (hence, its description as a 'single complete whole').

The other remaining references to disorder in the *Timaeus* concern *un*natural disorder, in other words, to disorder in the world that does not help maintain its equilibrium. Plato suggests that this kind of disorder should and can be avoided (e.g. by education or temperance), but avoidance is a difficult task, involving as it does the unbridled human 'will.' The idea of 'freedom of will,' e.g. to choose to be unruly or to be disciplined, is not explicitly argued by Plato, but is nonetheless, at least to some degree, thought possible for the soul and is consequently assumed by him throughout the text.[382] He does this by his several references to unbalanced appetite, unfettered motions of the soul, ignorance, and stupidity, any or all of which according to Plato bring about such disorder and are to be avoided, the idea of avoidance involving choice, choice in turn involving the idea of freedom, and the agent of this freedom understood as the human will (within the soul).[383] Finally, therefore, although he cannot argue that the Demiurge's ordering is at all disordered (this would be a contradiction), Plato must nevertheless account for the disorder or imperfection he observes in the finished Cosmos and this he does in a two-fold fashion: First, by including certain kinds of disorder as part of the balancing process and as necessary in achieving a good end[384] and second, by never claiming that the Demiurge completely subdued Ananke, but rather, did so only to the degree that the Cosmos is and remains ordered, in which absence all remaining instances of disorder or as I have collectively named them, 'unnatural disorder,' fall.[385]

382. Although Plato explicitly states that we acquire our faults owing to two causes outside our wills (i.e. a bad physical make-up and failure to be educated properly), he enjoins the reader to "try with all our might . . . to avoid evil and grasp its contrary" (see *Ti.* 87a7-b9). He does not go into a fuller explanation concerning this in the *Timaeus* stating, "that, however, is another story."

383. See *Ti.* 81e6-92c3 (end), where Plato deals with diseases of the body, mind, and soul.

384. This type of disorder is understood by Plato as being part of the ordering process taking place in the world and thus, forms part of the good end achieved.

385. It is contrary, I suggest, to Plato's understanding of the complete goodness of the Cosmos to grant Ananke any sort of role in supplying unwanted disorder to it. According to Plato, the world as generated is wholly good, wherefore the Demiurge is assigned as the cause of this goodness and thus, as cause, is itself wholly good. Any unwanted disorder in the Cosmos is simply a fact, ungenerated by the Demiurge, but which Plato must nonetheless account for and for which he must also, according to his metaphysics, assign a cause, which he does in the form of the disordering cause, Ananke. In the *Timaeus*, however, he does not fully account for the disordering powers lying *within* the Cosmos, acknowledging only the disordering cause, Ananke, in its relation to the initial ordering of the Cosmos and with as little note as possible. As a consequence, disorder within the Cosmos is left largely unexplained and within its mythical context, i.e. as that residual part of the Cosmos wherein Ananke was not fully "persuaded" by the Demiurge so as to allow the Cosmos to be fully ordered (see *Ti.* 47e5-48a5). In *Laws X*, however, through his further development of the notion of 'soul,' Plato is finally able to incorporate into his

3.3.3 'All that was visible'/was 'naturally visible'

The second issue in this passage which I would like to raise and which has not been addressed previously concerns the idea of the Demiurge having taken in 'all that was visible,' and further, his taking into account of what was 'naturally visible.' It is interesting that Plato should make this distinction between the visible Cosmos or further still, the *naturally* visible Cosmos and that which is not visible or naturally so. In some respects this language is clearly metaphorical or more properly speaking, anthropomorphic, since the Demiurge is depicted looking over what he is about to change or arrange, and yet eyes, with their function of vision, will only have existence later, as a part of that generation. The perspective of the Demiurge, moreover, is unabashedly anachronistic (he seems to already have a grasp of the difference between things that are rational over things that are not, even before the generation of the Cosmos takes place), much in the same way as a person might think of what is not yet there, when imagining something. The above, however, as regards both its anthropomorphic and anachronistic style has nothing to do with the Demiurge therefore being a character from what might be interpreted as a 'mythic' portrayal of the early Cosmos

---

metaphysics the idea of unwanted disorder or 'evil' as an active power in the world, but not until then. It was not thought possible by Plato for either the Demiurge or Ananke to be regarded as active powers within the generated world in the *Timaeus*, since according to its metaphysics he does not regard either of these forces to be souls or to be self-generated, generated, or a part of that which is generated in any way whatsoever. They are metaphysical powers lying entirely outwith the generated Cosmos, whose effects appear within the Cosmos as ordered or disordered aspects of it and which effects are a result of their respectively fair or foul play prior to, as well as at the time of, the act of generation. According to Plato's metaphysics, moreover, the World Soul was generated to maintain the order that the Demiurge wrought at the outset of the Cosmos, while any residual disorder simply remained from before the Cosmos was generated. How this disorder is maintained within the Cosmos is not addressed in the *Timaeus*, except to say that it remains in spite of generation having taken place. Plato only once refers to the maintenance of order or disorder by the primal causal powers, but this is with respect to the whole of the Cosmos and not just a part of it, when he states that the 'will' of the Demiurge was greater than those powers which had control before its generation and is why he (the Demiurge) would never destroy it (see *Ti.* 41a3-41b6). Consequently, while there is reference here to the Demiurge and to the maintenance of order, it is only with respect to the initiation of order from outwith the universe as it is being generated or as it might be dissolved. Similarly, Ananke is to be understood as the source of whatever disorder remained in the universe at its outset. In either case, the initiation of order or furtherance of disorder in the universe has no need to be understood as the result of these powers remaining active within the generated Cosmos, the latter coming to be, as it were, after the fact of their having acted. For an explanation of metaphysical activity that takes place within the Cosmos, Plato must further develop his notion of 'soul,' but which development takes place later on in *Laws X*.

and its generation,[386] but rather, is Plato's chosen method of describing what is very difficult to describe and therefore, of including, when reference is given to it or is required (of it or other early players and their activities), what is equally "indeed a hard task" (*Ti.* 28c3-5).

3.3.3.1 The Demiurge and the visible Cosmos

Plato has not yet completed his description of the Demiurge at this stage, but what will emerge is a god who is wholly good and wholly perfect. The Demiurge, moreover, is a tremendously powerful god, and whereas we are told the degree to which he used his powers, we are not told whether this use represented the full extent of his powers. This argues against the position that he is not omnipotent. We do not know. It seems plausible, however, that a god who is portrayed as being completely good and wholly perfect, would have powers to match, whether he chooses to use these powers or not. Further, with time only coming into existence with the generation of the heavenly bodies, the Demiurge is logically left to stand outside of time and all considerations relative thereto. The Demiurge, moreover, by definition 'the best' (*Ti.* 30a7), is a god having the highest degree of 'nous,' wherein his faculty of intellection has as its object the full complement of Forms.[387] To describe the Demiurge, therefore, as 'having taken in all that was visible' need not imply that it was visible to him, since he clearly stands outside of time and has no eyes, but need only imply that that which the Demiurge set out to order would be visible to someone who had eyes if that person were there. The Demiurge, however, being privy to all the Forms and subject neither to time nor to what he has generated, nor seemingly to the ungenerated universe, could quite plausibly envision this without eyes. Plato does not ask his readers to take him literally—on the contrary, he reminds them of how difficult it is to write about such subjects and why any considerations regarding these can only be approximate—but he does demand from his readers that they acknowledge the truth behind his writing and its images and goes into great detail to ensure that if one image does not quite capture what he is trying to

386. I strongly disagree with Cornford, who, in rejecting the metaphysics in Plato's cosmology, proposes that the Demiurge and chaos are merely symbols, further suggesting that there was never a moment of creation before which the chaos existed and whereupon he concludes "and this part of the mythical imagery is not to be taken at its face value" (*PC*, 37). This extensive reinterpretation of Plato's cosmology as myth opposes at every turn what Plato actually wrote and by this literary dismissal simply ignores the logic and complexity behind this cosmology. In Cornford's attempt to extricate Plato from the grip of the theologian, he casts him into the mould of the modern man of science, both of which, when simply imposed and then taken to the extreme, are in error.

387. The Forms relative to this work are referred to throughout, but are dealt with specifically at §3.3.4 of Part I on the *Timaeus*, as well as at §2.3.2.1.4 of Part II on *Laws X*. See also fn. 500.

explain, another will or will aid in that understanding. This happens time and time again in the course of the *Timaeus*, ultimately building up images—the Demiurge, Ananke (Necessity), the eternal Living Being, the Nurse of Becoming, the Forms—as a kind of meta-language in the supra-cosmic scheme of things for the eternally existent, inside and outside of time, within and outwith the generated universe, where they, as core players and features of generation, are and remain faceless except for these images.

3.3.4 Non-visible/invisible substrata and Plato's theory of triangles

Another consideration I would like to put forward is the possibility that behind the decision of the Demiurge to order 'all that was visible' is the suggestion by Plato of a non-visible substratum. Although Plato says very little about such a substratum, addressing for the most part only the visible world, he nonetheless does mention particles too small to be seen at *Ti.* 56b7-c3 (as discussed earlier at §2.3.5.2). These, he explains, only become visible once enough of them have joined together, like with like, so as to *become* visible, as one of the four elements. At this microcosmic level, therefore, there is, according to Plato, a physical world composed of triangles already ordered into their particular elemental shape, but which cannot be seen.

Early on in the *Timaeus* Plato's astronomer states, "anything that has come to be must be corporeal, visible, and tangible, but nothing can be visible without fire, nor tangible without solidity, and nothing can be solid without earth."[388] A little further on he explains that "fire, earth, water, and air are bodies, and all bodies are solids."[389] Continuing, he states, "all solids again are bounded by surfaces, and all rectangular surfaces are composed of triangles." However, if so, what can be said of these triangles, themselves? It is unlikely that Plato would be comfortable with the idea that the triangles which make up the basic elemental particles have acquired their triangular shape through either chance or Ananke, since his notion of chaos (and chance, but to a lesser degree) is irregularity, which is incapable of achieving ends of a rational nature or only rarely, as in the case of chance, but always without consistency.[390] When he first mentions the triangles in the *Timaeus*, it is not in respect of *their* provenance. Rather, his interest at this point is with a later stage of the development of the Cosmos, when these triangles have come together sufficiently to form specific elemental types, i.e. earth, water, air, and fire. According to Plato's description of these primitive triangles, they are completely regular and form the basis of what eventually becomes the fully generated, and being generated, rational and hence, meaningful (with respect to ends) Cosmos. Accordingly, these triangles most likely repre-

388. *Ti.* 31b4-6.
389. *Ti.* 53c4-8.
390. *Ti.* 69b5-8.

sent an early stage of the organization of the universe, with Plato's cosmology beginning, in turn, at a somewhat later stage, when some of these triangles have come together by chance and then design by the Demiurge to form the elements. Plato's depiction of the process of generation is, therefore, I suggest, an evolutionary one, involving forces that exist side by side, the Demiurge being a force for order and Ananke, for disorder, as the Cosmos is being formed. As metaphysical forces, moreover, the Demiurge and Ananke always exist and act outside of time. Accordingly, the initiation and maintenance of whatever order or disorder that is wrought by these forces during the generation of the Cosmos would be one and the same thing, the appearance of objects extended in time being an *effect* of such metaphysical forces at work and never, however construed, to be mistaken with their identity. The only exception to the principle of these forces acting outside of time is in how their acts are perceived and not with the principle itself, i.e. when one power overrides the work of the other, as in the case of the Demiurge "persuading" Ananke in the *Timaeus* during the generation of the Cosmos, this act is perceived as the beginning of the Cosmos and similarly, in *Laws X*, where, when redrawn as non-generated Souls, these causative powers interact with the world bringing about good or evil within it, displacing the acts of one with the other in a world of change,[391] these acts are perceived as good or evil events, the latter as disorder that diminishes the world and the former as order that sustains it.

There is, moreover, an *in*visible aspect also, relative to two different sources, which is evident in Plato's understanding of the universe. The first source, not being physical this time, in other words, not being due to the elemental particles being too small to be seen, is purely noetic and has to do with the role which the Forms have to play in the organization of the triangles into elemental particles (at least as far as Plato draws his cosmology). When the Demiurge sets about to order the Cosmos, the elements out of which the universe is constructed already contain a certain degree of order.[392] I have argued that at least in part this is due to the fact that Plato's cosmology does not appear to commence at the very beginning of time, but starts later in the process when the Demiurge has already begun his ordering. It is also possible, therefore, that outside of time, that is prior to generation proper, Plato would be fully aware of the potentiality which the Forms, present to the Demiurge as objects of his own subjective reflection, would have for all present and future ordering and thus hints, when referring to "all that was visible," to what is *in*visible or the Form behind the object and not just to the *non*-visible Cosmos (i.e. that which is too small to be seen) in contrast to the visible. I would suggest, therefore, that when Plato refers to "all that was visible," he may in part be contrasting this with the *in*visible substratum or Forms, which he postulates are responsible for the formal organization of all

391. *Lg.* 896e4-6.

392. See *Ti.* 52d2-53c3 and also *Ti.* 69b2-c3.

objects within the universe, while when he next cites "all that was *naturally* visible," he is at this point referring to the microcosmic world or to the physical particles themselves, some perhaps already elements, but as yet too small and too dissipated for us to see, but which will eventually comprise the full visible Cosmos.

The second source of *in*visibility, I would suggest, comes in the form of the primal triangles themselves, which only when amassed together become one of the four visible elements and after that, particular objects within the Cosmos. We learn in the *Timaeus* that at the beginning of the ordering of the Cosmos there existed triangles, ordered but non-visible particles as yet incapable of amassing together into particular elements until they were more complexly ordered into this or that kind of sided triangular shape. The question remains, however, whether or not they were in fact *in*visible prior to being *non*-visible, since they existed at this stage only on one plane, incapable of being fully perceived *ever* until further shaping (with other triangles forming sides) took place. I would suggest that the answer to this question is yes. These ordered triangles would have existed at first 'invisibly' (they could never be consistently visible without further dimensional contours added) and eventually 'non-visibly' (even with further dimensions, until amassed together like with like, being too small, would remain non-visible), eventually making up the suggested ordered substratum of the pre- or semi-ordered state of the Cosmos before its full generation, depicting, moreover, its evolutionary nature. How this substratum first came to be ordered and the degree and extent of its further ordering, if indeed this state was achieved over time as a part of initial generation or had always been ordered from the outset of generation, is not discussed or even suggested as being known by Plato. It is curious, however, that the Demiurge could 'pluck' so efficiently from the disorder of Ananke a whole universe without recourse to a stabilizing medium. Although one could reason that the Forms inhabit these invisible/non-visible spheres (substrata), they could never account for the stability of the objects within the Cosmos, only for their form, being themselves purely non-physical, noetic objects of 'nous,' nor could the Forms account for the mysterious[393] provenance of the triangles, only for their later organization when they are numbered and ordered by the Demiurge.[394] Plato never states that order and visibility are one and the same, and in fact, the opposite is proposed when in his description of the original chaos, the disorder that is found is described as com-

393. See *Ti.* 53d4-7.

394. Plato explains at *Ti.* 52d2-53c3 that before the God touched them, the basic constituents—fire, water, earth, and air—were all without proportion or measure (although they bore some traces of their proper nature), but the Demiurge reduced them to order by giving them a definite pattern of shape and number, which he did via the triangles (their substructure), whereby different sized triangles were put together in specific ways. See *Ti.* 56a6-c7. The details of how this was carried out are not discussed, rather, only that this is how order was achieved. For a scientific account, see Cornford, *PC*, 230–39.

prising "all that was visible." It is worth considering, therefore, that these primal triangles, out of which Plato later gives the elements their full visible form and stability following upon the Demiurge's further engineering, naturally inhabit this invisible and later non-visible, but logically suggested sub-stratum.[395] The degree to which this substratum is ordered or unordered, moreover, would be the degree to which the Demiurge or Ananke, respectively, actively participate in it. The upshot of the above considerations is that they reveal Plato to be a realist of profound depth and breadth, whose realism includes both supra- and non-physical notions, as well as a physical universe of both microcosmic and macro-cosmic proportions.

3.3.5 A 'likely' account

The third matter, a theme that constantly recurs throughout the *Timaeus* and shows up initially in this early passage, is the idea of the account being 'likely.' The idea of 'likelihood' is a twin, but not identical, idea with that of 'possibility.' This latter idea has been discussed elsewhere and most extensively within the discussion of '*as far as he could*' or '*as far as possible*' (τὸ . . . δυνατόν) at §2.3.4. However, in the present passage 'likeliness' is at issue, not 'possibility.' The difference lies not just in meaning, in that the Greek word which Plato uses in this early passage is τὸν εἰκότα, a participle with the nominal meaning of 'what is likely' or 'probable,' deriving from the verb ἔοικα, a perfect with a present sense, which in turn is from the verb εἴκω, 'to be like,' rather, the notion of careful deliberation or the choosing of what is likely or probable amongst possibilities is also introduced. In other words, what Plato is describing is an account of the Cosmos, which *according to reason* (κατὰ λόγον) is likely or probable. Thus, the context itself at the very outset of Plato's cosmology adds the stricture of reasonableness to the idea of what is 'likely,' with no hint of myth or fancy or suggestion that the "*Timaeus* is a poem."[396]

Later, at *Ti.* 48d1-4, Plato will again refer to his original intention as outlined in this initial passage, making it clear that what he is trying to undertake is as likely an account as possible of both particular things and the totality of things, and recommencing at that juncture from an even earlier origin. Thus, he writes:

> While preserving the meaning of the likely account which was laid down from the beginning[397] concerning all things individually and as

395. See *Ti.* 56b7-c3; see also 43a1-3.

396. Cornford, *PC*, 31.

397. Whereas at *Ti.* 48d1 of this passage τὸ δὲ κατ' ἀρχὰς carries the meaning of 'from the beginning,' this clearly refers to Plato's account, i.e. from the beginning of Timaeus's cosmology, and not to the beginning of the totality of things or the Cosmos itself, which totality has yet to be described in detail.

> a whole, I will undertake nothing of less likelihood, rather more, recounting an even earlier beginning.

As in the passage at *Ti.* 30b6-c1, in this later passage at 48d1-4 Plato again uses the participle εἰκότα:

> τὸ δὲ κατ' ἀρχὰς ῥηθὲν διαφυλάττων, τὴν τῶν εἰκότων λόγων δύναμιν, πειράσομαι μηδενὸς ἧττον εἰκότα, μᾶλλον δέ, καὶ ἔμπροσθεν ἀπ' ἀρχῆς περὶ ἑκάστων καὶ συμπάντων λέγειν.

Moreover, at *Ti.* 29c4-d3, prefacing 30b6-c1, while Timaeus and Socrates are still discussing what Timaeus is about to expound upon, Plato, in Timaeus's dialogue, makes it very clear what he means to impart by his use of the term εἰκότα, namely, that he will not be able to give accounts that are "always in all respects self-consistent and perfectly exact." This is a disclaimer, not an announcement that the language of myth or poetry is about to be used. He further explains that any lack of self-consistency or exactness in his treatment of the Gods (the heavens) or the generation of the universe is due to their being such huge topics, concerning many matters and *not* that these matters cannot be discussed or must be encrypted. Thus, Plato, by this disclaimer, not yet having embarked on his exposition at this stage, allows for a certain dynamism and approximation in the account as it unfolds. The necessity for this is soon made evident as Plato repeatedly begins his account anew, layered by the introduction of new key players within his cosmology. In this way Plato is able to build a degree of sophistication into his presentation that would otherwise appear convoluted and 'mythic' by the sheer crowdedness of the stage. To the contrary, it is precisely these poetic accounts and overcrowded cosmogonies, which he wishes to correct.[398]

There is a sense in which if Plato's cosmology were to be misunderstood as attempting to describe the actual physical beginning of the Cosmos that it could, with reason, be called poetry and not a true cosmology, containing as it does a fairly high degree of mythic language, too much to be called anything but a cosmogony. However, that this would be offensive to Plato is extremely likely, given both his low opinion of poets generally[399] and his distaste for cosmogonies, as well as due to the fact that he states very clearly in the *Timaeus* that what he is about to do is give a 'likely' (not '*un*likely') account of the generation

---

398. See *Ti.* 19c8-20a5 and 27a3-6.

399. See *Socrates' Defense (Apology)*, 22a6-c9, trans. by Hugh Tredennick; also, *The Republic*, 377d3-379a5, trans. by Paul Shorey, in the *Collected Dialogues of Plato*. Although both passages contain speeches purported to be by Socrates concerning the poets, Plato no doubt shared a similar view. See also *Laws X*, 885c7-e6, trans. by A. E. Taylor, where Plato again voices caution with regard to the poets, although this time through his central character, the Athenian.

of the Cosmos, repeating this mantra of 'likeliness' numerous times throughout his cosmology. Thus, in order to avoid an imputation of 'poetry' I think it is important to point out that Plato's cosmology does not attempt to describe 'a beginning' either before or at the time of the Cosmos's generation, although he does state that the universe *had* a beginning.[400] As the passage at *Ti.* 48c2-d1 clearly shows he eschews describing first principles of the universe altogether, on the basis that they are too difficult to explain in the context of the present discussion, leaving the earliest stages of both the chaos and the generated Cosmos alone. An "earlier stage" is the earliest he dares to go.[401] Plato's cosmology, therefore, consciously and consistently avoids what cannot be presented as 'likely,' whereupon he must begin anew with each new stage of the Cosmos he wishes to explore, but never at its very beginning and rarely in exact chronological order.

400. *Ti.* 28b2-c2.

401. See my translation of *Ti.* 48d1-4 at Part I, §1.2. At 48d3-4 Plato writes καὶ ἔμπροσθεν ἀπ' ἀρχῆς περὶ ἑκάστων καὶ συμπάντων λέγειν, where he introduces the adverb ἔμπροσθεν with the temporal meaning of 'earlier.' The use of this word brings into focus Plato's intention to start afresh and to delve even further back in time than he had done previously. Thus, it is not to the very beginning of the Cosmos he is referring, but rather, to an earlier stage in its formation.

Chapter 4

4 ***TI.* 41A3-B6**
***TI.* 42E7-43A3**
***TI.* 68C7-D7**

4.1/2 Original Greek/Translation

(i) *Ti.* 41a3-b6

( ) ἐπεὶ δ' οὖν πάντες ὅσοι τε περιπολοῦσιν φανερῶς καὶ ὅσοι φαίνονται καθ' ὅσον ἂν ἐθέλωσιν θεοὶ γένεσιν ἔσχον, λέγει πρὸς αὐτοὺς ὁ τόδε τὸ πᾶν γεννήσας τάδε—

"Θεοὶ θεῶν, ὧν ἐγὼ δημιουργὸς πατήρ τε ἔργων, δι' ἐμοῦ γενόμενα ἄλυτα ἐμοῦ γε μὴ ἐθέλοντος. τὸ μὲν οὖν δὴ δεθὲν πᾶν λυτόν, τό γε μὴν καλῶς ἁρμοσθὲν καὶ ἔχον εὖ λύειν ἐθέλειν κακοῦ· δι' ἃ καὶ ἐπείπερ γεγένησθε, ἀθάνατοι μὲν οὐκ ἐστὲ οὐδ' ἄλυτοι τὸ πάμπαν, οὔτι μὲν δὴ λυθήσεσθέ γε οὐδὲ τεύξεσθε θανάτου μοίρας, τῆς ἐμῆς βουλήσεως μείζονος ἔτι δεσμοῦ καὶ κυριωτέρου λαχόντες ἐκείνων οἷς ὅτ' ἐγίγνεσθε συνεδεῖσθε. ( )

Now when all the gods, both those that wander about manifestly and those that appear only as they will, had come to birth, He who brought forth this universe said to them the following: "Gods, gods, and works of whom I am Maker and

Father, being created by me, are indissoluble except by my will.[402] For although everything that is bound can be dissolved, yet to loosen that which is beautifully harmonized and well formed is the will of that which is evil. Wherefore you also, seeing that you have been generated, are neither immortal nor entirely indissoluble. Yet, by no means will you be dissolved or participate in death, possessing in my will a greater and more sovereign bond than those to which you were bound when you came into being.

(ii) *Ti.* 42e7-43a3

( ), καὶ λαβόντες ἀθάνατον ἀρχὴν θνητοῦ ζῴου, μιμούμενοι τὸν σφέτερον δημιουργόν, πυρὸς καὶ γῆς ὕδατός τε καὶ ἀέρος ἀπὸ τοῦ κόσμου δανειζόμενοι μόρια ὡς ἀποδοθησόμενα πάλιν, εἰς ταὐτὸν τὰ λαμβανόμενα συνεκόλλων, οὐ τοῖς ἀλύτοις οἷς αὐτοὶ συνείχοντο δεσμοῖς, ἀλλὰ διὰ σμικρότητα ἀοράτοις πυκνοῖς γόμφοις συντήκοντες,

And having taken hold of an [the][403] immortal principle of a mortal living being, imitating their own Maker, borrowing portions of fire and earth, and also water and air from the cosmos, as though to be given back, cemented together what they took, not by the indissoluble bonds with which they were held together, but rather, fused into one mass through numerous pegs, invisible for smallness.

(iii) *Ti.* 68c7-d7

( ). τὰ δὲ ἄλλα ἀπὸ τούτων σχεδὸν δῆλα αἷς ἂν ἀφομοιούμενα μείξεσιν διασῴζοι τὸν εἰκότα μῦθον. εἰ δέ τις τούτων ἔργῳ σκοπούμενος βάσανον λαμβάνοι, τὸ τῆς ἀνθρωπίνης καὶ θείας φύσεως ἠγνοηκὼς ἂν εἴη διάφορον, ὅτι θεὸς μὲν τὰ πολλὰ εἰς ἓν συγκεραννύναι καὶ πάλιν ἐξ ἑνὸς εἰς πολλὰ διαλύειν ἱκανῶς ἐπιστάμενος ἅμα καὶ δυνατός, ἀνθρώπων δὲ οὐδεὶς οὐδέτερα τούτων ἱκανὸς οὔτε ἔστι νῦν οὔτε εἰς αὖθίς ποτε ἔσται.

---

402. The text is corrupt; see Cornford, *PC,* 137 and his Appendix 3, 367–70; see also Bury, 88.

403. No definite article is present in the Greek.

And as to the others, from these examples it is quite clear which combinations should be brought together so as to preserve the likely account. But if someone were to seize upon a practical test to ascertain these things, the difference between the human and divine nature would not be known, because God, being both wise and powerful, is able to mix the many into one and again, to break up the one into many. On the contrary, of mankind no one is able, neither now nor in the future, to do either of these tasks.

4.3 Commentary and Notes

4.3.1 A division of powers

These three short passages taken together articulate what for Plato was a clear divide between the creative ability of the Demiurge and the creative element within the natural order, which natural order at its top or highest end included the divine, but not the Demiurge, the divine or gods in the heavens having been created or ordered by the Demiurge. Plato's cosmology also appeals to two 'wills' present at the outset of this ordering process, not just one, that of the Demiurge, as well as a prior, non-ordering 'will,'[404] identifiable in other passages as that of 'Ananke' or the 'Wandering Cause.' Thus, turning first to this latter point, in the initial passage of the three listed above, reference is made to an original bonding together of primordial matter by a 'will' which did not take harmonization or beauty of form into account. Plato further asserts, through a declaration made by the Demiurge, that his bond (that of the Demiurge) is a stronger and more sovereign will than that of the other non-ordering will (that of Ananke). Plato's appeal therefore is not only to a sharp division of powers between the Demiurge and what is created or ordered, but also between a primal 'will' that chooses and chose to order and harmonize, and a primal 'will' that does and did not. Further, while the position in these passages supports the view that the universe existed in some non-ordered form before its ordering by the

404. The first passage quoted (*Ti.* 41a3-b6) does not commit Plato to the notion of a morally 'evil' or negative 'will,' which is a more developed idea than simply a metaphysical power whose will it is to maintain disorder in a pre-generated universe or potentially to bring disorder into a post-generated one. However, for Plato, once the universe has been ordered, unnatural disorder, i.e. disorder within nature that does not contribute to any of its rational ends, is evil, since it would be against the will of the Demiurge, whose will with respect to the Cosmos is always one of organization or order, which order is composed of reasonable or noetic ends throughout. Accordingly, not only would unnatural disorder within the generated Cosmos be considered evil, the source or cause of this disorder would also be considered evil. Plato will develop this idea of good and evil further in *Laws X* where he introduces the idea of at least two primal Souls, a Soul that is good and which brings about good works in the world and one that is evil and brings about bad. See Part I, §3.3.2. See also Part II (on *Laws X*), §3.3.1–3.

Demiurge, Plato refrains from giving greater detail, keeping in line with his constantly reiterated stricture that the account be likely and thus refraining from giving either a generally speculative or mythic cosmology. Plato can reason back as far as this point, but not earlier without recourse to either speculation or mythic symbolism and because of this, refrains from doing so. Thus, to refer to him as a realist in this regard captures accurately his perspective.

4.3.2 The creative powers and their consequent metaphysics

With regard now to the first point, that is to the sharp division which Plato draws between the creative powers of the Demiurge and the creative element which exists within the natural order, the question of the robustness of the consequent metaphysics can now be raised. It has been argued by Cornford[405] and suggested by others that the Demiurge is a mythic figure, in other words, the metaphysics that would necessarily follow upon the Demiurge not being mythic must somehow fail. However, to depict the Demiurge as a causal power outwith the natural order, whose being is not generated and whose powers are wholly noetic, as Plato has done, is to deny outright that it is simply a mythic figure representing some aspect of the natural world, since the metaphysics given in relation to this cause, in fact to either cause, the Demiurge or Ananke, are ultimately not reducible to physical phenomena. Thus, a description of any mythic configuration would have to include an explanation of Plato's metaphysics with regard to these causes, and not just his physics or implied physics, to remain true to the text and thus, viable. However, in doing this, the explanation would inevitably lead back to where it began, with two primary causes standing in need of an adequate explanation, physical causation proving too reductive to be satisfactory.

From the passages under discussion and from other passages in the *Timaeus*, it is clear that Plato wishes to articulate within his metaphysical scheme a causal power, in fact two causal powers, one which orders, the Demiurge and one which does not, Ananke, whose ontology and powers are entirely distinct from physical causation and its effects. I have earlier denied a mythic interpretation for either of these causal powers, arguing on several occasions for the position that neither the Demiurge nor Ananke is simply a mythic figure. In other words, they are not imaginary beings introduced by Plato to explain difficult and poorly or not understood scientific phenomena.

Granting this earlier position has been sufficiently argued, a look, nonetheless, at the supposed failure of a metaphysical position that posits causal powers outside of the generated universe is still in order. For the sake of argument, moreover, a further look at the question of whether or not a more sophisticated interpretation of the Demiurge and Ananke as mythic figures, and not simply as

405. Cornford, *PC*, 31–32, 34.

literary figures acting as substitutes for physical phenomena, is also in order and of whether or not it provides an adequate understanding of what Plato intended. With regard to the first concern, namely, that of the failure of Plato's metaphysics should it be taken at face value, there is little to substantiate this claim other than a skeptical approach generally to *a priori* ideas of this sort and especially to a metaphysics involving supra-human type or divine beings, and therefore a denial outright of these causes and the metaphysics which support them. However, the intuition that Plato's move is weak and thus fails, is a different matter. What would be of concern here, therefore, would be if there were available by way of possible replacement, a stronger and equally applicable proposal as the one which Plato has offered. The supposed failure of a *prima facie* interpretation seemingly lies in the fact that known physical phenomena can now be substituted for Plato's supposed metaphysical players—the 'meta' becomes physical or primordial matter and the players become phenomena—with the result that an adequate description of the physical world is deployed without recourse to fantasy. However, Plato's concern was not merely a description of the physical world, although indeed that was part of his plan and a culminating part, but rather, he also wanted to show just how the physical world would have come about, not exactly perhaps but plausibly, given his knowledge of first principles and his understanding of their powers from a reflection on the effect of those powers. The one would not be valid without the other. The first principles intimated by the existence of the Cosmos would logically have had to exist prior to the Cosmos itself, as cause (the primal powers) is to effect (the Cosmos). In order to give a likely account, moreover, as was Plato's explicit intention, he required to show just how these primary causes existed in relation to the Cosmos (not apart from the generated Cosmos, which he acknowledges would be very difficult, if not impossible to achieve)[406] and this he painstakingly sets out to do. Indicative of where his interest in the *Timaeus* began, Plato will ultimately describe the effect or in other words, the Cosmos, itself, at length. However, while this could be described in detail, the primary causes, whose striving brought about the Cosmos, could only be hypothesized, their provenance denying further appeal beyond what an understanding of the existent Cosmos required of them. It is noteworthy that Plato does not appeal to either a theory of Forms or to religious belief in postulating his primary causes—any references to the Forms and to religious belief appear separately—rather, he is moved directly by considerations of the Cosmos and by further logical deductions which he draws when hypothesizing the existence of these causal powers. In this way, Plato is able to bridge his core belief in such powers with a rational structure relative to the

406. *Ti.* 28c3-5. It is clear from this passage that Plato's hesitancy to discuss the Demiurge separately from consideration of the generated Cosmos is primarily because of the limitations of his audience, i.e. "all men," and not because he did not believe himself to have such knowledge.

Cosmos, each supporting the other and confirmed by the inner working of the Cosmos.

4.3.3 An axiology emerges

Plato's pivotal consideration was his observation of the ordering and hence, goodness of the Cosmos, which led him to an axiology or valuing of everything he saw or hypothesized, both of the physical world and of the metaphysical, which latter he envisioned to have preceded the physical. Plato's Cosmos, therefore, was not *just* a cosmos, it was a *moral* Cosmos. It was good because it was ordered, harmonious, reasonable, intelligible, beautiful, and balanced. Thus, so, too, was its maker, the Demiurge, necessarily (like begetting like) a supremely good 'will' and Ananke, its opposite, a wicked or supremely bad 'will,' ever ready to return the Cosmos back to its original disorder or chaos, which, lacking in form, was inharmonious and unintelligible.[407] The possible charge, therefore, directed at Plato of undue anthropomorphisms is vacuous. What is really at issue here are scholarly 'divine-isms' or 'primal cause-isms' misapplied to the generated Cosmos and not the other way round, since Plato argues consistently and repeatedly that it is the divine or primary cause which was the source of order within the generated Cosmos and which Cosmos the God tried "to make as like himself as possible,"[408] never that the Demiurge and Ananke were somehow reflections of the existing world.

4.3.4 Order from a disordered cosmos

A difficulty does arise, however, when consideration is given to Plato's notion that the Cosmos pre-existed in a disordered fashion before its ordering.[409] The universe seemingly had always existed, but not always in an ordered state. The argument could then follow that Plato, relying on an ancient taxonomy and a science still in its infancy and which had not yet divested itself of other concerns, i.e. religious and moral, is really presenting an ancient version of, for instance, the Darwinian scheme or the Big Bang Theory, or some other modern theory yet to be informed by the world Plato was trying to describe (i.e. the conclusions of science today that do not require primal powers) and at the same time was burdened by the past (i.e. by the accepted, but transient religious and moral views influencing his belief in the gods and primal powers, as well as the standards of 'good' and 'bad' that followed thereon). Plato's metaphysical causes, it could then be argued, are unnecessary; he just did not know it. A First Mover is not required—infinity is a very long time. I would concede that al-

---

407. *Ti.* 41b2-6.
408. *Ti.* 29e1-30a2.
409. *Ti.* 30a4-5.

though there is myth going on here in the sense of presenting what cannot be fully known or be depicted in any other way, nonetheless when Plato attempts to give a likely account and incorporates myth, it is an unconscious and necessary use of mythology which Plato's draws upon and not a conscious or optional one. In other words, he is not presenting myth for myth's sake within his cosmology, but because there is no other way to describe its metaphysical and to a lesser degree, its physical components. The Demiurge, although most assuredly not a father, is all of this and much more. The Nurse of Becoming, although not a nurse, serves this function in the physical realm. However large the mythic portrayals emerge, it is their minimum reality and not their maximum that Plato is trying to capture.

4.3.5 Myth and Plato's cosmology

While the possible argument that Plato is really presenting a pre-scientific scenario in mythic form is in some ways compelling, it breaks down and fairly quickly when the claims Plato actually makes are revisited and are consequently subjected to the scrutiny of cross-examination. I would like to suggest, first, that Plato's metaphysics would have failed had he not proposed at least two causal powers, that of the ordering Cause and that of the disordering Cause. One causal power was not enough to satisfy Plato's claims with regard to the original disorder in the Cosmos. The importance of Plato's axiological assumptions cannot be underestimated or dismissed in this regard. The world was not just there, it was there in a good, beautiful, and perfectly harmonized way, except, of course, where it was not and where it was not, unnatural disorder resulted and this, too, had to be explained. A cosmological interpretation that does not take into consideration the valuing which Plato placed upon the Cosmos with regard to the good (or ill) of the universe and its contents radically alters the metaphysical requirements, since such a position does not necessarily require a metaphysical cause, let alone two. Physical causes could alone account for the occurrence of all events. However, such a position presents a cosmology radically different from the one Plato envisioned and would, given his understanding and assessment of the generated Cosmos and its generator(s),[410] be viewed as incomplete and as it stood, false. For Plato, physical causes were not true causes, as they were entirely dependent on other causes and ultimately on higher-order causation outwith the generated world itself or a primary cause.[411] Even the World

410. I wish to reiterate my position that Ananke did not have a direct 'hand,' so-to-speak, in the ordering of the universe as did the Demiurge, but rather, was the cause of its disordered state, pre-generation and was also the cause of any unnatural disorder remaining in the Cosmos after its generation.

411. In the *Timaeus* Plato does not argue specifically in his cosmology for a 'First Cause,' but rather, for his present purposes, is satisfied with positing two primary causes,

Soul, which Plato introduced to account for ordered movement within the generated world, could not account for either its own inception or for the generation of the physical world, since generation required not only movement, but also 'nous' or intelligence prior to the existence of the World Soul, nor could the World Soul account for the apparent goodness or overall perfection of the generated world of which it formed only a part, except in a purely functional way as means to an end, the end being the world's own perfection and existence. The physical world for Plato offered causes only of a strictly physical and accessory nature and because of this, represented for him an incomplete, although important, part of the cosmic assay. This required for Plato to look elsewhere, outwith the generated cosmos, to find causes of a non-physical nature that could account not only for the generation of the Cosmos, but also for its goodness and perfection, as well as for its residual disorder and disharmony and for the source of its ongoing ordered movement once generated. These causes emerged as the Demiurge and Ananke outwith the generated Cosmos and as 'soul' later infused within it.

4.3.6 'Like generating like' – a causal hierarchy

One of the principles at work in Plato's thought was that 'like generated like'—goodness comes from goodness,[412] evil from evil,[413] soul from Soul,[414] the physical from the physical[415]—which idea leads to the next point, namely, that a

---

with the primal 'stuff' of the universe also described as being present in a disordered form at the beginning point being described, its provenance left unexplained.

412. 'Goodness' was equated with the perfection of the Demiurge, who because of such goodness generated all things, which things in turn also participated in this goodness (*Ti.* 29e2-5).

413. *Ti.* 29e2-5. Just as the Demiurge, the ordering cause, was considered by Plato to be the source of the ordered universe (i.e. generated Cosmos) and its goodness, so, too, Ananke was its opposite, which, as the disordering cause and the antithesis of perfection, was considered to be the source of 'evil' or disorder in the Cosmos.

414. *Ti.* 41c2-3.

415. Plato's adherence to the principle of 'like from like' is again evident in his requirement that the Cosmos was generated from pre-existing material (*Ti.* 52d4-53a9). The idea of creating something from nothing would have contravened that principle. I would suggest, too, therefore, that it is precisely because of the firmness with which Plato held this principle that he understood the Demiurge as having generated, in the sense of having 'arranged,' the original cosmic material, rather than having created it from nothing. In other words, the Demiurge's act of generating or arranging the Cosmos from pre-existing chaos, rather than having created it from nothing, was not, as has been argued, because the Demiurge lacked any sort of power or was not possibly omnipotent—Plato already declared that the Demiurge was perfect, which perfection included every good—rather, the Demiurge's course of action would have been because of the very power Plato declared it wielded. Thus, just as omnipotence or perfection carries with it the notion and requirement of perfect ordering, so, too, a hierarchy of being and action would have been

cosmological explanation for Plato had to take into consideration not only that the universe existed (the physical from the physical), moved (soul from Soul), and was beautiful and perfectly harmonized (evidence of a perfectly good Demiurge or maker) for the most part (which when lacking, was evidence of the disordering cause, Ananke), but that it was also reasonable. It is here that Plato's axiological assumptions again come into play and help determine the reach of his metaphysics. What is reasonable[416] is deemed by Plato to always be better than that which is not, leading to a hierarchy of perfection according to the degree to which 'nous' is or is not present in a particular object, thereby determining its place within the overarching hierarchy. At the lowest end of the causal,[417] non-physical, hierarchy is Ananke, a pure disordering force without 'nous,' which presence Plato postulates by the physical chaos he deems was requisite for an ordered Cosmos to have taken place and which alone is attributable by Plato for all unnatural disorder. At the highest end of this hierarchy is the Demiurge, a pure ordering force, from which all 'nous' originates and which alone is designated as the source of all natural order. Natural disorder, it should be noted, is distinguished by Plato from unnatural disorder in that the former was determined by Plato to be part of the ordered pattern of generation and decay, and not brought about by Ananke. Likewise, unnatural order is distinguished by Plato from natural order by being solely attributable to chance, whereas the latter is attributable to design.[418] Further, whereas Plato does not strictly identify the Demiurge with nous, 'nous' is the most singular characteristic attributable to the Demiurge and acts as an umbrella for the other qualities or modes attributed to this cause. Thus, whereas the Demiurge is described as being good, it is in the degree to which the Demiurge is identified with nous that he is judged to be good and which principle of identification also holds for the other qualities assigned to the Demiurge.

In summary, any understanding of Plato's cosmology which chooses to do away with either or both of the Demiurge and Ananke as genuine primary causes has first to abandon Plato's axiology or at the very least, greatly reduce its im-

---

put in place to reflect this, which Plato adhered to by the principle of 'like from like' and which is subsequently carried out within his cosmology by the Demiurge, as Nous par excellence, in generating the world as a visible god, whereby, in the final words of the *Timaeus*, it was declared to be supreme in greatness and excellence (*Ti.* 92c7-9).

416. *Ti.* 30a6-b3.

417. A causal hierarchy is meant here in the sense of constituting a force for order (the Demiurge) at one end of the hierarchy and a force for disorder (Ananke) at the other. Thus, while both are forces, they are causal opposites with regard to the ends they bring about. Although in the *Timaeus* Plato argues that the Demiurge is superior to Ananke for having ordered the Cosmos and for that order prevailing, the causal powers presented in *Laws X*, namely, the Beneficent and Non-beneficent Souls, are considered to be on a more equal footing by Plato. Thus, in this later dialogue the Non-beneficent Soul is presented as having at least as much power as the Beneficent Soul.

418. *Ti.* 28b2-7.

portance, but in so doing, and in both cases, radically reduce its richness and with that, the overall robustness which a moral universe, such as Plato's, offers. Accordingly, to understand the Demiurge as representing 'intelligibility' in the world and/or the 'Reason' in man and nothing more, and hence, as not existing prior to the Cosmos as the *source* of its intelligibility and reason, but rather, having existence only in the abstract as 'intelligibility' itself in an already existing universe (and hence, the need for mythic characterization) is mistaken. Equally mistaken, and again grossly reductive, is the view that Ananke is either mechanical motion or disorder in a Cosmos that has always existed. Plato flatly denies the latter, arguing from the start for a generated Cosmos (γέγονεν) and for the pre-existence of the powers that such generation entailed.[419] Given the above and prior arguments, I again strongly suggest that choosing to interpret either the Demiurge or Ananke as mythic or representational is clearly not what Plato had in mind. For in so doing, much of what Plato wrote concerning the Demiurge and Ananke as strictly first order causes within his metaphysics and physics, and so too, within his axiological account, must be abandoned and to the degree that it is, such an interpretation becomes misinterpretation.

4.3.7 Mythic names versus descriptive names

Finally, and importantly where myth is concerned, as part of its literary aspect neither the Demiurge nor Ananke are given proper names as one might expect in a mythic account, but rather, these first-order causes are given names which are always descriptive of and are limited to their function and/or nature. An example of Plato's regular use of myth and associated naming appears at the beginning of the *Timaeus* (20d-25d), where he relates a myth through the ancient voice of the poet Solon, in which the lost continent is named 'Atlantis.' It is important, moreover, that this myth be told through the eyes of a poet, which for Plato is indicative of its status as myth or story-telling and is in contrast to the body of *Timaeus* proper where the story of the Cosmos, in being stressed as a 'likely account,' is told through the eyes of an astronomer and hence, its basis as being derived from a well-developed science. Another example of mythic naming as used by Plato is in a myth told at the very end of *The Republic, Bk. X* (614b-621d), where the hero is a warrior named 'Er.' Also within *The Republic* is the well-known myth of the Ring of Gyges in *Bk. II* (359d-360d), where once again, the main mythic character, Gyges, is given a proper name. These examples are typical of the myth genre as used by Plato in his giving of proper names to the main characters or in the case of 'Atlantis,' to the key mythic feature.

This is not the case, however, as regards either the Demiurge or Ananke. The word 'Demiurge' in the *Timaeus* is essentially a descriptive term (ὁ δημιουργός) meaning 'craftsman,' 'maker,' or 'author' and is not a proper name

419. *Ti.* 28b3-7.

when used by Plato, although functionally the name is apt and is likely the reason why it has been widely adopted by scholars when referring to this primary cause. What actually takes place in the *Timaeus* is that the primary ordering cause or 'Demiurge' is referred to in several different ways by Plato, but none of which are proper names, for example, as 'the God' (*Ti.* 30a1, ὁ θεὸς), as the Constructor (*Ti.* 29e1, 'He that constructed it': ὁ συνιστὰς), in predicative terms, "he was good" (*Ti.* 29e1, ἀγαθὸς ἦν), and then more directly identified with the 'good,' as in 'for Him who is most good' (*Ti.* 30a7, οὔτ' ἔστι τῷ ἀρίστῳ), and as 'the Father' (*Ti.* 37c7, ὁ πατήρ), with similar varied references continuing to the end of this work.

This cause is also referred to three times in the *Timaeus* as the 'Demiurge.'[420] The first time 'Demiurge' is pronounced is by the Demiurge himself, when he announces in a markedly dramatic passage that he is 'ἐγὼ δημιουργός' (*Ti.* 41a7-8), immediately qualifying this by co-joining it with 'father' (πατήρ) and thus further refining its meaning. It is clear from this passage, however, that the God is putting this descriptive forward for himself as an example to the other gods who are being instructed to carry out similar 'demiurgic'[421] activity in their fashioning of the remaining mortal beings. Thus, it is purely as a mode of activity that this term is employed and to this extent it is also used in reference to the lesser gods (*Ti.* 41c4-5, τὴν τῶν ζώων δημιουργίαν). What is relevant here is that besides being a 'Constructor,' the Demiurge is equally referred to as a 'God' and as being 'very good' or 'the best.' Thus, the name 'Demiurge' by itself would have been far too narrow a term to have properly named or fully designated this primary cause. In strikingly similar fashion, the term 'δημιουργός' also appears in the later passage at *Ti.* 68e2, followed closely a third time at 69c3,[422] where once again the role of the Demiurge as divine craftsman is being emphasized and where the lesser gods are likewise enjoined (τῶν δὲ θνητῶν τὴν γένεσιν τοῖς ἑαυτοῦ γεννήμασιν δημιουργεῖν προσέταξεν). Here, too, the Demiurge is referred to as ὁ θεὸς (*Ti.* 69b3) when his role as craftsman is being singled out, reaffirming Plato's use of the term 'δημιουργός' as a descriptive, rather than as a name.

Ananke (from ἀνάγκη, with the meaning of a 'force' or more literally, a 'violence' which causes disorder), on the other hand, while equally a primary cause with the Demiurge, but not its equal, is captured even less with a descriptive by Plato than the Demiurge, being offered neither exemplary role-model status in the *Timaeus*, as was the Demiurge, nor moral praise. Ananke as a primary cause is referred to by Plato as a 'force' (ἀνάγκη) or 'necessity' (e.g. *Ti.* 47e6, δι' ἀνάγκης) when he wishes to refer generally to its causal power or alternatively, as the 'errant' or 'wandering cause' (*Ti.* 48a6-7) when he wants to

---

420. See *Ti.* 41a7-8 and the two subsequent instances at *Ti.* 68e2 and 69c3.

421. *Ti.* 41c3-6.

422. See Chapter 5 of Part I.

refer specifically to its disordering power. Again, in a previous passage where there is discussion of an evil 'will' (*Ti.* 41b2, κακοῦ), I have argued that this also refers to Ananke. However, references to Ananke are always brief or are only hinted at. Like the Demiurge, no single term fully captures the identity or essence of Ananke. Thus, another atypical feature not supportive of a mythic interpretation for either the Demiurge or Ananke in the *Timaeus* is that not only were these causes not given proper names, as was Plato's usual treatment for mythic characters (and with it, a granting of imaginary stability), their designations constantly changed. To wit, what descriptive terms Plato did use when capturing their function or essence *changed* as he began anew a fuller or further description of his cosmology. A mythic interpretation, therefore, on any of the suggested accounts cannot be easily or consistently sustained. The metaphysical divide, therefore, that exists between the Demiurge and the generated Cosmos, and Ananke and the pre-cosmic chaos, remains intact.

Chapter 5

5 *TI.* 68E1-69C6

5.1 Original Greek

Ταῦτα δὴ πάντα τότε ταύτῃ πεφυκότα ἐξ ἀνάγκης ὁ τοῦ καλλίστου τε καὶ ἀρίστου δημιουργὸς ἐν τοῖς γιγνομένοις παρελάμβανεν, ἡνίκα τὸν αὐτάρκη τε καὶ τὸν τελεώτατον θεὸν ἐγέννα, χρώμενος μὲν ταῖς περὶ ταῦτα αἰτίαις ὑπηρετούσαις, τὸ δὲ εὖ τεκταινόμενος ἐν πᾶσιν τοῖς γιγνομένοις αὐτός. διὸ δὴ χρὴ δύ' αἰτίας εἴδη διορίζεσθαι, τὸ μὲν ἀναγκαῖον, τὸ δὲ θεῖον, καὶ τὸ μὲν θεῖον ἐν ἅπασιν ζητεῖν κτήσεως ἕνεκα εὐδαίμονος βίου, καθ' ὅσον ἡμῶν ἡ φύσις ἐνδέχεται, τὸ δὲ ἀναγκαῖον ἐκείνων χάριν, λογιζόμενον ὡς ἄνευ τούτων οὐ δυνατὰ αὐτὰ ἐκεῖνα ἐφ' οἷς σπουδάζομεν μόνα κατανοεῖν οὐδ' αὖ λαβεῖν οὐδ' ἄλλως πως μετασχεῖν.

Ὅτ' οὖν δὴ τὰ νῦν οἷα τέκτοσιν ἡμῖν ὕλη παράκειται τὰ τῶν αἰτίων γένη διυλισμένα, ἐξ ὧν τὸν ἐπίλοιπον λόγον δεῖ συνυφανθῆναι, πάλιν ἐπ' ἀρχὴν ἐπανέλθωμεν διὰ βραχέων, ταχύ τε εἰς ταὐτὸν πορευθῶμεν ὅθεν δεῦρο ἀφικόμεθα, καὶ τελευτὴν ἤδη κεφαλήν τε τῷ μύθῳ πειρώμεθα ἁρμόττουσαν ἐπιθεῖναι τοῖς πρόσθεν. ὥσπερ γὰρ οὖν καὶ κατ' ἀρχὰς ἐλέχθη, ταῦτα ἀτάκτως ἔχοντα ὁ θεὸς ἐν ἑκάστῳ τε αὑτῷ πρὸς αὑτὸ καὶ πρὸς ἄλληλα συμμετρίας ἐνεποίησεν, ὅσας τε καὶ ὅπῃ δυνατὸν ἦν ἀνάλογα καὶ σύμμετρα εἶναι. τότε γὰρ οὔτε τούτων, ὅσον μὴ τύχῃ, τι μετεῖχεν, οὔτε τὸ παράπαν ὀνομάσαι τῶν νῦν ὀνομαζομένων ἀξιόλογον ἦν οὐδέν, οἷον πῦρ καὶ ὕδωρ καὶ εἴ τι τῶν ἄλλων· ἀλλὰ πάντα ταῦτα πρῶτον διεκόσμησεν, ἔπειτ' ἐκ τούτων πᾶν τόδε συνεστήσατο, ζῷον ἓν ζῷα ἔχον τὰ πάντα ἐν ἑαυτῷ θνητὰ ἀθάνατά τε. καὶ τῶν μὲν θείων αὐτὸς γίγνεται δημιουργός, τῶν δὲ θνητῶν τὴν γένεσιν τοῖς ἑαυτοῦ γεννήμασιν δημιουργεῖν προσέταξεν. οἱ δὲ μιμούμενοι, παραλαβόντες ἀρχὴν ψυχῆς ἀθάνατον, ( )

5.2 Translation

All these things, then, which had been brought forth in this manner out of Necessity, the Maker [Demiurge] of both the most fair and the best amongst the things generated took them over when he was begetting the self-sufficient and most perfect God, using them[423] as subordinate causes of all things, while He himself formed the good in all things generated. On which account, then, it is necessary to distinguish two kinds of cause, the necessary and the divine, and to seek the divine, on the one hand, in all things for the sake of the acquisition of a happy life to the extent that our nature allows, while the necessary, for the sake of those things [in which we seek the divine], discerning how, without the former, it is impossible to observe by themselves alone those divine things after which we make haste, neither, moreover, to understand them nor to grasp them otherwise at all.

Now that the kinds of causes lie sorted, all ready before us like wood to joiners, out of which it is necessary to weave together the remaining speech, let us

423. In this passage, it is the fairest and best amongst things generated which Plato describes the Demiurge as using for secondary causes with respect to other things. What makes them best in Plato's overall scheme is their participation in rationality. This is fully in line with his high regard for and understanding of rationality and its ends. Accordingly, the more integral something is to a specific rational end (and even these are valued, e.g. the pursuit of philosophy is seen as the highest end with respect to eyesight as a secondary cause), the higher it will be in the hierarchy of generated things, since, with respect to purpose (which is to bring about the particular rational ends which constitute a fully rational universe), it is closer in kind to the Demiurge, who, as 'nous' and intelligence par excellence, is the ultimate genitor of all rational ends.

return once again briefly to the beginning and in quick review go over where, up until now, we have arrived. At this time it is fitting that we should make an attempt to add to the story an ending and climax for what has gone before. For indeed, just as was also said at the beginning, all things being in a state of disorder, God worked in symmetries, both in each thing with respect to itself and with respect to one another, in as many ways it was thought possible to be proportionate and symmetrical. For at that time neither did any of these partake of this, except by chance, nor was there anything at all worthy to be called like those now named as 'fire' and 'water' or as any of the others: But first of all He set all of these in order, and then out of them He generated this universe, one living being having all living beings within itself, both mortal and immortal. The divine, the Maker [Demiurge] crafted himself, but He appointed his children to work out the birth of the mortal, and they, in imitation, having received the immortal principle of soul, . . .

5.3 Commentary and Notes

5.3.1 The generation of man – a causal proem

This passage forms a final and major transition point in the *Timaeus*, where once again Plato moves forward to a new and this time culminating object of his inquiry, the generation of man. Before doing this, however, Plato recapitulates his present position regarding *kinds* of causes and in so doing slightly revamps his previous position regarding 'Ananke' and 'the Demiurge.' This is followed by a brief summary of what has gone before. It is now 'the necessary' (implying, for the most part, the physically necessary, although occasionally the logically necessary) to which Plato will refer and not 'Necessity' or the primal cause, Ananke. Likewise, except for the two references to the Demiurge as δημιουργὸς at the opening and close of this passage (which positioning and reiteration reaffirms this God's place always as the ultimate source of generation), it is no longer the Demiurge to which Plato will refer in the subsequent dialogue (with the exception of divine praise noted in his closing statement), but instead, his references will be in more general terms of 'the divine.' Moreover, the Divinities, οἱ θεοί, the gods generated by the Demiurge (his "children"), will likewise drop out of the picture once the generation of man is complete.

This passage, therefore, is essentially divisible into two parts (shown both in the Greek and in its translation as two paragraphs). The first part represents Plato's position regarding the kinds of causes operative within the generated cosmos, while the second is a summary of his previous position up to this point regarding the primordial causes, Ananke and the Demiurge, and their role in the generation of the Cosmos. The first passage, however, is prefaced by a brief citing of the primary causes, while the second, by an equally brief reference (now that all the causes have been named) to the totality of causes.

5.3.2 Change, but not a material change

The change that is effected at this juncture is not a material change. Plato does not alter his original understanding of Ananke as a 'wandering' or 'wayward' cause or of the Demiurge as the 'generative' or 'constructor' god. These two powers remain for him the primary causes. In fact, as noted above, Plato reiterates his position with regard to the latter, describing now for the second time the Maker of the Cosmos as the δημιουργὸς.[424] The task undertaken by the Demiurge of generating the Cosmos is now complete. What are then subsequently presented as causes within the generated universe Plato describes as *kinds* of causes or causes of a general sort, since both the Demiurge and Ananke as metaphysical causes exist outside of all physical causation occurring in place and time. Plato will now therefore speak of *divine* causes when referring to the good in the Cosmos. As such, these causes do not refer to either the Demiurge or the God, itself, or to the divine or the gods (the Heavens), whose own task set by the Demiurge will be complete with the generation of man. Rather, these divine *ends* are those things in which man will find happiness or the good, and therefore are causal in respect of directing or having directed an individual's actions.

5.3.3 Ananke poses a problem

Ananke, on the other hand, poses a problem for Plato. As a primordial disordering cause it has been rendered essentially powerless by the Demiurge due to the ordering of the Cosmos. Yet, as a true primal cause, it remains a force, albeit, one whose role is greatly altered with respect to the generated Cosmos. However, if the Demiurge alone brings about the good in the universe and as a whole, this living, visible God (the Cosmos) is completely good, the question remains of where Ananke fits into the finished cosmic picture. As a true causal power for disorder, as long as there is order, the answer can only be nowhere. However, having been subdued by the power and workings of the Demiurge, the idea of a disempowered Ananke (or mostly, since individual instances of disorder are still possible), or at least with respect to its ability to maintain or bring about global disorder, is nevertheless borrowed by Plato and its role with regard to the finished Cosmos, transformed. He now incorporates the name of this disordering cause into the ordered Cosmos, but as a term only, symbolic of its submission, using it to describe in general terms the accessory or physical causes within the generated Cosmos, causes which, not being unmoved movers (i.e. the primary causes) or self-moved movers ('soul'), have no causal power of their own. Up to this point in Plato's cosmology, although the accessory causes had been previously described,[425] he did not at that time associate these causes di-

424. See my discussion in Part I at §4.3.7.
425. See *Ti.* 46c3-d7.

rectly with Ananke, since as a power it had not yet been subdued by the Demiurge.[426] However, now, with its power relinquished, the force alone of Ananke has been borrowed by Plato, thus, its submission acknowledged, and as a final consequence, all future references to the primal disordering cause in the *Timaeus* are abandoned.

5.3.4 A transition – from the metaphysical to physical realm

With the Cosmos now fully formed and with only the ordering of man left for him to describe, Plato chooses to leave the metaphysical realm behind. Accordingly, with the exceptions of brief mentions of 'soul' in relation to the body, followed by a call to the philosophical life lest one should become a woman or an animal and a final pause for divine exaltation of the Cosmos as a "visible god,"[427] again prefaced by a warning to choose the philosophical life, the remaining dialogue, to the end of the *Timaeus*, is written almost exclusively in terms of the physical universe. Thus, with the earliest stages of the ordering of the Cosmos over, so, too, is Plato's mention of them. This move solely to the realm of the physical world, however, the generation of the Cosmos now being complete, requires from Plato a statement of the chief kinds of causes active within it in order that he can give a further description of man as fully formed, the culminating goal of the *Timaeus*. This he carries out, but not without some difficulty as to the meaning of his terms, leaving room for equivocation by the reader. What Plato does is borrow and adapt previously used causal terms, namely, those used for his primary metaphysical causes, 'Ananke' (or 'Necessity') and the 'God' (which in the above passage is referred to as the '*Demiourgos*' or 'Demiurge'), re-introducing these same terms as 'the necessary' and 'the divine' respectively, but which now become causes within the physical realm. From Plato's current perspective, therefore, which is to focus on the physical world as it now is, combined with his ultimate goal, which is to end with the story of man, it is essential that all subsequent discussions leave out further mention of the metaphysical causes, except in summary passages. The reason for this is that these causes exist entirely outwith the physical realm due to their provenance and role as primary generative agents and hence, have no direct bearing on the day-to-day workings of the world, nor does 'soul' with its separate psychic function, while the secondary or auxiliary causes, as regards both their provenance and function, become the causal crux of the physical world, being responsible for and participating in the carrying out of all its processes.

426. *Ti.* 47e3-48a5.
427. *Ti.* 92c4-9.

5.3.5 Primary causes versus *kinds* of causes

The problem for Plato is to discover and retain a relationship between these causes without compromising their integrity both as separate causal powers and as separate *kinds* of causes. To achieve this, he chooses to adapt the causal terms as much as possible. Accordingly, his primary causal powers, the Demiurge and Ananke, now beget *kinds* of causes in the generated Cosmos. Ananke, having been subdued by the Demiurge and left essentially powerless as a causal force in light of the actual generation of the Cosmos, allows Plato to then safely incorporate its name for the secondary causes within that Cosmos, having already declared these auxiliary causes not to be true causes, i.e. unmoved movers (the primary causes) or self-moved movers ('soul')[428] in any event. Thus, as a true causal power for disorder, Ananke (Necessity) has been ultimately overridden by the ordering cause, the Demiurge, by the very fact of the ordering of the Cosmos. However, the further idea closely bound to Ananke, that of brute force, is nonetheless retained by Plato and is later resurrected as a *kind* of cause in the Cosmos, namely, *carte blanche*[429] the secondary or mechanical causes which he now identifies as 'the necessary.' The Demiurge, on the other hand, remains the principal metaphysical causal power in the generation of the Cosmos and therefore cannot be used by Plato as a *kind* of cause in the physical world. He consequently decides to show its relationship to the Cosmos by referring instead, not to the Demiurge, but to 'the divine' as causes or good ends in the world, which man chooses for the sake of his own happiness and which ends in man, according to Plato, are produced only by the proper functioning or ordering of the human soul, just as the world about him (man) is ordered. These good ends, moreover, whether in relation to man's soul or the world's soul, have been generated solely by the Demiurge or God for their optimum well-being and continuance, as reflections and in imitation of its own perfection and immortality. The divine causes *in* the Cosmos, therefore, correspond like-to-like with the Demiurge's generative handiwork *of* the Cosmos.

It is essential to note, however, that although the terms 'necessity' and 'divine' are used in earlier passages of the *Timaeus*, this present passage does not

---

428. The notion of 'soul' as a self-moved mover does not fully emerge until *Laws X*. Rather, in the *Timaeus* the role of the 'World Soul' is more that of a 'moved mover,' where its noetic movements have been designed by the Demiurge to bring about particular ends within the generated Cosmos. With regard to the human soul, however, also designed, the idea of self-movement, although not yet fully articulated or developed (which will not happen until *Laws X*), is nonetheless assumed in this work to the degree that Plato recognizes or assumes choices involving the human 'will.'

429. The 'necessary' now acts as a kind of genus for the secondary causes, under which every physical process or descriptive category when understood as either a process, e.g. 'eyesight' or as the cause of something else, e.g. 'earth,' becomes a kind of subspecies.

offer a restating of any previous position. This would have been fruitless, given that according to Plato's cosmology the metaphysical causes described earlier are entirely outwith the Cosmos or physical world they have shaped. Rather, as a preface to his description of man, what Plato does is describe in familiar terms certain aspects of the Cosmos already in place. Thus, he now introduces the 'necessary' and the 'divine' as *kinds* of causes. In so doing, he chooses first to acknowledge the larger picture, of which man forms a part and within which he moves, in terms of some of the processes of the physical world and a description of their causes. Only then will Plato embark on the final and principal aim of his work, a description of man himself, which he then carries out.

5.3.6 The finished universe – a new causal statement

Thus, a completely new statement coincident with the finished universe is now given, albeit in familiar terms. Plato writes: "on which account, then, it is necessary to distinguish two kinds of causes."[430] The account to which Plato is referring is that of the *finished* Cosmos, a world capable of being inhabited by man and whom he is about to describe in detail. The two *kinds* of causes referred to by Plato as now being in place in the universe are indeed reflective of and dependent upon the two primary causes responsible for the ordering of the Cosmos in the first place—and this is no doubt the reason why Plato borrowed these previously used causal terms, namely, to retain some carry-over of meaning. Thus, 'the necessary' and 'the divine' become *kinds* of causal movements linked to, but not the same as, their causal predecessors, namely, Ananke (Necessity),[431] the disordering force and the Demiurge or God, the ordering force, Plato's two primal and pre-generative causal players.

430. *Ti.* 68e6.

431. As explained earlier, although I have retained 'Necessity' at certain places as an alternative English translation of the Greek word ἀνάγκη or Ananke, I have done this chiefly for ease of reference, 'Necessity' having been used by all three aforementioned commentators extensively, regardless of whether primary or secondary causes were being considered. The reason I suggest for such broad usage of this word by these commentators is its mistaken identity solely with the secondary causes of this passage or τὸ μὲν ἀναγκαῖον, Ananke, as a primary cause, having been dismissed early on as myth. As suggested at fn. 135, 'Anarchy' is a better English translation for the disordering cause Ananke, with its inherent meaning of 'lawlessness' or 'disorder,' than 'Necessity,' the term 'the necessary' on the other hand being a suitable translation for the secondary causes. The word 'Anarchy,' however, like 'Necessity,' remains somewhat misleading as a name for Ananke, inasmuch as it could mistakenly be identified with the effect that such a force produces and not with the causal force itself.

5.3.7 πεφυκότα ἐξ ἀνάγκης – produced out of Ananke/necessity

However, not only is there a carryover or association of prior meaning taking place within this passage, there is also a change of meaning occurring here with respect to these terms. When introducing this passage, Plato allows for the possibility of some play on the meaning of his primal causal term 'Ananke' (Necessity). Although he states that "all these things" were "produced out of Ananke," by choosing to use the common prepositional phrase ἐξ ἀνάγκης, which in all its subsequent occurrences to the very end of the *Timaeus* is used exclusively with the adverbial meaning of 'necessarily,' thereby attaching a kind of 'causal' necessity (i.e. a certain action must take place or else violate a natural law) to the physical processes being referred to, the meaning of ἀνάγκη in this opening remark can be read in one of two ways, either as the primal cause 'Ananke,' i.e. on account of Ananke having been "persuaded" (subdued) or as the adverb 'necessarily.'[432] In either case, the phrase 'all these things' refers to the preceding processes that have just been discussed by Plato. They are presented, depending on which reading, as either a result of the Demiurge having taken over the visible universe on account of ('out of') Ananke having been first subdued or alternatively, when ἐξ ἀνάγκης is read adverbially (as with the remainder of the account), so as to stress their dependent or mechanical nature.

Further, while these necessary processes are described earlier by Plato as causes (here referred to as "all these things . . . being produced out of Ananke"), such causes are not considered by Plato to be true causes, in the sense of directing their own movement, but rather, are physical objects moved by and moving other physical objects. Thus, besides a change taking place with regard to the meaning of a familiar term, a thematic change has also occurred. Accordingly, as regards change of meaning, the 'wayward cause,' Ananke, in being subdued by the Demiurge, has become its opposite, the 'necessarily caused' (the events of process), which in both cases can be expressed by the prepositional phrase ἐξ ἀνάγκης. As to the suggested change of theme, Plato's vision is likewise now earthward. Dropped is his interest in the primary causes, Ananke and the Demiurge, and stressed are the secondary causes or those that serve another end, namely, the happiness of man. These subordinate causes are therefore classed as *kinds* of causes and named respectively 'the necessary' and 'the divine,' thereby

432. The meaning of 'necessarily' or 'out of necessity' for the Greek phrase ἐξ ἀνάγκης arises from the following line of reasoning: The idea of brute force follows from the idea of a disempowered Ananke. The universe was ordered and the power of Ananke overridden, but not destroyed. Consequently, this previously disordering force, now anchored to and overridden by the natural logic or presence of 'nous' inherent within the ordered universe, becomes a passive tool or accessory causal force for that very ordering. Thus, Ananke, in being subdued by the Demiurge, ultimately becomes a passive causal tool of Nous within the ordered universe and its power or force appropriated to causal ends, which flow 'necessarily' by virtue of the end envisioned.

becoming at once vestigial reminders of their ancient primogeniture in the primary causes, Ananke (Necessity) and the Demiurge (the God).

5.3.8 τὸ μὲν ἀναγκαῖον, τὸ δὲ θεῖον – the necessary and the divine

Plato presents this thematic change subtly, as is his style in the *Timaeus*, relying on the reader's previous understanding thus far assumed and then proceeds anew from a streamlined conceptual platform towards a new vision within his cosmic landscape. He initiates this move, moreover, as he has done before in the *Timaeus*, by dropping previously depicted cosmological features or players no longer required. Ananke, for instance, as a brute causal force is not required for the present discussion and is therefore dropped. The reason for this is that in his subsequent discussion of man, Plato assumes a post-generative Cosmos, in other words, a Cosmos in which Ananke has already been subdued or overpowered by the Demiurge. Assuming any earlier state of the Cosmos other than a fully realized one would have made the task of describing man a much more difficult one. Consequently, the idea of Ananke with its original and full meaning of a disordering metaphysical causal power has been laid aside. Its associated meanings, however, of chaotic particular movement (which movement has now been ordered to a good end)—keeping in mind that initially this disordered movement of mostly incoherent mass was all there was to the universe—and the subsequent causal force of that movement (now overpowered) have been broadly retained. Thus, with this shift of emphasis from an emerging Cosmos to a fully emerged one, and with only man left to fully describe, Plato re-assigns his vocabulary in order to accommodate this move. Hence, he now talks of "all these things . . . produced of necessity," i.e. the effects of a subdued Ananke or Necessity or alternatively, the physical processes which occur *necessarily* after a certain manner, and not specifically of 'Necessity,' the Wayward Cause, itself. Likewise, the Demiurge as a primal cause can only be reminisced upon. So instead, it is now 'the divine' which Plato will seek to find in the finished Cosmos or in other words, that which points to the good and which brings about man's happiness. Thus, this is the first passage in the *Timaeus* where Plato now refers to 'the necessary' or τὸ μὲν ἀναγκαῖον as a *kind* of cause, while contrasting it with another *kind* of cause, the divine or τὸ δὲ θεῖον: διὸ δὴ χρὴ δύ' αἰτίας εἴδη.

5.3.9 Ταῦτα δὴ πάντα – 'Now all these things'

It is essential, moreover, when investigating this passage and Plato's understanding of it that his use of the plural "all these things . . . produced of necessity" in the opening phrase is not ignored, nor his choice of mere association with Ananke, rather than direct reference to Ananke itself. Likewise, his reference to the divine as a *kind* of cause, rather than to the Demiurge itself as a cause, is also important to note. These are significant markers indicative of the causal aspect which Plato's investigation has now taken, namely, that of the physical realm and, as indicators of aspect or perspective, are not themselves to be taken as objects of it. To make 'all these things' objects of his investigation results in this passage being taken as a summary passage, which it is not. The summary passage is clearly in the next paragraph beginning: Ὅτ' οὖν δὴ τὰ νῦν τέκτοσιν. Rather, this first paragraph is a continuation of and conclusion to his previous ongoing description of the physical processes in a fully working and viable Cosmos, set now as the platform from which Plato will soon describe man.

Similarly, Plato's declaration that 'all these things' form the subordinate causes must be explained in light of what he has already proposed regarding such causes. His previous explanation concerning the auxiliary causes, therefore, must be taken into consideration. What potentially happens when this passage is glossed over is that the entire pericope comprised of both paragraphs is then understood as and paraphrased into the singular, with the consequence that 'all these things' are then wrongly re-interpreted as the effects of a single cause, Ananke. As a further result, Ananke is then understood as a subordinate cause to the Demiurge, and 'all these things' are likewise subsumed as relatively unimportant effects of Ananke. This new physical understanding of Ananke as a subordinate cause is then wrongly applied to all of the *Timaeus* by way of the summary in the second paragraph, with the outcome that Plato's metaphysics is essentially abandoned and is now replaced with a purely physical explanation of the Cosmos, the exact position that he angrily condemns in the *Laws*.[433] To avoid such a misreading, this passage, comprised of two very distinct parts, must not be taken in isolation, but must be seen in the light of both the content and style of what has gone before, the first paragraph in light of the physical processes described over the last several pages and the second paragraph, as a summary reminder of Plato's original metaphysical explanation. Moreover, Plato's lack of reference to Ananke or Necessity as a primary cause in this second paragraph, referring now only to the Demiurge as a causal power, is not indicative of the non-existence of Ananke. Rather, I suggest, it is a further reminder of Plato's revulsion of evil, with as little consideration given to it as possible, since, as he later expresses, πᾶν δὴ τὸ ἀγαθὸν καλόν, τὸ δὲ καλὸν οὐκ ἄμετρον, "it is better to devote our attention to good than to evil."[434]

Thus, although there is a change taking place here, the above-noted mistaken interpretation wrongly assumes that rather than initiating change at this juncture, Plato has simply been sloppy in his writing, not only generally, but also in the particulars of his earlier cosmology, as well as with regard to his referents. The consequence that ensues, however, is that upon scrutiny new and quite ingenious problems are now presented. What exactly are "all these things" constituted of

433. See *Lg.* 889b1-890a9.
434. *Ti.* 87c4-5.

Ananke? Moreover, how can Ananke be a subordinate cause within the Cosmos, when it has already been presented as a primary cause outwith the generated universe? Further, how is it that without the so-called 'necessary' causes, we cannot perceive, comprehend, or in any way attain our objective?[435] These questions, however, are inevitably raised by the above interpretation. Plato has already discussed perception and sight in some depth under the topic of the subordinate or assistant causes, which subordinate causes are quite separate from the primary causes, namely, the Demiurge and Ananke. Quite clearly, therefore, there is a problem when the above reading is assumed in which Ananke is deemed to be the subordinate cause of the Demiurge, rather than the second of the two primary causes.

Read as written, however, and not glossed over, where in fact 'all these things' and *not* the metaphysical cause, Ananke, are deemed to be the subordinate causes of the Demiurge, this passage requires no such interpretation to be fully comprehensible both *in situ* and in relation to the *Timaeus* as a whole. A closer inspection of the relevant passages supports this view. Very early on in the *Timaeus* Plato presents two primary causes, the Demiurge and Ananke, whereby he gives a likely account of the metaphysical basis for the generation of the Cosmos. He gives further summary accounts of these causal figures when he plies layer upon layer to the cosmic story as he fleshes out its many implications. Although Ananke is described as having been subdued by the Demiurge by virtue of the fact that the ordering of the universe had taken place, at no place and in none of the summaries is Ananke ever described as being a subordinate cause within the Cosmos. To the contrary, Plato clearly distinguishes what he describes in some detail as being the subordinate or assistance causes within the generated cosmos. To this end, he first provides an example of an accessory cause beginning at *Ti.* 45b2. Here he gives a detailed account of the eyes, presenting them as the physical organs for the mechanism of vision, and latterly, a little further on (*Ti.* 46c7), as a prime example of the subordinate or accessory causes, which assistant causes, however, he notes are often mistaken for true or primary causes. Prefacing this passage at *Ti.* 45a6, Plato further writes:

> They [the gods] set the face in the front thereof and bound within it organs for all for the forethought of the Soul.

The eyes in this passage, soon after described as accessory causes, are positioned on the body for the benefit of the Soul, man's divine or best part, and the source of his beneficence or happiness. Thus, in physical terms, Plato has already stated what he will later restate, but now in freshly re-drawn causal terms:

435. Lee, 96.

> On which account, then, it is necessary to distinguish two kinds of causes, the necessary and the divine, and the divine, on the one hand, to be sought in all things for the sake of the acquisition of a happy life to the extent that our nature allows, while the necessary for the sake of those things [in which we seek the divine].

Plato explains that these 'organs for the sake of the Soul' or latterly, these 'necessary causes for the sake of the divine' are not true causal powers, since they (and their parts) are purely physical in nature, their causal efficacy depending entirely on movement from without, upon which the 'divine' and irrational powers do not depend, being noetic and anoetic respectively.

To summarize, it must be ruled out categorically that 'all these things' in this passage refer to Ananke itself, as nowhere is there reference to Ananke in the long section which precedes this opening remark, which remark alludes to and gives closure to what has just preceded. Given the text, it is clear that reference is indeed being made to the previous section. In this section, beginning at *Ti.* 65b5 and comprised of several smaller passages, Plato gives an account of the sensations that occur in particular organs of the body and their causative agents. As in his earlier description of the eyes and their related mechanism of vision, these new descriptions follow with similar appeal. He therefore first deals with the tongue and the sensation of taste, followed by the nose and the sensation of smell, the ears and the sensation of hearing, and finally, the eyes and this time round, the sensation, as Plato understands it, of color. He prefaces this entire pericope with an explanation of the physical basis of sensation (*Ti.* 64a2) and within this, offers a further explanation of how bodily sensations as a whole give rise to pleasurable and painful feelings. Thus, what is to hand is a physical account of bodily sensations as a whole, of their relation to pleasurable and painful feelings, and then more specifically, an account of what happens within the physical organs themselves to cause these sensations and the feelings consequent upon them. How far one should go back within the dialogue to encompass the full scope of 'all these things' is left up to the reader, the point being that the content of 'all these things' is in fact all of the physical processes accruing to enable the formation of man and certain of those—the senses—proceeding therefrom. This description, I suggest, therefore, follows from and comprises the whole of the organization of the primordial triangles into the basic elements by the Demiurge, up to and including man's ongoing experience of physical sensations. A further description of the mortal parts of the soul (emotion and appetite) will follow later, as will the functioning of various internal organs, i.e. the heart, lung, liver, spleen, and intestines.

5.3.10 The markers

As noted above, there are a number of markers showing Plato's change of perspective in causal aspect as he moves from the metaphysical to the physical realm. He has also put into place literary markers indicative of this same thematic move. First, this is the last time the primal cause 'Necessity' is referred to by Plato as a cause, and the first time, it is so directly addressed. Instead, the prepositional phrase ἐξ ἀνάγκης, with its adverbial meaning of 'necessarily,' will now be solely used for the remainder of this work to its end. Second, and similarly, other than *Ti.* 41a7-8, this passage contains the only two remaining direct references to the Demiurge (δημιουργὸς), with all future references dropped. Instead, Plato, in naming his two causes, will now refer exclusively to two *kinds* of causes, namely, 'the necessary' (τὸ μὲν ἀναγκαῖον) and 'the divine' (τὸ δὲ θεῖον). In so doing, Plato moves away from the larger metaphysical inquiry of his earlier writing and through to the end of the *Timaeus* now addresses exclusively a physical description of man, with the above characteristics of the metaphysical powers borrowed by Plato and consequently refocused and reshaped in general terms as kinds of causes relative to the physical sphere. Third, it is also noteworthy that along with this move from the metaphysical to the physical, Plato, except for his account of the generation of the sexes, is able to carry out the remainder of his discussion primarily in the present, whereas earlier his writing has been largely in the historic tenses, indicative of the past. The primary significance of this well-marked move from the metaphysical to the physical world of man is twofold. Firstly, it cautions against dismissing out of hand Plato's primary causes, the Demiurge and Ananke, as merely mythic entities symbolic of their supposed physical counterparts. As his writing clearly shows, Plato is more than capable of providing physical explanations for phenomena according to the knowledge of his time and given this, he has no need to resort to fantastic ones. Equally important is the fact that his metaphysical explanations have often so little bearing on physical processes that if they are nonetheless so construed, they have to be interpreted as mythic and hence, symbolic, to make any sense of them at all. Secondly, and singularly important as regards the overall integrity of the work as a literary entity, by this move from the metaphysical to the physical, from primary causes to secondary causes, from pre-Cosmic chaos to man, Plato is able to complete and thus accommodate perfectly the task Timaeus was asked to carry out by Critias, namely, πρῶτον λέγειν ἀρχόμενον ἀπὸ τῆς τοῦ κόσμου γενέσεως, τελευτᾶν δὲ εἰς ἀνθρώπων φύσιν,[436] which is to give an account beginning first with the origin of the Cosmos and ending with the nature of man. Thus, there is no need now but to believe that Plato achieved the full complement of his undertaking.

---

436. *Ti.* 27a5-6.

5.3.11 Causal conclusions – the commentators

5.3.11.1 Cornford

Of the several passages studied from the *Timaeus*, this passage in particular (*Ti.* 68e1-69c6) brings to completion the causal views concerning this work of the classical commentators chosen for comparison, namely, those of Cornford, Archer-Hind, and Taylor. Cornford quite rightly divides this passage into a "*Conclusion*" and "*Recapitulation*,"[437] which division I have shown above as two paragraphs. The main problem with the conclusion that Cornford subsequently draws, however, is that he fails to negotiate the several markers set by Plato in the *Timaeus* with regard to its theme and as a consequence, oversteps the parameters of this work's conclusion. Thus, with regard to the first of these two paragraphs (*Ti.* 69e-69a) Cornford refers to "necessary causes" or "what comes about of Necessity," explaining that according to Plato we must study these as the only way of approaching the manifestations of rational purpose in nature. However, as I have argued above, Plato does not refer to 'necessary causes' in his conclusion here, but rather, to '*secondary* causes' and notably, to those which are of particular interest to man, several of which Plato had been discussing over the prior few pages, namely, the tongue and taste, the nose and smell, the ears and hearing, the eyes and sensation of color, prefaced in turn by an explanation of the physical basis of sensation, with a further explanation of pleasurable and painful feelings. Although Cornford correctly observes that the subsidiary causes are the ones being discussed, he obfuscates, without sufficient clarification, the 'necessary' with the purely 'secondary,' which latter according to Plato are not true causes at all. I have argued above, however, that Plato does in fact introduce a decisive change of theme at this juncture, wherein 'Necessity' as a primary cause drops out and 'the necessary' as in the 'logically necessary' or conditions of process become the focus through to the end of the *Timaeus*, but given this, it is not a theme which Plato has visited earlier or one with which he has been "concerned throughout."[438] To the contrary, up to this point Plato has been concerned primarily with the metaphysics of his cosmological position, interspersing metaphor with natural accounts to capture the breadth of a universe about which he can only speculate (hence, his overall driving vision of a 'likely' account). Moreover, it is a metaphysics that postulates quite candidly at least two primal causal figures, Necessity (Ananke*)* and the Demiurge. Thus, except for his earlier discussion of eyesight,[439] in which he specifically denies that its process is truly causal in any way (lacking in self-motion) and subsequently separates this type of cause unequivocally from the sort of cause that he consid-

437. See Cornford, *PC*, 279–80.
438. Cornford, *PC*, 279.
439. *Ti.* 45b2-46d3.

ers to be a true cause, Plato has shown minimal interest in the secondary causes. Moreover, the two kinds of causes which he then contrasts one with the other in this earlier passage concerning eyesight, distinguishing "those which, with Mind, are makers of the beautiful and good" from "those which, being devoid of thought, bring about random disorder each time they produce,"[440] have nothing in common with the secondary causes. Rather, conversely, the secondary or ancillary causes that Plato subsequently goes into great detail about later on in the *Timaeus* through to its end are *never* random and produce *nothing* of themselves. In the context of this passage concerning eyesight, therefore, it and the other secondary causes, whose discussion take up the latter part of the *Timaeus*, form part of 'the beautiful and the good' and hence, as argued by Plato, are themselves not causal, but rather, are the caused.

A definitive argument against identifying the secondary causes with those that "bring about random disorder each time they produce" is that they are ultimately traceable by Plato back to the Demiurge as their source, the cause of order and not of random disorder. In this regard Plato cites eyesight and the good which the God brings mankind[441] as being exemplary of the secondary causes, further enumerating other important secondary causes in relation to man, i.e. speech, hearing, and the sensing of rhythm, each of which according to Plato are given to man as "heaven sent ally in reducing to order and harmony any disharmony in the revolutions within us."[442] These, he explains, are given to us by the divinities, who in turn have been generated and instructed by the Demiurge to carry out their tasks. A consequence, therefore, of this two-fold tier of generation, traceable first to the gods and then back to the Demiurge, is that both they (the divinities) and their works (the soul and body of man, and all the secondary causes in which they participate) are necessarily wholly good, which goodness, in being founded upon the twin notions of reason and intelligence, can in no way be the cause of randomness or disorder, its contrary. Hence, Plato writes:

> And with regard to the things before these, concerning bodies according to parts of their generation and concerning soul, how, through the agency and providence of the gods they were wrought, we must put forth the best possible account and thus proceed for these to accordingly be given [*Ti.* 44c7-d2]. . . . Of the organs, they first constructed the light-bearing eyes, fixing them in place for the following reason

440. *Ti.* 46e3-6. These two causal powers are also translated as "those that operate intelligently and produce results that are good" and "those that operate without reason and produce effects which are casual and random" (Lee, 64–65); see also Cornford, *PC*, 157.

441. Plato explains that the God (the Demiurge) invented and gave us (mankind) sight (gifted to us through the divinities) in order that we could observe the ordered revolutions of the heavens and in this way guide the troubled revolutions in our own understanding (*Ti.* 47b5-c3).

442. Lee, 65 (*Ti.* 47d4-7).

> [45b3-5]. . . . Accordingly, these are all assistant causes which serve God who uses them in carrying out, as far as it is possible, the idea of the best [46c8-d2].

As far as what is *not* possible, the primordial force, Ananke, with its residual effects, is the only cause of unnatural or true disorder within Plato's cosmology.[443]

The brunt of Plato's prior discussions up to the beginning of this present speech concerning the various processes has been almost entirely theoretical or speculative in nature, which, to list some of the topics, include: 'being' and 'becoming'; whether the Cosmos began or always was; the nature of its maker, 'the Demiurge'; the reason for its generation; the extant materials and existent power, 'Ananke,' both of which the Demiurge had to work and contend with; the formation of the body of the world; the soul of the world; time; the formation of the first of the living creatures, the celestial gods; the traditional Greek gods; the remaining three forms of living creatures, which include birds, fish, and land animals; the formation and destiny of the human soul; the human body; and plants.[444] Some of these topics Plato discusses only briefly, while others, he re-

---

443. For a discussion of the different kinds of order and disorder, see Part I, §4.3.5–6.

444. Plato's later discussion at *Ti.* 90e1-92c3 of the formation of the sexes, birds, fish, and certain land animals resulting from the inept behavior on the part of individual men (i.e. they were not generated distinctly by the gods, but instead *de*-generated into their species from generated man) is, I suggest, not a serious discussion by Plato, but rather, is a tongue and cheek account, given the literary markers at the beginning of this particular pericope (*Ti.* 90e1-6) and the lack of argument from a philosophical perspective. Although a reasonable account is given within this later pericope of the process of reproduction, this is a separate matter from that of the generation of the sexes, which latter is largely an historical consideration. Plato's much earlier statement at *Ti.* 41a7-d3 states that the celestial gods generated three distinct kinds of mortal creatures *so that the cosmos might be perfect*. The idea of the perfection of the cosmos as the reason for the generation of these creatures (which includes both men and women) against this later closing account, a plausibly inserted aside, of the imperfections of men as part of a degenerative process to account for the appearance of women, are wholly incompatible. Whilst the former is in keeping with the overall tenor of the *Timaeus* in its description of the world as a visible god, supreme in excellence and perfection, the latter is not. Moreover, the implicit assumption in the last sentence of the earlier passage (*Ti.* 41d2-3) is that these creatures were capable of proliferating at the outset of their generation, in that it is decreed that they should partake of a cyclical pattern, which without division into the sexes would have been impossible. Plato thus presumes that there is a physical process which takes place between the sexes in order for a full human being to be born, which process he later describes in his passage at *Ti.* 90e6-91d6. This is clearly indicative of the separation of such creatures into the two separate sexes at the generative stage, but beyond this assumed fact, Plato remains silent. The reason I would suggest for the derisive but fairly jocular banter at the end of the *Timaeus* regarding the degeneration of men into the female sex and other creatures is an unabashed call to the philosophical life, which for Plato is a masculine and idealistic venture, and in this respect remains Platonic, but

visits in greater detail. He stops rather abruptly after discussing vision to begin anew with a further metaphysical description in which he separates very sharply the secondary causes, such as that just discussed of vision, from the true primal causes already mentioned, the Demiurge and Ananke. Thus, acknowledging that a body must have a receptacle to receive it, Plato thereby resumes his prior metaphysical description, describing now the receptacle of becoming,[445] the nature of becoming and all change, i.e. the elements understood as qualities, the difference in kind between intelligence and true opinion as a basis for postulating the existence of 'being' and 'becoming' respectively, followed by a discussion of being and becoming as 'form' and 'copy.' Only at this juncture does Plato begin to move into the predominantly physical or natural realm, starting first with a discussion of the primitive chaos, followed by a long description of the formation of solid bodies from basic triangles, from which he then turns to his description of the physical basis of sensation and the organs of sensation, bringing the reader full forward to the present passage, soon to be crowned by the formation of man, and which physical processes Plato will remain primarily vested to the end of the *Timaeus*.[446]

---

clearly, given what has been stated in the *Timaeus* regarding creatures generally, is not part of his cosmology *per se*. Although Plato does briefly discuss the generation of the sexes earlier at *Ti.* 41d4-42d2, he does so in very similar terms as his later account, first announcing humankind as being of two sexes (in the later account this is assumed in his discussion of the reproductive processes) and only then, as at *Ti.* 90e1-92c3, instead of attempting a more detailed physical explanation, offers instead a brief explanation of the *de*-generation of men into the two sexes and certain animals in the context of the theory of the transmigration of souls. This, therefore, like the later passage, is a philosophical and didactic account of the generation of man, rather than a physical or historical one, which non-physical and a-historical accounts, following upon the heels of the generation of the soul and body respectively, are intended to call the whole of man, both body and soul, to the philosophical life.

With regard to the literary markers, there are two. The first is Plato's mention that his account of the universe is now complete (*Ti.* 90e1-3). This is decisive and there is no need to question this statement. The second marker lies in what immediately follows (*Ti.* 90e3-6), which passage Lee has captured well in his translation "a brief account along the following lines seems more in keeping with the subject" (Lee, 122); see also Cornford, *PC*, 356. In other words, if Plato is going to give any further account, it will be more in keeping with his interest in philosophy, rather than that of metaphysics or the natural world, which accounts are now complete, thus bringing Timaeus's discussion full round to that of philosophy. Plato's discussion of plants, on the other hand, is reasonable given his account thus far and therefore is to be taken at face value.

445. The closest to this concept would be our modern notion of 'space.'

446. Although Plato will still refer to God (the Demiurge), his 'sons' (the celestial divinities), 'things divine,' and the human soul, notably at *Ti.* 68e1-72d8, he does so specifically with reference to what he will describe as the 'mortal' parts of the tripartite soul, namely, the emotive and appetitive or sensual parts and how these are placed and interact with the organs of the body, while also giving brief mention to the divine part of the soul in comparison. The dualism which Plato's account espouses in embracing both a meta-

'The necessary' in this passage (τὸ μὲν ἀναγκαῖον), therefore, whose meaning is consequently summed up adverbially within the context of the prepositional phrase ἐξ ἀνάγκης ('necessarily' = the logically necessary or end result of process) through to the end of the *Timaeus*, while far removed from the original causal figure 'Necessity' or more precisely, 'Ananke,' with its inherent meaning of a force for disorder (or 'anarchy') from the earlier passages dealing with Plato's metaphysical theory, is equally distanced from Cornford's understanding of it as a 'necessary cause.'

5.3.11.2 Archer-Hind

Archer-Hind's understanding of this passage with regard to 'the necessary' is similar to that of Cornford.[447] Thus, although Archer-Hind correctly identifies τὸ μὲν ἀναγκαῖον as referring to the class of secondary causes, rather than to the 'necessary causes' of Cornford's reading, like Cornford, he fails to explain how this is so. In other words, Archer-Hind fails to explain how ἀνάγκη completely gives way to the idea of τὸ . . . ἀναγκαῖον given Plato's earlier metaphysical description and strong injunction against understanding these secondary causes as truly causal. The consequence of this reading is that Archer-Hind, like Cornford, assumes the same meaning for the causal force ἀνάγκη as it appears or is referred to in the whole of the preceding text of the *Timaeus* for what is now understood as τὸ . . . ἀναγκαῖον in this latter, quite different closing section of the *Timaeus*, where metaphysics has given way to physics or the study of the natural world. What then follows is that Archer-Hind, along with Cornford, greatly extenuate Plato's grounding metaphysics in the light of this understanding.[448] Moreover, Archer-Hind's reading of τὰ τῶν αἰτίων γένη at *Ti.* 69a6 as 'secondary causes,' rather than 'kinds of causes,' is not a correct translation and further exacerbates the problem of retaining Plato's full cosmological causal account, which account includes not only secondary causes, but also the primary causes, Ananke and the Demiurge, which latter are specifically recalled at the beginning of this passage.[449] The main problem with failing to include these previously drawn metaphysical causal powers, and not just the secondary causes, into the summary passage which next follows, is that a mythical or metaphorical reading of the true causal powers, the Demiurge and Ananke, in-

---

physical and physical basis is not philosophically problematic for Plato. Rather, the latter requires the former in Plato's theory, in other words, the physical Cosmos is incumbent upon such a metaphysical basis and therefore both are philosophically *necessary* regardless of the attendant difficulties for human understanding.

447. A-H, *TP*, 252–53.

448. Ibid., 253. See Archer-Hind's alternative translation of *Ti.* 69a-70d in his running notes. Here he fails to distinguish *kinds* of causes generally from the specifically secondary kind in this further recounting of the early generation of the Cosmos.

449. See *Ti.* 68e1-2.

evitably follows, since these causal powers are no longer included as 'wood for the joiner' in the retelling and final telling of the story of the Cosmos. Thus, in the wake of such a reading, the whole of the *Timaeus* is then reinterpreted in this light and the secondary causes become inured with a causal importance which Plato has already not only soundly rejected, but explicitly denied.[450] Although Archer-Hind is quite right in noting the importance and interest which Plato now gives to the secondary causes or the "investigation of phenomena"[451] as a means of attaining to higher truths or 'the divine,' this is not of causal importance, but rather, it is one of means to an end. In assigning importance to the world of becoming, Plato, for the first time, is giving to the world of phenomena a purpose and hence, value, in the overall scheme of human happiness, which hitherto has been absent. It is noteworthy in this respect, however, that contrary to Archer-Hind's further explanation that this is an injunction to study the copy in order to get to the form or in Aristotelian terms, to study particulars in order to gain knowledge of the ideal, Plato does not refer here to the Forms or to any of his previous theory in this respect, but rather, he stops short of this, referring now only to 'the divine,' τὸ θεῖον. Thus, he states that for mankind to attain 'the divine,' wherein consists the happy life, and to the degree that it is possible for him to do so, there is no other way for him to achieve this goal except through the study of the world of change or physical process, in other words, he must study the natural world with its plethora of secondary causes. What the 'divine' specifically refers to in this passage is not further explained by Plato, but what is clear is that it is something which can be observed apart from the phenomenal world by some sort of mental abstraction and can also be understood or apprehended through such study. Moreover, the divine is desirable, as something after which 'we make haste.' In this respect, Plato appears to move beyond earlier passages in the *Timaeus*, where the divine is drawn up in more restricted terms as having a direct correlation with the 'Soul' or the best part of a man. In the light of Plato's consideration in this present passage of the secondary causes as the sole medium through which we attain to the divine, it is not surprising that at the end of the *Timaeus*, immediately following his explication of the various secondary causes and their effects, he should remind his readers, through his allegory of how the sexes[452] and creatures other than man came into being, that

---

450. *Ti.* 46c7-e6.

451. A-H, *TP*, 253.

452. Within this second explanation of how the sexes came about (*Ti.* 90e6-91d6; see also 41d4-42d2), Plato includes this time round a reasonable account of the process of physical reproduction. However, his comment upon entering into this explanation that "it is reasonable to suppose" (κατὰ λόγον) that the first generation of men who lived cowardly or immoral lives were reborn in the second generation as women, is not reasonable, especially given what he has assumed and said earlier in the *Timaeus* (see my previous comments at fn. 444 on the sexes). Its obvious *un*reasonableness and lack of historicity is the reason why, I suggest, Plato has said this, namely, as a marker to indicate that what he

the philosophical life must be the end of such study and nothing short of this, lest the hapless student degenerates into a lesser being.

5.3.11.3 Taylor

Taylor's approach to the passage at *Ti.* 68e1-69c6 is largely dependent upon his understanding of the earlier passage at *Ti.* 48a1ff. Although he correctly understands the Demiurge in this previous passage as the true cause of the οὐρανός or Cosmos, his interpretation of Ananke as

> a collective name for the sum-total of the 'conjunctions' or 'collocations' which we have to accept as so much ultimate and unexplained bare fact—as conjunctions, to use Hume's phrase, in which we can see no 'connexion'—because *we*[453] cannot see what they are good *for.*[454]

is highly interpretative with regard to the text, which at best leaves Ananke unexplained and at worst, is wrong. I have argued with regard to this prior passage that Ananke is as truly causal as the Demiurge, but that it is a *disordering* cause, not an ordering one. As such, Ananke is the cause of all disorder before the universe was ordered, is the cause of all remaining disorder within the ordered Cosmos and is also the cause of any unnatural disorder thereafter. Given this, Taylor is quite correct in stating that God (the Demiurge) is the "true 'cause' of the οὐρανός,"[455] since Ananke has no ordering power whatsoever, leaving the Demiurge alone to order the Cosmos. The Demiurge has effectively overridden, by the very fact of the existence of the ordered Cosmos, the disordering power—for the most part—of Ananke. Unlike Cornford and Archer-Hind, moreover, Taylor has not obfuscated Ananke with the secondary causes, τὸ . . . ἀναγκαῖον, in the later passage at *Ti.* 68e1-69c6 now under discussion. Thus, the secondary causes in this passage, at least initially, remain distinct from Taylor's "ultimate and unexplained bare fact."

However, this does not long remain the case. Taylor is very quick in assuming identification between this earlier 'unexplained bare fact' and the later subsidiary or secondary causes at *Ti.* 68e1-69c6. The former is assumed to be the latter. A problem with such an identification is that in many, if not all, of the cases where the 'secondary causes' are so identified by Taylor, these same accessory causes are explained quite well in the *Timaeus*. In other words, mankind

---

is saying is *not* to be taken as a reasonable part of his explanation, but rather, is a tongue-in-cheek aside given for other purposes than for its purported rationale of explication, i.e. is a part, the final part, of his extended exhortation to follow the philosophical life.

453. The italics are Taylor's own.

454. Taylor, *PT*, 491.

455. Ibid.

has an immediate and quite controlling command and understanding of secondary causes, so much so that Plato complains that these causes are often attributed with powers that they do not have.[456] Thus, Taylor, I suggest, has made an illicit 'leap of faith' in his understanding here, namely, from "unexplained bare fact" to the idea of the subsidiary cause or 'the necessary,' which move, moreover, is not sustained by his initial understanding of Ananke.

Taylor's reason for doing so, nonetheless, remains curious. He suggests that Plato's character, Timaeus, recognizes that "faith has its legitimate place in our view of the world, no less than demonstration."[457] Thus, such a move is a call to 'faith.' Further, by immediately suggesting that there is no ultimate dualism between God and the 'subsidiary causes' and also, that the οὐρανός or Cosmos is imperfect by way of it being generated and not "from any 'radical evil' in its ingredients,"[458] it is clear that Taylor has now ultimately identified the secondary causes with Ananke, wherein he is rejecting certain interpretations of the latter and is at the same time assuming identification of the causal power Ananke with 'the necessary,' τὸ . . . ἀναγκαῖον, or secondary causes of this later passage.[459]

Finally, with regard to the ὕλη or 'wood' at *Ti.* 69a6, following upon the above-noted identification of Ananke with 'the necessary,' Taylor affirms this identification when he writes:

> The ὕλη we have before us is the two kinds of cause, the primary, God, and the subsidiary, ἀνάγκη.[460]

Although this metaphor describes the requisite causes as now lying before us as 'wood to the joiner,' at this juncture Plato does not move immediately back into his ongoing description of 'the necessary,' but offers first this brief reference to the whole of his causal position, followed by a summary survey of the metaphysical or pre-Cosmic state. I would suggest, therefore, given Plato's previous treatment of all the causes up to this point, that, if he had wanted to distinguish not merely kind*s* of causes but specifically *two* kinds of causes, he would have

456. *Ti.* 46c7-d4.

457. Taylor, *PT*, 491.

458. See Taylor, *PT*, 499. It is highly unlikely that at *Ti.* 69d1, when Plato refers to 'pleasure' as being the "strongest lure to evil," what he really means here is that pleasure is the "strongest lure to secondary causes." However, Taylor's ultimate interpretation of ἀνάγκη as merely an early reference to 'the necessary' or the secondary causes, leaves him with this very odd reading. Moreover, Taylor avoids answering this call, preferring to refer to κακοῦ as 'sin' or 'moral mischief,' rather than to its basic meaning of 'evil,' but again, the question of the connection of 'moral mischief' to secondary causes is left unasked and unanswered.

459. Taylor, *PT*, 492.

460. Ibid., 493.

done so,[461] enumerating and specifying these causes here also. Rather, what Plato appears to be doing here is making a general statement concerning the various causes before introducing his final summary passage and then resuming his explanation of the secondary causes. Moreover, Plato's choice in the prior paragraph (of this pericope) of referring to the secondary cause as 'the necessary,' τὸ . . . ἀναγκαῖον and not as a συναιτία or subsidiary cause as he had done earlier,[462] is significant, but not as a signal to amalgamate this later meaning with the previously discussed primal cause, Ananke, rendering the latter as non-existent, but rather, to distinguish it *from* Ananke, which disordering cause has, for the most part, been subdued with respect to the generated Cosmos. Thus, Plato has now placed his readers entirely within the realm of that completed Cosmos, where neither primal cause, not the Demiurge or Ananke, have an active role. Instead, what now become significant to man in Plato's cosmology are 'the necessary' and 'the divine,' vestiges each of their more powerful primal source, through and to which man must learn and journey to find his present and ultimate happiness.

461. See *Ti.* 46e3: λεκτέα μὲν ἀμφότερα τὰ τῶν αἰτιῶν γένη; see also the above passage at *Ti.* 68e6: διὸ δὴ χρὴ δύ' αἰτίας εἴδη διορίζεσθαι.

462. See *Ti.* 46c7-e6; in the last line at 46e6 Plato also uses the alternate Greek term συμμεταίτια for the secondary cause.

PART II

COMMENTARY ON *LAWS X*

The 'Ανάγκη / 'Ananke' (Necessity) and 'Demiurge' Dichotomy and a Journey into 'Soul'

Introduction

The passages in *Book X* of Plato's *Laws* relating to ἀνάγκη or as I argue, relating to a thoroughly overhauled, but at the same time derivative understanding of ἀνάγκη, will be presented in the order in which they appear. This is in keeping with Plato's written intention at the outset of *Laws X* to build an explicit argument against impiety[463] and in so doing, present a polemic[464] that might lead to the conversion of or at least reflection by the impious. The passages chosen are at least indirectly related, either individually or together, to the notion of ἀνάγκη as it appears in 47e3-48d6 of the *Timaeus* and as argued for in Part I. However, should any ideas mentioned in *Laws X* be found to relate directly to the above notion of ἀνάγκη, it and their relevance will be noted and explained.

The basic translation of the text from *Laws X* is that of A. E. Taylor,[465] with comparisons also made with those of Trevor Saunders[466] and Thomas Pangle[467] where their translations diverge. These ultimately present no significant difficulties. An original translation of the whole of these passages, therefore, will not be given. Should a specific word or section of a passage come into question, however, the original Greek and its translation will be examined with respect to that particular pericope.

In *Book X* of the *Laws* ἀνάγκη, or a close cognate or derivative of it, appears 17 times. All of these instances represent the ordinary, conversational use of this word as it had developed over time and do not refer to the primal disordering cause, Ananke. Accordingly, ἀνάγκη is variously translated and understood from the different contexts and in its various forms in this work as indicating some degree of necessity or force, but not in any significant metaphysical sense and rarely in a causal one. The form and meaning, therefore, of the word 'ananke' as

463. *Laws X*, 887 (*Laws*, hereafter abbreviated *Lg.*).
464. *Lg.* 907.
465. Taylor, *The Laws of Plato, Book X*, 274–304 passim.
466. Saunders, *(Plato) The Laws, Book 10*, 410–47 passim.
467. Pangle, *The Laws of Plato, Book X*, 280–311 passim.

it appears in *Book X* can be listed as follows: Its nominal form (ἀνάγκη) used as a modifier, 'to be sure,' 'bound,' 'must' (= 'necessarily'), 'is sure' (*Lg.* 887c8, 896e1, 898e4, 899a5); in a nominal phrase, meaning 'they cannot be dispensed with' (νῦν δὲ ἀνάγκη = 'are now necessary') (891b4); a verb form (ἀναγκάζουσιν), 'are forcing' (887e9); in a prepositional phrase (ἐξ ἀνάγκης), 'inevitable' (889c1/2); 'must' (892a7, 895a8), '(is) bound' (898c3); the nominal form (ἀνάγκη) used as a one-word indication of strong and necessary agreement, 'Why, necessarily,' 'inevitably so,' 'No doubt, we must' (892b2, 896d4, 900e5); as an adverb (ἀναγκαίως), 'necessarily' (895b5); and finally, as an adjective (ἀναγκαῖον), 'in the consequence' (= 'unavoidable'), 'to be sure,' 'inevitably' (896d5, 905e2, 906c8). These passages will not be explicated with respect to the Greek word ἀνάγκη due to the fact that at this later stage of Plato's writing any metaphysical rendering of this word has either significantly diminished or disappeared altogether.[468] What now becomes metaphysically important in *Laws X* with regard to 'order' and 'disorder' and the primal forces behind them, which in the *Timaeus* took the form of the Demiurge and Ananke, is Plato's understanding of ψυχή or 'soul.' It is within the developed notion of 'soul' as a self-moved mover found in *Laws X* that Plato passes on the idea of causal power. Thus, primary causal power now takes the form of Beneficent and Non-beneficent Soul(s) and human causal power takes the form of individual souls either departing from or aligning themselves with the Beneficent Soul (the ordering Soul), which by the end of this work is referred to as 'God' or alternatively, with the Non-beneficent or disordering Soul. It is this elevated and transformed notion of 'soul' that Part II will explore and discuss.

468. As already seen in the *Timaeus*, when Plato leaves aside the purely metaphysical background of the generation of the Cosmos, concentrating thereafter solely on the physical description of certain of its aspects well after generation, the meaning of 'Ananke' subsequently used is without metaphysical significance.

Chapter 1

1 *LG.* 887C5-888A2

1.1 Original Greek

ΑΘ. Εὐχὴν μοι δοκεῖ παρακαλεῖν ὁ λεγόμενος ὑπὸ σοῦ νῦν λόγος, ἐπειδὴ προθύμως συντείνεις· μέλλειν δὲ οὐκέτι ἐγχωρεῖ λέγειν. φέρε δή, πῶς ἄν τις μὴ θυμῷ λέγοι περὶ θεῶν ὡς εἰσίν; ἀνάγκη γὰρ δὴ χαλεπῶς φέρειν καὶ μισεῖν ἐκείνους οἳ τούτων ἡμῖν αἴτιοι τῶν λόγων γεγένηνται καὶ γίγνονται νῦν, οὐ πειθόμενοι τοῖς μύθοις οὓς ἐκ νέων παίδων ἔτι ἐν γάλαξι τρεφόμενοι τροφῶν τε ἤκουον καὶ μητέρων, οἷον ἐν ἐπῳδαῖς μετά τε παιδιᾶς καὶ μετὰ σπουδῆς λεγομένων καὶ μετά θυσιῶν ἐν εὐχαῖς αὐτοὺς ἀκούοντές τε, καὶ ὄψεις ὁρῶντες ἑπομένας αὐτοῖς ἃς ἥδιστα ὅ γε νέος ὁρᾷ τε καὶ ἀκούει πραττομένας θυόντων, ἐν σπουδῇ <τε> τῇ μεγίστῃ τοὺς αὐτῶν γονέας ὑπὲρ αὐτῶν τε καὶ ἐκείνων ἐσπουδακότας, <καὶ> ὡς ὅτι μάλιστα οὖσιν θεοῖς εὐχαῖς προσδιαλεγομένους καὶ ἱκετείαις, ἀνατέλλοντός τε ἡλίου καὶ σελήνης καὶ πρὸς δυσμὰς ἰόντων προκυλίσεις ἅμα καὶ προσκυνήσεις ἀκούοντές τε καὶ ὁρῶντες ῾Ελλήνων τε καὶ βαρβάρων πάντων ἐν συμφοραῖς παντοίαις ἐχομένων καὶ ἐν εὐπραγίαις, οὐχ ὡς οὐκ ὄντων ἀλλ' ὡς ὅτι μάλιστα ὄντων καὶ οὐδαμῇ ὑποψίαν ἐνδιδόντων ὡς οὐκ εἰσὶν θεοί—τούτων δὴ πάντων ὅσοι καταφρονήσαντες οὐδὲ ἐξ ἑνὸς ἱκανοῦ λόγου, ὡς φαῖεν ἂν ὅσοι καὶ σμικρὸν νοῦ κέκτηνται, νῦν ἀναγκάζουσιν ἡμᾶς λέγειν ἃ λέγομεν, πῶς τούτους ἄν τις ἐν πρᾳέσι λόγοις δύναιτο νουθετῶν ἅμα διδάσκειν περὶ θεῶν πρῶτον ὡς εἰσίν; τολμητέον δέ· . . .

1.2 Translation

The earnestness and passion of your speech are, I feel, an invitation to prayer; they leave no further room for postponement of the argument. Come then; how shall we plead for **the existence of gods** dispassionately? To be sure, no man can help feeling some resentment and disgust with the parties who now, as in the past, impose the burden of the argument on us by **their want of faith** in the stories heard so often in earliest infancy, while still at the breast, from their mothers and nurses—stories you may say, crooned over them, in sport and in earnest, like spells—and heard again in prayers offered over sacrifices, in conjunction with the spectacle which gives such intense delight to the eye and ear of children, as it is enacted at a sacrifice, the spectacle of **our parents addressing their gods, with assured belief in their existence,** in earnest prayer and supplication for themselves and their children. Then, again, at the rising and setting of the sun and moon, they have heard and seen the **universal prostrations and devotions of mankind**, Greeks and non-Greeks alike, in all the varied circumstances of **evil** fortune and **good**, with their implication that **gods are not fictions, but the most certain of realities, and their being beyond the remotest shadow of a doubt**. When we see **all this evidence** treated with contempt by the persons who are **forcing us into our present argument**, and that, as any man with a grain of intelligence will admit, without a single respectable reason, how, I ask, is a man to find gentle language in which to combine reproof with instruction in the initial truth about the gods—that of their existence? Still, the task is to be faced; . . .

1.3 Commentary and Notes

1.3.1 An argument against impiety

This passage vividly portrays Plato's impassioned plea against impiety, which theme offsets *Laws X.* This work bears little resemblance to the *Timaeus* except in being a later and possibly his last work. Whereas in the *Timaeus* the existence of the gods is assumed as a premise at the outset of his cosmology and throughout that work, in *Laws X* Plato positions himself to argue for the existence of the gods against all "naysayers" or those who do not believe in them. However, the decision to enter into rational debate against the contentious arguments of non-believers requires from Plato a more developed notion of 'soul.' As a consequence, the idea of soul as originally presented in the *Timaeus* now appears greatly transformed.[469] Plato has moved away from the notion of 'soul *as generated*' to the idea of 'soul *that generates*.' This will be discussed in some depth later. Moreover, this move in *Laws X* to the idea of soul as a causal force makes no amends for its impingement on other metaphysical features in the cosmic scheme of things, notably Plato's earlier understanding of the primary causes, the Demiurge and Ananke, as the primary causal forces. As I have previously argued with respect to the *Timaeus*, only these two causes have any true causal power for Plato. Thus, any implications arising from the developed notion of soul now taking place in *Laws X* must be carefully examined.

The causal scope in *Laws X*, therefore, appears to have broadened. Causal powers previously given only to the primary powers, the Demiurge and Ananke, in the *Timaeus*, and only secondarily to 'soul' through its generation and sharing in 'nous,' are now delegated exclusively to soul, with no mention given to either its source or to generation. In order to avoid confusion, therefore, 'soul' in *Laws X* must be understood in the light of Plato's earlier metaphysical assumptions regarding causality found in the *Timaeus*.

1.3.2 The Cosmotheogony of *Laws X*

Plato sets out from the beginning of *Laws X* to defend the gods in whom the legislators would have the world believe. He attempts to prove that there really are gods, that they are not detached from human concerns, and that they are "too good to be diverted from the path of justice by the attraction of gifts,"[470] which beliefs, Plato contends, have led to the willful choosing of "unhallowed deeds and lawless discourse." In particular, Plato seems to be directing his arguments against the opinions of the "first-rate poets, orators, prophets, and priests, and countless others," who encourage and incite such moral disorder.[471] Thus, to a large extent *Laws X* is a polemic, centered round a philosophical discussion. A necessary question to raise, therefore, is to ask in what ways this cosmotheological discussion, written perhaps up to thirty years later than the *Timaeus*, presents new philosophical material, specifically metaphysical material, beyond that of the *Timaeus*. It is also necessary to assess the quality of this material and its impact on our understanding of the causal powers, the Demiurge and Ananke.

---

469. This understanding of soul differs not only from that found in the *Timaeus*, but from other earlier works as well, e.g. *The Republic*.

470. *Lg*. 885c5-e6.

471. Ibid.

1.3.3 A comparison with the *Timaeus*

In the *Timaeus* Plato offers no arguments for the existence of God or the Demiurge. This is a given upon which his cosmology rests, derived, albeit, from the logical assumption that all organized objects of becoming, including the Cosmos, must have a maker. Likewise, whereas Plato acknowledges the presence of Ananke as a disordering power before the generation of the Cosmos, its existence is assumed and not argued for. Thus, according to Plato's metaphysical scheme, not only ordered objects, but also disordered objects of becoming must have a power behind their existence. In one sense, therefore, the *Timaeus* is more 'theological' than *Laws X*, taking as its basic premise that there are two primal powers, first, a powerful, designing god or Demiurge, supreme over all, including both the lesser gods[472] and the universe which it has generated and second, a less powerful, non-generative, disordering primal force, Ananke, responsible for all disorder both before and after generation has taken place. In contrast to the *Timaeus*, however, Plato does not take the gods for granted in *Laws X*, but rather, sets out, as one of his three primary goals, to *prove* that they (at least two) exist.[473] The number of gods is not at issue in *Laws X*. In this dialogue it is the fact of the gods' existence, all gods, which has been questioned and accordingly, it is their rejection that Plato wishes to adamantly oppose as being irrational.

To this end, in *Laws X* Plato offers four preliminary arguments for the existence of the gods, beginning with (1) the cosmological (*Lg.* 886a2-5): observance of the earth, the sun, and the planets, and the ordering and beauty of the seasons, the implication being that there is design and hence, intelligence and care behind this. He also offers three further dialectical arguments,[474] namely: (2) the belief that all mankind, Greeks and non-Greeks alike, believe in the gods,

---

472. Although Plato discusses in some detail the generation by the Demiurge of the Heavenly Gods (the sun, moon, stars, and planets, etc.), he also acknowledges quite separately in the *Timaeus* the existence of the traditional gods, but offers little explanation or comment about them.

473. *Lg.* 896e4-7. Reference is made in this passage to at least two *souls* ("we must not assume fewer than two"), rather than to two *gods*. From the ensuing argument, however, it is clear that these souls, because they are self-moving, are identified by Plato as 'gods.'

474. As distinguished by Aristotle, this is reasoning which proceeds syllogistically from opinions generally accepted, whereas demonstrative reasoning begins with primary and true causes. However, for Plato, dialectic, strictly speaking, is the science of first principles, differing from the other sciences by dispensing with hypotheses. It forms the basis from which the other sciences proceed and for which there is no proof. True opinions as objects of belief or faith, however weak (with respect to evidence) or scanty (as to content), would for Plato point to the truth of that which they profess. Thus, arguments (2), (3), and (4) above remain as arguments, however unsatisfactory, in favor of the existence of the gods, given this understanding.

thus, an argument from number; here, true belief, in this case the observance that all mankind believes in the gods, is allowed to function as a dialectical axiom or unproved, single proposition argument—not everyone can be wrong (ibid.); (3) an argument from tradition: "the spectacle of our parents addressing their gods, . . . they have heard the universal prostrations and devotions of mankind" (*Lg.* 887d-e), similar to argument (2) above; and (4) an argument from faith: non-believers lack faith (*Lg.* 887e); 'faith' or true belief itself (in the gods), again, an unproved proposition (dialectical axiom), is accepted as the basis of this argument. In this way Plato leads up to what he considers his most persuasive argument in support of the "old traditional belief of the being of the gods [*Lg.* 890d]," namely, (5) his argument for the primacy of soul. He soon begins to present this fifth and final argument, taking him through to the end of *Laws X*. Thus, it is argued, as "elder-born than all bodies [that is, 'soul'] and prime source of all their changes and transformations [*Lg.* 892a]," all that is akin to soul must therefore be prior to all that is proper to bodies. Plato supports his position by defining 'soul' in terms of 'being alive' or 'self-movement' (*Lg.* 896a4). By this identification 'soul' is then assumed to be the source of all movement and therefore prior to the bodies being so moved (*Lg.* 896b1-4). The soul, moreover, which has control over the heaven and earth and their whole circuit is described as the 'supremely good soul' (*Lg.* 897c7). Hence, Plato argues that this soul and other soul(s) which direct the universe by inhabiting bodies like animated beings, give evidence that all things are "full of gods" (*Lg.* 899b2-8).

1.3.4 Comment on *Lg.* 887c5-888a2

This passage is important as an introduction to the ideas and their context in Plato's later discussion of 'soul' in *Laws X*. Although proof of the existence of the 'wayward cause' (Ananke) as depicted in the *Timaeus* is not specifically enjoined—proof *generally* of the existence of the gods is—it is within the subsequent key discussion of soul that Plato will make reference to a 'non-beneficent soul,' thereby re-introducing the notion of evil and providing a link to its predecessor (Ananke) from the *Timaeus*. The nature and extent of this link in relation to the individual ontologies of each of these sources of 'evil' (i.e. Ananke and the 'Non-beneficent Soul') or as regards a possible sharing of identity will become crucial. Initially, however, *Laws X* gives little mention to the notion of evil save indirectly in terms of 'evil fortune and good' (ἐν συμφοραῖς . . . και ἐν εὐπραγίαις, *Lg.* 887e4-5). It remains for later passages, beginning at *Lg.* 896d5-8, as the concept of soul is presented and then developed to include the notion of evil that the sources of misfortunate and good luck are brought to bear on Plato's metaphysics.

Chapter 2

2 *LG.* 895E10-897B5 (omitting Clinias's perfunctory responses)

2.1/2 Original Greek/Translation

(i) *Lg.* 895e10-896a2

ΑΘ. Ὧι δὴ ψυχὴ τοὔνομα, τίς τούτου λόγος; ἔχομεν ἄλλον πλὴν τὸν νυνδὴ ῥηθέντα, τὴν δυναμένην αὐτὴν αὐτὴν κινεῖν κίνησιν;

Well, then, what is the definition of the thing for which *soul* is the name? Can we find any but the phrase we have just used, 'the motion which can set itself moving'?

(ii) *Lg.* 896a5-b1

ΑΘ. Φημί γε· εἰ δ' ἔστι τοῦτο οὕτως ἔχον, ἆρα ἔτι ποθοῦμεν μὴ ἱκανῶς δεδεῖχθαι ψυχὴν ταὐτὸν ὂν καὶ τὴν πρώτην γένεσιν καὶ κίνησιν τῶν τε ὄντων καὶ γεγονότων καὶ ἐσομένων καὶ πάντων αὖ τῶν ἐναντίων τούτοις, ἐπειδή γε ἀνεφάνη μεταβολῆς τε καὶ κινήσεως ἁπάσης αἰτία ἅπασιν;

(Clinias has just sought affirmation that what the Athenian means is that *soul* has *self-movement* as its definition:)

I do. But if this is indeed so, is there anything we can desiderate,[475] anything further towards complete demonstration of the identity of soul with the primal becoming and movement of all that is, has been, or shall be, and of all their contraries, seeing it has disclosed itself as the universal cause of all change and motion?

(iii) *Lg.* 896b4-8

ΑΘ. Ἆρ' οὖν οὐχ ἡ δι' ἕτερον ἐν ἄλλῳ γιγνομένη κίνησις, αὐτὸ δὲ ἐν αὐτῷ μηδέποτε παρέχουσα κινεῖσθαι μηδέν, δευτέρα τε, καὶ ὁπόσων ἀριθμῶν βούλοιτο ἄν τις ἀριθμεῖν αὐτὴν πολλοστήν, τοσούτων, σώματος οὖσα ὄντως ἀψύχου μεταβολή;

Then must not the motion which, wherever it arises, is induced by something else, but never confers the power of self-motion on anything, come second in the scale, or as low down as you please to put it, being, in fact, change in a truly soulless body?

(iv) *Lg.* 896b10-c3

ΑΘ. Ὀρθῶς ἄρα καὶ κυρίως ἀληθέστατά τε καὶ τελεώτατα εἰρηκότες ἂν εἶμεν ψυχὴν μὲν προτέραν γεγονέναι σώματος ἡμῖν, σῶμα δὲ δεύτερόν τε καὶ ὕστερον, ψυχῆς ἀρχούσης, ἀρχόμενον κατὰ φύσιν.

Consequently it will be a right, decisive, true, and final statement to assert, as we did, that soul is prior to body, body secondary and derivative, soul governing in the real order of things, and body being subject to governance.

---

475. 'Desiderate' has the meaning of 'to still want.' Thus, this passage might read, "I do. But if this is indeed so, is there anything we can still want . . . ?" Saunders translates this passage: "I do. And if this is true, are we still dissatisfied? Haven't we got ourselves a satisfactory proof that soul is identical with the original source of the generation and motion of all past, present and future things and their contraries? After all, it has been shown to be the cause of all change and motion in everything" (Saunders, 426).

(v) *Lg.* 896c5-7

ΑΘ. Μεμνήμεθά γε μὴν ὁμολογήσαντες ἐν τοῖς πρόσθεν ὡς, εἰ ψυχὴ φανείη πρεσβυτέρα σώματος οὖσα, καὶ τὰ ψυχῆς τῶν τοῦ σώματος ἔσοιτο πρεσβύτερα.

But we have not, I imagine, forgotten our earlier agreement that if soul could be proved older than body, the characters of soul must also be older than those of body.

(vi) *Lg.* 896c9-d3

ΑΘ. Τρόποι δὲ καὶ ἤθη καὶ βουλήσεις καὶ λογισμοὶ καὶ δόξαι ἀληθεῖς ἐπιμέλειαί τε καὶ μνῆμαι πρότερα μήκους σωμάτων καὶ πλάτους καὶ βάθους καὶ ῥώμης εἴη γεγονότα ἄν, εἴπερ καὶ ψυχὴ σώματος.

And so moods and habits of mind, wishes, calculations and true judgments, purposes and memories, will all be prior to physical lengths, breadths, and depths, in virtue of the priority of the soul itself to body.

(vii) *Lg.* 896d5-8

ΑΘ. Ἆρ' οὖν τὸ μετὰ τοῦτο ὁμολογεῖν ἀναγκαῖον τῶν τε ἀγαθῶν αἰτίαν εἶναι ψυχὴν καὶ τῶν κακῶν καὶ καλῶν καὶ αἰσχρῶν δικαίων τε καὶ ἀδίκων καὶ πάντων τῶν ἐναντίων, εἴπερ τῶν πάντων γε αὐτὴν θήσομεν αἰτίαν;

Hence we are driven, are we not, to agree in the consequence that soul is the cause of good and evil, fair and foul, right and wrong; in fact of all contraries, if we mean to assert it as the *universal* cause?

(viii) *Lg.* 896d10-e2

ΑΘ. Ψυχὴν δὴ διοικοῦσαν καὶ ἐνοικοῦσαν ἐν ἅπασιν τοῖς πάντῃ κινουμένοις μῶν οὐ καὶ τὸν οὐρανὸν ἀνάγκη διοικεῖν φάναι;

Well then, if indwelling soul thus controls all things universally that move anywhere, are we not bound to say it controls heaven itself?

(ix) *Lg.* 896e4-6

ΑΘ. Μίαν ἢ πλείους; πλείους· ἐγὼ ὑπὲρ σφῷν ἀποκρινοῦμαι. δυοῖν μέν γέ που ἔλαττον μηδὲν τιθῶμεν, τῆς τε εὐεργέτιδος καὶ τῆς τἀναντία δυναμένης ἐξεργάζεσθαι.

And is this done by one single soul, or by more than one? I will give the answer for both of you, 'by more than one.' At least we must assume not fewer than two, one beneficent, the other capable of the contrary effect?

(x) *Lg.* 896e8-897b5

ΑΘ. Εἶεν. ἄγει μὲν δὴ ψυχὴ πάντα τὰ κατ' οὐρανὸν καὶ γῆν καὶ θάλατταν ταῖς αὑτῆς κινήσεσιν, αἷς ὀνόματά ἐστιν βούλεσθαι, σκοπεῖσθαι, ἐπιμελεῖσθαι, βουλεύεσθαι, δοξάζειν ὀρθῶς ἐψευσμένως, χαίρουσαν λυπουμένην, θαρροῦσαν φοβουμένην, μισοῦσαν στέργουσαν, καὶ πάσαις ὅσαι τούτων συγγενεῖς ἢ πρωτουργοὶ κινήσεις τὰς δευτερουργοὺς αὖ παραλαμβάνουσαι κινήσεις σωμάτων ἄγουσι πάντα εἰς αὔξησιν καὶ φθίσιν καὶ διάκρισιν καὶ σύγκρισιν καὶ τούτοις ἑπομένας θερμότητας ψύξεις, βαρύτητας κουφότητας, σκληρὸν καὶ μαλακόν, λευκὸν καὶ μέλαν, αὐστηρὸν καὶ γλυκύ, καὶ πᾶσιν οἷς ψυχὴ χρωμένη, νοῦν μὲν προσλαβοῦσα ἀεὶ θεὸν ὀρθῶς θεοῖς, ὀρθὰ καὶ εὐδαίμονα παιδαγωγεῖ πάντα, ἀνοίᾳ δὲ συγγενομένη πάντα αὖ τἀναντία τούτοις ἀπεργάζεται. τιθῶμεν ταῦτα οὕτως ἔχειν, ἢ ἔτι διστάζομεν εἰ ἑτέρως πως ἔχει;

So far, so good. Soul, then, by her own motions stirs all things in sky, earth, or sea (and the names of these motions are wish, reflection, foresight, counsel, judgment—true or false—pleasure, pain, hope, fear, hate, love), stirs them, I say, by these and whatever other kindred, or primary, motions there may be. They, in turn, bring in their train secondary and corporeal movements, and so guide all things to increase and decrease, segregation and integration,[476] with their attendant characters of heat and cold, weight and lightness, hardness and softness,

476. Also translated "separate or combine"; see Saunders, 428.

white and black, dry [bitter] and sweet. By these and all her instruments, when wisdom is her helper, . . . she conducts all things to the right and happy issue, whereas when she companies with folly, the effect is clean contrary. Shall we set it down that this is so, or have we still our doubts that it may be otherwise?

2.3 Commentary and Notes

2.3.1 Soul as the source of all change and movement

These 10 passages from *Laws X* together form the core and substance of Plato's views on 'soul' in this work. Important to the idea of disorder, they also incorporate a view on the nature of evil not hitherto visited in the *Timaeus*, connecting evil to when the soul 'companies with folly' and also to the movements of a particular 'non-beneficent soul.' Moreover, Plato's understanding of soul in *Laws X* has changed considerably from that held in the *Timaeus*.[477] Not only is soul now presented as that which generates, rather than as that which was generated, the tripartite understanding of soul from his earlier works gives way to the idea of soul as a complex, but whole entity, with clear divisions of the soul noticeably absent.

2.3.2 The Demiurge and soul

2.3.2.1 The *Timaeus*

2.3.2.1.1 Soul as generated and not begotten

In the *Timaeus* the Demiurge is not equated with 'soul,' although one might mistakenly draw the conclusion that it is, since the World Soul, although wrought by the Demiurge, is wrought directly and as a consequence is described as being both divine and immortal.[478] Likewise, the initial part of the human soul is made by the Demiurge, although wrought from less pure ingredients than the World Soul, and is the one immortal, divine part that the Demiurge bestows to mankind.[479] Thus, there are similarities between the Demiurge and 'soul,' whether

---

477. Due to similarities in Plato's understanding of the soul in the *Timaeus* and *The Republic*, the divergence in *Laws X* away from his earlier view in the *Timaeus* also applies to *The Republic*.

478. Thus, I agree with R. D. Mohr in asserting that "the *prima facie* evidence is that the Demiurge is not [a soul]. . . . Rather souls, and especially the World Soul and what rationality souls have are viewed as products of the Demiurge (*Timaeus*, 35a, 36d-e, *Philebus* 30c-d, *Statesman* 269c-d)" in *The Platonic Cosmology* (Leiden: E. J. Brill, 1985). For his argument, see Chapter 10, 178–83.

479. *Ti.* 41c6-d1. However, in the *Timaeus*, as in *The Republic*, the soul is described as being tripartite. Thus, in addition to the immortal part of the soul described here, there

speaking of the World Soul or the human soul, namely, of being both divine and immortal.

However, this is where further identity between the Demiurge and 'soul' ends. First, not only is 'soul' described in the *Timaeus* as having been generated, whereas the Demiurge is clearly not generated, 'soul' is cited as having been generated *by* the Demiurge. Thus, the Demiurge is depicted as the causal agent of the soul's generation, but the reverse is not true. Soul is never described in the *Timaeus* as generating anything, its causal power being limited solely to that of providing noetic (psychic) movement within the generated universe. Thus, as a causal source, its powers are exclusive of any others (e.g. the power of generation). Consequently, soul is presented in Plato's cosmology as being the primary source of noetic movement within the Cosmos, but never of generation or of movement outwith the generated Cosmos.[480] The Demiurge, on the other hand, is inclusive with respect to its possession of the causal powers of nous, delegating its noetic powers as it sees fit and in particular, the power of movement within the generated Cosmos, which it delegates to 'soul.'

The ontology of soul, moreover, as it is presented in the *Timaeus* is complex, where identity with the Demiurge would belie the latter's supreme simplicity[481] and causal priority with respect to soul. Soul is described in this work as being wrought by the Demiurge and enjoined to the world, acting as the causal agent for the world's movement. However, although soul is described by Plato as though it were the causal agent for the physical movement of the universe, like the Demiurge, soul itself does not *physically* move the universe as, for instance, a first mover in a chain of events. As Plato explains,[482] the universe existed in chaotic flux before its generation and therefore, was never in a static state before

---

were also two additional *mortal* parts added later by the gods when they took over the immortal principle of soul given to them by the Demiurge, and after generating the body, joined it with the human soul (*Ti.* 69c5-d6). The immortal part of the human soul was consequently encased in a globe (the 'head'). These additional *mortal* parts of the soul, however, emotion and appetite, were wrought by the gods as "indispensable equipment" for the whole of the body, which in turn acted as a vehicle for 'immortal soul' and accordingly were placed appropriately elsewhere in the body (*Ti.* 69d6-72d8). Not having been directly generated by the Demiurge, however, they were therefore neither divine nor eternal.

480. It would appear that movement outwith the generated Cosmos, being disordered, was due not to the causal power of nous ('nous' depicting the *kind* of causal power) associated with the Demiurge, but to the disordering causal power (lacking in 'nous') of Ananke. Thus, the only movement outwith the Cosmos of which the Demiurge could be the causal agent would be generation itself (whether of the Cosmos or some other generated object).

481. The Demiurge is ungenerated and perfect, neither proceeding from nor being a part of anything greater, nor does it have any parts. Moreover, its movement, as I argue, is entirely noetic.

482. *Ti.* 52d2-53a7.

being organized by the Demiurge. Rather, this chaotic movement was *ordered* by the Demiurge through its powers of reason or 'nous' and which ordering—not physical movement itself—was then delegated to the generated soul, either to the World Soul or the human soul, whereby through its purely noetic[483] movements soul impacted the generated world, bringing about an ordered physical equivalent with respect to its ordered noetic movement.[484]

However, depending on whether the World Soul was the cause of the movement or the human soul was, this movement was one of either pure reasoning and intelligence or it could also involve the emotions and lower appetitive motions (i.e. as with movement initiated by the human soul).[485] The less pure mixture[486] given to the human soul was possibly meant to accommodate the impact that the emotions and senses could have on the soul, while its immortal, divine part, unless so impacted, like the World Soul, remained purely rational. Thus, the type of movement assigned to soul, while aligning with a particular physical movement, is not itself physical, but is entirely noetic, just as the Demiurge's agency is noetic. The only exception to this pure noetic movement is where the human soul participates in its lower emotive and appetitive motions or its mortal parts,[487] which 'lower' noetic motions are due to and dependent on the human soul's less pure origins. The description, therefore, which Plato gives to the soul as being compounded of the Same, Different, and Being so as to embrace the complexities of the world in its own parallel or sympathetic noetic form, was not so that soul could be a source of physical motion (this being impossible), but rather, in order that soul could provide "a divine source of unending and rational life for all time."[488]

---

483. By 'noetic' is meant the character some entities have due to their resulting from 'nous' or reason alone; thus, they are non-sensual and non-empirical in nature. According to Plato's cosmology these entities are also non-physical in nature, being either ungenerated or divinely generated in origin. Thus, the movements of immortal soul are noetic and so are those of the Demiurge. I refer to the movements of the mortal parts of the soul as being "lower" noetic movements, inasmuch as like noetic entities they are non-physical in nature and account for movement and can be directed by noetic movements to rational ends, but are 'lower' inasmuch as they do not proceed from 'nous' or reason themselves, but from the emotions and the senses and thus, are impacted by the physical, empirical world. As a result, they are not by nature rational entities, but can only be directed to rational ends by the 'higher,' rational part of soul (e.g. see *Ti.* 70a2-7).

484. *Ti.* 34a1-3. In this passage the noetic movement of the Demiurge is being aligned with its physical counterpart. Thus, of the seven physical motions, uniform circular motion on the same spot "most properly belongs to intelligence and reason."

485. See *Ti.* 69d6-e2.

486. *Ti.* 41d4-7.

487. *Ti.* 69c5-d6.

488. *Ti.* 36e2-5.

Plato does not address in the *Timaeus* any concerns over the actual source of the seven physical motions[489] (i.e. by demanding a first mover) or suggest any further alignment of the remaining six motions with similar movements of the soul, as he did for the "uniform circular motion in the same place" with the Demiurge's reason and intelligence. In his ontology the soul was generated before the body of the universe.[490] This preempts the idea that 'soul' is some descriptive feature of the physical world. In Plato's cosmology, moreover, physical movement existed as part of the chaotic flux, ruled at that time only by Ananke, before either the generation of the soul or the ordering of the body of the world out of that pre-existing chaos had taken place. This further preempts any idea that 'soul' has the role of first mover for any of the physical motions with respect to the world. Soul[491] is entirely noetic in nature and was fashioned by the Demiurge for an already dynamic universe, so that its chaotic nature might be transformed and become informed by the principle of rationality, of which soul consisted, and thus, become an organized universe and sustainable as an ordered entity. Movement would now have ends towards which it was focused, but this would be the only way in which soul 'moved' the physical universe. Physically speaking, soul had no input whatsoever, nor does Plato suggest as much.

When Plato next addresses the physical motions after constructing the body of the universe, the movement discussed is in relation to the 'divine form' (made of fire and surrounding the visual Cosmos), described as "universal cosmic embroidery" and also to the heavenly gods (planets and stars) or divine beings. Again, this movement is part of the generated Cosmos and thus, is ordered. The movement of the 'divine form' is described as following that of the universe and highest intelligence, namely, uniform circular motion in the same place. In this same passage,[492] he also gives the heavenly bodies two motions at the time of their generation, namely, uniform motion in the same place and forward motion, again imitating noetic or thought functions. Uniform motion is connected to "always thinking the same thoughts about the same things" and forward motion as being subject to movement (of soul) which calculates "the Same and uniform."[493] Thus, not only has the universe become organized, so has its motion,

489. The remaining six physical motions are: up and down, backwards and forwards, and right and left.

490. *Ti.* 34c4-35a1.

491. Here, 'immortal' soul is meant. The two mortal parts of the human soul, fashioned not by the Demiurge, but by the Divinities when they fashioned the mortal body, had a 'lower' noetic function with respect to the human body, providing "indispensable" or "necessary" equipment for man's mortal life (*Ti.* 69d4-6). See Cornford, *PC*, fn. 3, 281.

492. *Ti.* 40a2-b4.

493. For the idea of 'sameness' as a noetic movement of soul see *Ti.* 37a5-b3. "Whenever the soul comes in contact with anything whose being is either dispersed or indivisible, it is moved throughout and calculates similarity and difference . . ." Plato's

but not without a noetic counterpart imitating the Demiurge, its organizing force, to carry this out, namely, 'soul,' generated by the Demiurge for this express purpose. The remaining five motions, being less perfect, are unassigned, but reappear as normal physical movements within the earth, "our foster-mother."

2.3.2.1.2 Primary causal movement, self-movement, and secondary movement

When considering movement with respect to the generated universe, there are three kinds of movement at issue: primary causal movement, self-movement, and secondary movement. The idea of 'self-movement' is not raised until *Laws X* as Plato further develops his ontology of 'soul' and soul's ability to move (again, through its organizational ability with respect to movement and not through any innate physical power) whatever comes into contact with it. This leaves only primary movement and secondary movement as the two sources of motion within Plato's cosmology.

In the *Timaeus* the primary movement of the Demiurge is never equated with the movement of 'soul.' Soul, although participating in the divine nature and 'nous,' having been wrought by the divine, is nevertheless generated and is always attached to a body, whereas the causal force identified with the Demiurge is that of the *organizing* power of nous, unlimited, unembodied, and ungenerated. Thus, the causal movement of the Demiurge is identified neither with the 'self-movement' (due to the noetic powers of soul contained therein or however soul moves such bodies, as Plato himself was not sure) of the heavenly bodies nor the secondary movement—physical motion—of the organized universe. Rather, the movement of the Demiurge, having no physical nature, is understood by Plato as solely that of the organizing principle of generation—the workings of 'nous' unembodied and unlimited, carrying out the precept of 'like begetting like.' Having perceived the organization of the world as being wholly perfect from a rational perspective (i.e. with respect to ends) and thus, wholly good, Plato hypothesizes that it must have a maker and that its maker must be the same. The causal movement ascribed to the Demiurge, therefore, is that of the

---

discussion of the noetic movements of soul and their relation to the movements of the heavenly bodies is notoriously difficult, as is his equally difficult description of the heavenly bodies themselves. Cornford, for instance, does not translate the passage at *Ti.* 40a2-7 as referring to the bright aura surrounding the earth (which latter was thought to form the center of the universe and be stationary [see Bury, 35, fn. 6] and thus, fit well with the imagery), and above, wherein also reside the fixed stars, but rather, to the material makeup of the stars themselves (*PC*, 118). My reading of this passage, however, follows Lee's translation (55), where the 'divine form' in "τοῦ μὲν θείου τὴν πλείστην ἰδέαν . . . " is understood as referring to something specific (i.e. the fiery aura surrounding the earth) and hence, is not simply referring to what is generic (i.e. the material form of the stars).

invisible organizing force behind the Cosmos (evidenced by the organizing principle found within it), for which there is no connecting evidence, since only *ipso facto* does the world's existence require that of the Demiurge. The Demiurge, itself, however, is perceived by Plato as more than just a principle, it is the primal power *behind* that principle, entirely separate from the universe, and thus, is the power behind its organization and behind the organizing principle of 'nous' or Mind found within it. With regard to the generating power of 'nous' as a metaphysical principle, therefore, and as analogous to the organizing power of human thought, the world as generated is, I suggest, an eternal afterthought by the Demiurge (a noetic movement), which has a beginning but no end, whereas chaos was an eternal forethought by Ananke (an anoetic movement), which had an end, but no beginning. However, having now accounted for the organization of the universe as a whole on the principle of noetic movement by a primal, non-physical power(s), Plato must still account for the physical movement of the Cosmos itself, which he subsequently sets about doing.

2.3.2.1.3 Physical movement and the question of dualism

Plato achieves this in the *Timaeus*, but only in rough sketch form, where the soul (not described as self-moving until *Laws X*) is fabricated by the Demiurge as the closest analogue to itself. Although soul is made up of the rational powers of nous, and to this extent is divine and shares with the Demiurge its powers, being complex (having been fabricated), soul is nonetheless necessarily inferior to the Demiurge, since perfection, according to Plato's understanding, is singular ("residing happily in itself"). Soul, moreover, unlike the Demiurge, does not reside in itself, rather, it resides *in* and is formed *for* a body, whether the body of the world or a human body,[494] which latter, depending upon its aspiration or lack thereof, might ultimately reside in a heavenly body (i.e. its own star) or lower animal (see *Ti.* 41d4-42e4).[495] Although made prior to the body, soul is formed

494. According to Plato's cosmology, the idea of a soul residing in a non-human body (i.e. in its own star or other heavenly body) or without a body is understood by him as being in either a degraded secondary state or in its primary state ("its first and best form," *Ti.* 42b3-d2), whereby the soul has either been demoted from or promoted to its original state, respectively, according to the moral life which it has led following its first human and consequent later incarnations. The human state, although a secondary state and thus subject to deterioration, is considered a "necessary incarnation" for the sake of the existence of Man—not for the soul's sake—and thus is not a degraded state *per se*, but *becomes* degraded.

495. It is possible that for Plato a star represented a more rarefied and attenuated body than a human body, since, unlike the human body, a star (or other heavenly body, e.g. the moon) was capable of releasing a soul (to inhabit a living being) without itself being destroyed in the process. He does not discuss this, no doubt believing such knowledge to be either speculative in nature and therefore outwith the parameters of his 'likely' account or alternatively, closer to the mythic aspect of this account.

so as to be its dominating and controlling partner.[496] The Demiurge, moreover, is still or unchanging, in that it does not participate in the movement of becoming. The World Soul, on the other hand, once it has been intertwined with the world (the organized universe or Cosmos) is described by Plato as "revolving upon itself" (Lee) or "revolving within its own limit" (Cornford). This is not a description of a self-mover. The meaning of this passage is made clearer from the context of an earlier one, where at *Ti.* 34a1-5, in discussing the body of the Cosmos (the World Soul not yet having been considered) Plato writes, "For of the seven physical motions he [the Demiurge] allotted to it the one which most properly belongs to intelligence and reason, and made it move with a uniform circular motion on the same spot; any deviation into movement of the other six kinds he entirely precluded [Lee]." Thus, Plato's depiction now of the movement of the World Soul follows closely the metaphor already in place concerning the movement of the Cosmos, whereby Soul is imitated analogously by the movement of the world with which and for which it has been woven, not, however, as initiator and perpetuator of the latter's *physical* movement, but rather, as a psychic power, the source of 'nous' within the Cosmos, and the cause of its rational ends. As such, although there is physical movement of the world, moving as it does in uniform circular motion in the same place, the sympathetic movement of the soul is entirely psychical, carrying no physical motion of its own.

Physical motion, on the other hand, completely lacking in rational powers, is not only inferior, it is also always secondary and accessory. Plato makes no attempt in the *Timaeus* to directly connect the physical power or motion in the world with its noetic counterpart or to a first causal principle other than to introduce the notion of soul and souls as being present in the world, no doubt helping to impart to Plato (on an initially separate issue from that of the Forms[497]) the long-held and rightly (as I will argue) title of dualist.

Crucial for understanding Plato's account of causality in the world is the fact that there is no beginning to physical movement in his cosmology; there is only a beginning to *organized* physical movement, where previously the universe moved in chaotic disarray. Neither primary power, the Demiurge nor Ananke, bringing about generation or chaos respectively, enter into the sphere of physical causality directly. Theirs is a noetic and anoetic counterpart to this movement, impacting it with order or disorder by being co-aligned with it, but which counterparts are never identified with the movement itself.[498] Thus, as far as Plato's

496. *Ti.* 34c4-35a1.

497. That "extreme dualism" with regard to the body-soul relationship can be found in Plato is also noted by T. M. Robinson (see *Plato's Psychology*, 159).

498. Vlastos, in "Creation in the *Timaeus*: is it a Fiction?" from *Studies in Plato's Metaphysics*, R. E. Allen, ed. (401–19), grapples with the problem of the pre-existence of motion as it existed in the chaotic pre-organized universe in relation to Plato's understanding of 'soul' as self-generating motion and being the first cause of all motion (as

description of the world is concerned, to describe the first organized movement, even if he cared to (which he did not, no doubt believing such guesswork to be outwith the sights of probability or a "likely account"), would not lead to pure causality or to an unmoved mover. Rather, it would lead only to further movement, albeit chaotic, disorganized movement.

What the *Timaeus* offers, therefore, is a *parallel* account of causality, a dualistic understanding introduced by a soul-body dichotomy, in which the organization of the physical world is completely dependent upon the invisible presence of a soul or soul infused within it. Moreover, it is the soul's participation in 'nous,' of which the Demiurge is the sole benefactor, which gives soul its organizing power or intelligence. Although Platonic dualism is traditionally associated with the Forms, as Plato develops his theory of causality in the *Timaeus*, the Forms acquire a second-order noetic reality in relation to the importance and first-order noetic reality of the soul.

The difficulty of a dualistic causal account which does not take into consideration the fact of pre-existing motion, but which at the same time takes into account Plato's claim that the soul is the cause of all motion, so as to mean physical motion, is that the physical and psychical (the soul) must then somehow meet up. Dualism becomes a paradox. The soul moves the body, literally. The enigma of dualism, however, is that this appears irrational and viewed as a paradox it is irrational. However, according to the logic of ancient Greek thought, like moves like. The sensible and intelligible worlds, therefore, are two independent and mutually irreducible domains or worlds. With body and soul sitting at opposite poles, both as to their ontology and metaphysics, such direct interaction would be impossible and therefore is impossible.

As I have argued, a reading of dualism involving the direct interaction between body and soul is wrong. According to Plato's cosmology movement has always existed, although not always in an organized form. Moreover, once the Cosmos has been generated, things, too, as accessory causes, move other things, but at no point is an unmoved mover introduced to take over or carry out this

---

quoted from and understood by Cherniss, 417). Vlastos correctly approaches this problem by suggesting (although I do not agree with his understanding of the Craftsman as in any way *having* a soul) that "the self-motions of his soul [that of the Craftsman] then, no less than ours, supervene on material motions which he does not generate but only harnesses to the fulfillment of his creative purposes [418]." Vlastos does not go the one step further, however, which is to qualify *how* the Craftsman's soul can impact on material motions. This question is left unasked and unanswered. I argue, however, that Plato's cosmology assumes a dualistic position in this regard, whereby the Demiurge supervenes on material motions indirectly, impacting directly only on noetic or a-noetic motions. In other words, what the Demiurge impacts is either the pure noetic motion of 'nous,' whether directly (e.g. in an ungenerated soul) or harnessed within a generated 'soul' or alternatively, the anoetic motion maintained by Ananke or wrought by Plato's latest causal player (from *Laws X*), the Non-beneficent Soul.

task. What I have suggested above, therefore, to replace this reading is a 'parallel' account of body and soul, which is a dualistic understanding of causality in which the human soul does not physically move a body, although it is able to rationally impact upon it and thus, indirectly influence its movement towards ordered ends. As such, soul always exists parallel with a body as long as there is discernible intelligence or ends within that body. When there is not, it is simply 'soulless.' Thus, whereas soul is imitated by the physical movement of a body, in that intelligent design toward some end necessarily follows its presence, soul is not to be identified with the movement of that body itself, since, when movement is entirely irrational (as in the pre-generation chaotic state), although there is still movement, there is no longer a soul or an organized body. Thus, it is not a lack of movement that renders an object soulless, but rather, a lack of 'nous' or intelligence directing its movement so as to sustain it as that object. Accordingly, all physical movement in this reading has a noetic counterpart or movement of nous or soul to account for its intelligent end(s).

By the time Plato wrote *Laws X*, however, not only is soul (the 'World Soul' in the *Timaeus*) understood as being generally present within the Cosmos, impacting the functioning of all bodies towards ends supporting their individual functioning as bodies and the Cosmos as whole, but the human soul, as an individual noetic entity, now becomes closely associated with its own 'will' and of its ability to bring about secondary ends for itself that may or may not be good. Accordingly, Plato is fully aware that the human soul, in exercising its 'will,' can and does at times act irrationally, bringing about 'bad' ends for that individual, e.g. greed or sloth, which represent an imbalance with respect to its overall good. However, being originally generated by the Demiurge, who is all good (the *Timaeus*) and as such, the work of a God (*Laws X*), the human soul, itself, is, according to Plato, also wholly good. He therefore must account for the irrational behavior of an individual (soul), which he does by positing its alignment with a 'non-beneficent soul' or as I argue, an aspect of Ananke redrawn as immanent. This again is an example of parallel psychic movement, in which the individual human soul, impacted by one of two souls, either the Beneficent Soul (as with Ananke, an aspect of the Demiurge redrawn as immanent) or the Non-beneficent Soul, through its own 'will' directs the body's motions noetically towards secondary ends (those ends which can co-exist with the primary human ends keeping the body alive so as to house the human soul). Although Plato does not elucidate his causal landscape so as to hypothesize the interaction of body and soul more specifically in either the *Timaeus* or *Laws X* (leaving such details alone in the former and declaring them to be unknown in the latter) or yet, to allow his causal understanding to span over these two important works (which proved not his style to do so), the above scenario is close, I suggest, to what is ultimately presumed.

2.3.2.1.4 The Forms and the question of dualism

The Forms for Plato remain bare objects of thought, which objects are unchanging, uncreated, indestructible, admit of no modification, enter no combination, and are imperceptible to sight or the other senses.[499] It is noteworthy that beyond this description, Plato never discusses their metaphysics in the *Timaeus* or the *how* of their existence and with that, their connection to causality or the primary causal powers.[500] Moreover, although the Forms are a powerful tool in Plato's metaphysics in that they stabilize reality for the observer and allow a measure for discerning truth (intelligence) over true opinion, Plato does not provide support in his ontology for their separate existence[501] apart from the existence of the

---

499. *Ti.* 51e6-52a4.

500. I disagree, however, with W. Lutoslawski's perfunctory claim that "taking this for granted, we shall easily recognise that the ideas were nothing else for Plato when he wrote the *Timaeus* than God's thoughts," (*The Origin and Growth of Plato's Logic*, 477) or at least his initial claim, as he soon qualifies what he means by this. To wit, soon after Lutoslawski distinguishes between God's thought and of what it is understood to be comprised or its objects. Thus, the eternal models (the ἀίδια of *Ti.* 29a and 37d and the παράδειγμα of 28a, 37c, 39e, and 48e, etc.) are existent as objects of God's thought, their function to be eternal models of thought, which are reproduced later in the investigating souls of men. As with Plato's 'likely' account, it is difficult to remove all anthropomorphic material when describing the primary cause and is why, in trying to minimize this difficulty, I have described the Forms, as understood by Plato, as the result of God reflecting upon himself and when similarly qualified by Lutoslawski, as particularized in the reflective souls of men (i.e. models recognized by the soul's activity, 478). Essentially, therefore, I agree with Lutoslawski's more in depth understanding of the Forms, which is that they are the reflective objects of God's thought. His idea, moreover, that the Forms exist solely in the soul and can never exist as separate entities apart from the soul is particularly valuable in gaining a deeper understanding of Plato's extensive development of the notion of 'soul' in *Laws X.* For an excellent discussion of this view see W. Lutoslawski, "III. Latest Works: Laws" in *The Growth and Origin of Plato's Logic*, 491–516. Thus, he writes, "No conclusion can be drawn from the use or absence of the terms εἶδοσ or ἰδέα which Plato borrowed from earlier writers and used himself in many different meanings, 'very seldom with a pronounced metaphysical meaning' (294). Plato's philosophy is not a mere theory of knowledge, and his theory of knowledge is not limited to the conception of ideas. The soul is not an idea, and acts a more important part in later Platonism than all ideas of Middle Platonism. It is the soul, and not the ideas, which is the central point of Plato's later theory of knowledge" (494) and concludes, both because of Plato's argument for the priority of soul, as well as due to the close relationship drawn between soul and knowledge that "the world of eternal notions forming the system of human and divine knowledge . . . can only exist in souls" (495–96).

501. Lutoslawski writes: "This close relation between soul and knowledge is here [in the *Timaeus*] insisted upon as in the *Sophist*, and makes it evident that Plato no longer dreamed of separate ideas" (*The Origin and Growth of Plato's Logic*, 474). "Objectivity does not require substantial existence: anything that by its logical nature must be universally admitted is an objective truth," admitting that in the *Symposium* and the *Phaedo*, as

thinkers who perceive them, and in so doing, fails to present a full dualistic account of reality in which the Forms actually exist separately from the thinker.[502] A proper dualistic account of the Forms, however, demands this. Alternatively, a different understanding is required.

2.3.2.1.5 A parallel account

In a 'parallel' account of body and soul, the Forms do not require that they have a separate existence apart from the soul, since in this account, their totality, if enumerated, would simply comprise the parallel noetic counterparts of soul in relation to the organized body.[503] The Forms, as known, are the objects of nous as reflected in the mind[504] and accordingly, act as a function of soul, i.e. in its ability to impact physical objects with those of nous. The Forms, therefore, are how the intelligence of nous manifests itself, either in the mind as noetic objects or correspondingly, in the physical world as organized objects. Thus, an ontological position arguing for the separate existence of the Forms is not required, since they are dependent upon 'nous' for their existence, found only in the Demiurge and 'soul.' Accordingly, Plato's ontology need only account for the separate existence of the Demiurge and soul, and not for the Forms, which this

---

well as in *The Republic* and the *Phaedrus*, the difference between this objectivity and the objectivity of substances may not yet have been fully realized by Plato (521). By late Platonism, however, this was not the case, where the idea of soul and movement formed the ultimate explanation of everything that happened, brought forward by the *Sophist*, *Politicus*, and *Philebus* and maintained by the *Timaeus*, *Critias*, and *Laws*: "Knowledge is acquired by each soul through its own exertions, increased by constant exercise and imparted by teaching. Ideas exist only in souls—they are eternal and unchangeable because their first model is created by God in his own thought. Thus ideas are the patterns of reality, and their existence in souls is named true Being. But they are not now suddenly perceived in ecstatic visions, as in the period of Middle Platonism. They must be created and elaborated by each soul in its own turn, and sought for by the logical exercises of classification, generalisation, and division" (523–24).

502. Teloh argues for the radical separation of Forms from phenomena in the *Timaeus* where for the first time there are Forms for both attributive, relational, and incomplete characteristics, as well as for sortals (e.g. man, fish), which results, he suggests, in the "desubstantialization of phenomena" (*DPM*, 12–13). He further states that phenomena are *in toto* mere reflections of the Forms and are "totally dependent on the Forms and the receptacle for their existence." Although Plato does state that the Demiurge used the 'Living Creature' as a model for the Cosmos, there is no basis in Plato's cosmology for a reduced ontology for phenomena. What Plato hypothesizes are the Forms and the receptacle, and not the phenomena, the latter alone remaining substantial, if not fleeting, its matter made up of space encapsuled by polyhedra (see Vlastos, *PU*, 90) and its movement borrowed from chaos and transformed by 'nous.'

503. A doctrine of parallelism appears much later in Spinoza (*Ethics*, Bk. II; prop. 7 schol. and props. 11 and 12).

504. See *Ti.* 51d3-5.

'parallel' reading of Plato's metaphysics provides. What effectively takes place, therefore, is that when the noetic counterparts of soul, formed by the movement of soul as directed by nous in tandem with the body, are reflected upon by the mind and consequently conceived as separate entities, these become for Plato the Forms or pattern after which something is formed. Plato does not argue that the Forms existed before the generation of the Cosmos, only that all things which are good must have an intelligent pattern. Although Plato describes 'being' as existing before the generation of the Cosmos, he also describes 'becoming' and 'space' as existing before its generation.[505] Whereas the latter two can be accounted for within a chaotic universe, the former cannot. I would suggest, therefore, that 'being' here comprises 'nous' generally or *en masse* and only refers to separate distinctive Forms later, when 'nous' is disseminated by soul as it aligns with body and is reflected upon by the human mind.[506]

The singularity of the Demiurge, moreover, and hence, of 'Nous' in its totality, would preempt any notion of a pre-existing plurality of Forms. Before the generation of the Cosmos, the only pattern after which the Demiurge could reflect would be upon itself and therefore, of 'Nous' in all its perfection and goodness.[507] Thus, the individual Forms, in reality, are contingent on Plato's understanding of the Demiurge as a primal causal force and hence, on his theory of causality of the generated Cosmos and not the other way round, i.e. that the generated Cosmos is dependent upon the pre-existence of the Forms. The necessity of myth-making has confused the order. Plato, moreover, did not analyze the Forms as such, preferring in the *Timaeus* to refer only to pure being (having no movement) and to patterns (which presuppose noetic movement).[508] Their existence can be hypothesized, however, by abstraction, without their full story being told, as the workings of soul and nous are observed. This accounts for Plato's reference to the Forms outwith the confines of a cosmology and hence,

---

505. *Ti.* 52d2-4.

506. Similar to Lutoslawski, but without drawing on the close connection between 'soul' and 'nous' (and subsequently 'thought'), C. Ritter affirms that "I repeat, the Idea is nothing more than the designation of the objective basis of a conception; this objective basis assures validity to the conception and gives certitude to it," *The Essence of Plato's Philosophy*, trans. Adam Alles (London: George Allen & Unwin Ltd, 1933), 375. He further clarifies his position, stating that: "According to their logical content the Ideas are divine thoughts, a part of the content of God's way of doing things. Only by abstraction can we separate and distinguish the two; just as we can separate God from the world by abstraction. . . . The conclusion of the *Timaeus* confirms us in this conception. . . . In short, the world stands before us as 'the visible image of God whom we can conceive in thought alone.' From this statement we can see that God's will to give to this world the best possible form was completely realized. We may, therefore, say that God's power is adequate for the realization of his thoughts" (376).

507. See *Ti.* 29a2-b1.

508. See *Ti.* 28c5-29c3; see also *Ti.* 51b6-52d4.

before his writing of the *Timaeus*.[509] In modern terms, much in the way that language markers work, the Forms became by default ancient analytic tools. Thus, with the existence of the Forms having already been conceived by reflection upon the interplay of soul with nous, Plato was able to stabilize reality, while the function and form of causality and the fact of the physical universe were yet to be explained or having been explained, drop from the discussion. It was from this process, I suggest, that Plato's universals were born.[510]

2.3.2.1.6 The soul

In *Laws X* Plato tackles the problem anew of the relationship of the primary causes to the world and of causal power within the world, shifting his "likely account" now to a 'likelier' one. Thus, he moves away from the earliest stages of the formation of the Cosmos towards a later, more developed stage. Accordingly, Plato abandons his earlier metaphor of the primary causal powers, consisting of the supreme craftsman or Demiurge and its antithesis, Ananke, the work of these key metaphysical figures having been completed with the generation of the Cosmos. The entire backdrop of this metaphor therefore also disappears. With the metaphysics now greatly diminished, but with causation still requiring

509. Burnet's historical argument (see *Platonism*, passim, particularly Ch. III, 'The Theory of Ideas') that the 'Forms' were principally a belief of Socrates, which doctrine Plato faithfully reproduced in his early and middle dialogues, not as a philosopher presenting his own ideas, but as an artist and great admirer of Socrates, thereby bringing to life his mentor's teachings and person for a new generation, and only later presented his own ideas (e.g. in the *Timaeus* and *Laws X*, which, with the exception of a single reference in the *Timaeus*, do not mention the Forms) is convincing given the complete lack of serious discussion in either of these works concerning the Forms, where, in fact, such a discussion would have been warranted had they been a integral part of Plato's own belief system.

510. Vlastos, by way of improvement, offers an alternative to Plato's degrees-of-reality theory, by suggesting a 'kinds'-of-reality theory in which both particulars and universals are real in their different ways. I would suggest, however, that by the time Plato wrote the *Timaeus* (if not before, should the Forms, as Burnet argues, be Socratic in origin) it is possible that a kinds-of-reality theory was already to some degree assumed by Plato. There is no strong sense in the *Timaeus* in which the generated world is depicted as being in any real way inferior to a more perfect one. Its imperfections, rather, are understood as the remains or intrusion of *Ananke* and its own perfection is assumed in spite of this. Moreover, any fleetingness of appearance is ignored, with Plato indeed pointing out 'this' and 'that' while describing various objects within the world. Thus, contrary to the idea of a lesser reality, upon its completion the Cosmos is described as a visible god, "supreme in greatness and excellence, beauty and perfection" (*Ti.* 92c4-9). Its model, being exactly that, a model, no more and no less, becomes simply the intelligent pattern upon which the ordered universe, the Cosmos, was fashioned. See Gregory Vlastos, "Degrees of Reality in Plato," in *New Essays on Plato and Aristotle*, ed. R. Bambrough (London: Routledge & Kegan Paul, 1965), 1–19.

a source within the world, 'soul,' the only metaphysical entity generated for the Cosmos, becomes its only player, albeit greatly transformed and enhanced by new causal powers. Thus, in *Laws X* Plato re-introduces 'soul' as the "universal cause of all change and motion"[511] in an already generated world.

In the *Timaeus*, however, there is no argument or evidence to suggest that the causal movement of the Demiurge is identical with the movement of the soul or that soul is to be identified with the Demiurge. In this work, the generation of the soul (by the Demiurge, which by causal priority becomes an argument *against* identity between the soul and the Demiurge) is described in obscure, abstract terms (various combinations of the 'different,' the 'same,' and 'existence').[512] The soul's movement is just touched upon and how it apparently moves a body is not addressed. Similarly, however, neither is the movement of the Demiurge (or Ananke) discussed. The obscure terms used in the *Timaeus* to describe the soul's generation, moreover, have not been used before by Plato, nor are they used for any other purpose afterwards.[513] This "likely tale," therefore, argues for an opposite conclusion to that of identity, namely, that there can never be complete identity between the generated soul and the Demiurge or their respective movements.

The initial argument for this conclusion is the above-noted one of causal priority, which notion has two aspects. Not only is the Demiurge described as the agent who generates the soul, thus, is prior to the soul, but the Demiurge is also depicted as being ungenerated. We are therefore left with the intimation that the Demiurge is wholly simple, and its being and movement ultimately unknowable.[514] The generated soul, on the other hand, precisely because of its generation, is complex, and its being and movement, albeit with difficulty, are knowable. Unlike the Demiurge, the soul's ability to 'know' is through a complexity of noetic moves as it reflects upon itself impacted by the external world. The soul, moreover, must have an external world with which to interact. It is never bodiless. Its movement is dependent upon this, whether in a star or in a human person. Thus, because of the soul's very makeup, being made out of portions of

---

511. *Lg.* 896a5-b1.

512. *Ti.* 34b10-35b3; see also *Ti.* 41d4-8 with regard to the human soul.

513. There is no reason to believe that for Plato this noetic prescription of the generated soul from the *Timaeus* does not still hold true for the human souls or the World Soul in *Laws X*, although this is not explicitly stated. Where he differs in *Laws X* is in his introduction of two singularly powerful 'beneficent' and 'non-beneficent' souls, as well as in his introduction of divinities as souls, no longer maintaining an epistemology of soul as strictly generated, since clearly these souls are not generated.

514. 'Knowing' is an entirely noetic function of the human soul in the *Timaeus* (*Ti.* 46d5-6) involving movement between the knower and the thing known. Accordingly, knowledge is acquired through the senses with special emphasis on the role of the eyes. See *Ti.* 46e6-47e2.

existence,[515] difference, and sameness, as well as its interwovenness with the world and its objects, i.e. the heavens and human beings, 'soul' is able to observe what has existence (on the principle that 'like knows like') and the kind of existence it has, and then differentiate between what is different and what is the same, thereby enabling it to make judgments and move the world it impacts towards certain ends. The causal movement of the Demiurge, on the other hand, is not dependent upon the external world with which it interacts, i.e. on chaos, and therefore does not need a body. It reflects only upon itself in its totality and oneness, thus, only upon its own simplicity. In other words, the Demiurge reflects upon 'Nous' undifferentiated or as described by Plato, upon perfect 'Being.'[516] In so doing, the Demiurge is able to bring about its desired end, that is, to bring into existence the generated world, which at a perceived noetic level results in 'Nous' *differentiated* or the Forms. Again, the issue of priority is raised. The Demiurge not only constructed the soul, but also constructed the Cosmos for which 'soul' was made and whose purpose was to bring about constructive ends within it.

A second argument against identifying the Demiurge with the soul only becomes apparent in *Laws X*, where Plato re-introduces the noetic functioning and movement of the soul. Here we are introduced to the concept of the 'self-moved mover,' a completely different notion from the so-called 'unmoved mover' of the *Timaeus*. Although I have argued against either of these notions being actually present in the *Timaeus* with regard to physical motion, as explained above, a distinction can still be made between the Demiurge and the soul at the noetic level. Accordingly, to drop the idea of the 'unmoved mover' in *Laws X*, even at the noetic level, still requires an explanation to account for the organization of the world as it stands, as well as for the existence of 'soul.' As a consequence, either the idea of the world as having been generated must be abandoned or the existence of the primal causal powers must be assumed. I would suggest that the latter is more likely, given that the *Timaeus* was written with a metaphysical understanding of the generation of the Cosmos specifically in mind, while *Laws X*, which challenges its premises, is a polemic by Plato's own admission against impiety, addressing metaphysical issues in light of this need well after the generation of the Cosmos. In either case, the soul cannot be identified with the Demiurge (or Ananke).

Soul, moreover, according to the *Timaeus*, came into being with the generation of the world and yet it is both immortal and divine. Although there is a

515. 'Existence,' as described by Plato, can be one of three things, (1) the "indivisible, eternally unchanging existence" of perfect being, i.e. of 'nous' or the same reflected upon by the human mind and conceived of as the 'Forms,' (2) the "divisible, changing Existence of the external world," i.e. of 'becoming,' or (3) some intermediary form (*Ti.* 35a1-6), i.e. of generated 'soul.'

516. See *Ti.* 52d2.

sense in which the Demiurge and soul as subjects of 'immortality' and 'divinity' can be identified one with the other, i.e. they will never die and the whole of their activity is noetic (rational) and not physical, apart from these qualities, as players in the divine game they are very different. For example, in the *Timaeus*, whereas the Demiurge can generate the soul, the soul cannot generate itself or generate the Demiurge, nor can soul generate the world, but the Demiurge can and did. If this difference were not the case, Plato would have left the Demiurge out of his cosmology since all players within the *Timaeus* are essential to it. There are no exceptions. Therefore, although soul has been called immortal and divine, there are no grounds, other than the sharing of these two characteristics, for identification of 'soul' with the Demiurge.

Plato, moreover, never defines the Demiurge or God as a soul in the *Timaeus*. Rather, in this work the Demiurge is defined as the "best of causes" (*Ti.* 29a5-6) and the "best of intelligible and eternal things" (*Ti.* 37a1-2). There is also the intimation that the Demiurge is supremely one or simple. Soul, on the other hand, is never described in this way. Rather, it is fabricated and complex, which, in this same passage is described as "the best of things generated." Minimally, therefore, there is a strong case of priority of existence with regard to the Demiurge over 'soul,' arguing once more against identity. Moreover, while in the *Timaeus* the Demiurge is never defined in terms of self-movement, the soul is generated by the Demiurge in such a way so as to embody this very attribute, albeit, at a noetic level. Furthermore, Plato offers no epistemology in the *Timaeus* by which to understand the Demiurge, stating only that: "to discover the maker and father of this universe is indeed a hard task, and having found him it would be impossible to tell everyone about him" (*Ti.* 28c3-5). With regard to the epistemology of the soul, however, Plato readily draws upon the tripartite figure, which he has already set out in *The Republic*. As a consequence, therefore, of the clear separation which Plato had made between the Demiurge and soul in the *Timaeus*, his later reassignment in *Laws X* of all causal power and movement to soul alone provides a strong argument for understanding soul in this later work as a telescoping, at least to some extent, of Plato's key metaphysical players from the *Timaeus*. Thus, what is now presented in *Laws X* as 'soul,' comprising at least two souls, the Beneficent Soul and the Non-beneficent Soul, as well as the human soul(s), had been presented earlier (at the generation of the Cosmos) as separate metaphysical entities, namely, the Demiurge and Ananke, as well as 'soul,' which latter was comprised of the World Soul and the human soul(s).

2.3.2.2 *Laws X*

As stated above, *Laws X* moves on from Plato's earlier understanding of soul in the *Timaeus*. The Demiurge, which in this "likely story" depicted both the Cosmos and soul as his handicrafts, is no longer referred to. In this subsequent work, Plato's discussion of God is preempted by a discussion of 'the supremely good soul.'[517] This 'best of souls' is described as a self-moved mover, who is able to move and direct the whole of the universe, and in this sense is the 'Universal Cause.'[518] Thus, by apposition, the 'supremely good soul' and 'God,' as far as they are discussed, take on the same identity. What is clearly evident from *Laws X* is at no point is the movement that is ascribed to this 'supremely good soul' ever equated with physical movement. Rather, this self-movement, drawing from a similar image of 'soul' from the *Timaeus* and shared by souls generally, remains a movement of nous, now directing the generated world. Earlier, in the *Timaeus*, however, it was the World Soul, which directed the Cosmos, the Demiurge having first generated it to carry out this task. The generated soul, moreover, was never described as being "supremely good," a descriptive reserved only for the Demiurge or God. Thus, the quite separate notions of the primary cause or God (the Demiurge), the World Soul, and individual souls from the *Timaeus* collapse in *Laws X*, with the notion of 'soul' now subsuming these ideas and reformulating them anew as Plato further develops his understanding of soul.

2.3.3 The Demiurge and mechanical movement

There is no point in Plato's causal account, either in the *Timaeus* or *Laws X*, where he is concerned with *how* a seemingly non-material Demiurge or God (equated closely with 'nous' in the *Timaeus* and the 'supremely good soul' in *Laws X*) is able to interact with the material of the generated world and thus be the cause of any kind of physical movement, organized or otherwise.[519] This,

---

517. *Lg.* 897c6 (trans. by Taylor). Saunders translates this same phrase as "the best of souls." I prefer the reading by Taylor, as Plato abandons all mention of soul in *Laws X* shortly thereafter, referring directly to the gods, arguing by apposition for identity between the souls previously mentioned and the gods. As Plato begins to discuss the gods, he now also refers to "the God" (*Lg.* 902e4) and "the King" (*Lg.* 904a6), which together argue for a 'supremely good soul,' rather than the more generic 'best of souls,' which may be read as comprising more than one.

518. *Lg.* 895e10-896a2.

519. Vlastos's assertion that "by thinking and willing it [soul] can move the body to which it is attached and, through this, other bodies" (*Plato's Universe*, 31) is not one which Plato makes or one that he assumes. In *Laws X* Plato has the Athenian admit that he does not know how soul moves a body (*Lg.* 898e8-899a4). I would suggest, therefore, that if Plato had thought that by simply thinking or willing it 'soul' could effect physical

however, does not present a problem for Plato. From his perspective all physical movement in the universe is due to secondary causes and thus, is not the *kind* of movement attributable to the Demiurge or soul. Although it would still be possible to take a determinist reading of accessory movement, in which case all action necessarily relates back to a first cause, i.e. to the Demiurge in the case of the generated world as a whole or to 'soul' in the case of things *within* the generated world, I would suggest that this reading is wrong.

First, such a reading requires either an inexplicable dualist position in which the non-physical Demiurge or 'supremely good soul' somehow interacts with the physical world, thus pointing to the causal players, but providing little explanation as to what or how this actually happens and thereby remaining unsatisfactory as an answer or alternatively, the Demiurge or supremely good soul must be reduced to some aspect of the physical world, a position not supported by the "likely account" given in the *Timaeus*, as well as vehemently denied and explicitly argued against in *Laws X.* Plato is fastidious in clarifying physical connections, however erroneous, in the *Timaeus*. He does not attempt, however, to explain *ever* how the Demiurge 'pushed,' so to speak, the first dice or how the World Soul or individual souls give initial movement to things or individuals within the Cosmos. This is not carelessness on Plato's part, but an acute awareness of the metaphysical reality and philosophical problem he was dealing with, namely, that movement did not begin with the generated world, but continued within it, only now, with the generation of the Cosmos, this movement is impacted by the Nous of the Demiurge, toward organized ends.

The problem which causal determinism raises in Plato's account, given his understanding of physical movement as accessory, focuses upon *how* movement within the generated world can be ultimately connected to the Demiurge as the 'unmoved mover' or in other words, the problem of dualism. However, I would challenge the view that physical movement in Plato's Cosmos is determined by a non-physical cause, just as, at its opposite pole, I would challenge the verdict of physical determinism, akin to the atomism of Democritus.[520] The question

---

motion, Plato would have said so, but did not precisely because he has already claimed and deeply believed that only physical motion can effect physical motion. See *Lg.* 895a5-b7. In this passage Plato describes self-generating motion as the primary source in both *stationary* and moving objects, while all movement transmitted between objects is described as secondary. Clearly self-generating motion is a special *kind* of motion (i.e. I have argued that it is entirely noetic or a-noetic) and not simply a more complex notion of physical motion (or as argued by Vlastos, the unexplained linking of the psychic with the physical).

520. In countering Vlastos, Sorabji writes: "Aristotle does not postulate that every event conforms to high-level regularities. I doubt if Plato does either." (Sorabji, *Necessity, Cause and Blame*, 62.) His interesting argument follows upon an earlier one by G. E. M. Anscombe that events can be caused without being necessitated, although likely differing where Sorabji adds to his argument that desires and beliefs cause actions (ibid.,

central to this debate is to ask what metaphysical reality was Plato trying to describe? It is only in *Laws X*, however, with Plato's unreserved rejection of the strictly physicalist position, that the picture becomes clearer and a possible answer emerges.

We have already seen from the *Timaeus* that there is a considerable degree to which Plato agrees with the atomist understanding of the physical universe, differing with the atomists, however, with respect to his opposing causal views and their attendant metaphysics. Although Plato has a theory of soul by which he accounts for motion in the universe, he has no answer for *how* soul interacts with things or people or *how* the World Soul moves the universe, as the motions initiated by soul are not physical. Rather, their movements have been described by Plato as the various motions earlier associated with the different regions of the tripartite soul, namely, wish reflection, foresight, counsel, judgment—true or false—pleasure, pain, hope, fear, hate, love, and "whatever other kindred, or primary, motions there may be."[521] By his own admission, Plato makes it clear that he does not know how 'rational motion'[522] interacts with the universe, offering instead the proviso (as he similarly did in the *Timaeus*) that "we must not assume that mortal eyes will ever be able to look upon reason and get to know it adequately."[523]

What follows in this passage is important. Plato once again recalls the image of the sun and the need, due to the limitation of the human intellect, to look at these metaphysical realities obliquely or at their *image*, and not directly.[524] His subsequent discussion of the various physical motions and their parallel with the rational motions, accordingly, is meant to serve this purpose. His further discussion of *how* psychic movement (noetic or anoetic depending on which soul) interacts with the physical world soon follows, but again, obliquely. Consequently, Plato offers three possible ways in which soul interacts with the world. These remain, however, at a descriptive level, rather than providing the reader with a full causal explanation. Thus, at *Lg.* 898e8-899a4, using the sun as an example of a physical body, Plato suggests with regard to the motion of the soul ('rational motion') that either (a) the soul resides within this visible spherical body and carries it wherever it goes, just as our soul takes us around from one place to another, or (b) it acquires its own body of fire or air of some kind (as certain people maintain), and impels the sun by the external contact of body with body, or (c) it is entirely immaterial, but guides the sun along its path by virtue of pos-

---

28–29). Apart from Sorabji's argument against a deterministic position for these philosophers, this idea is potentially valuable in determining how a disordering cause, which I argue is completely without 'nous,' i.e. is a-noetic, might bring about and maintain disorder at a cosmic level.

521. *Lg.* 896e8-897b5.

522. *Lg.* 897e4.

523. *Lg.* 897d8-e2.

524. This is not an allusion to the Forms.

sessing some other prodigious and wonderful powers.[525] In (a) Plato does not provide an explanation of *how* the various psychic motions associated with the soul would be able to achieve this. In similar fashion, (b) does not mention *how* the soul would acquire its own physical body and even if it did, how the soul would interact with this body. Finally, (c), similar to (a) but with soul not residing *in* the object moved, similarly begs the question of how soul moves a body by offering a description of this unexplained dualist position in place of an answer. What this passage clearly enumerates, I suggest, are not causal explanations for the interface between the non-physical soul and a physical body, but rather, alternative descriptive guises for the 'parallel' causal relationship suggested earlier, which exists between them. Moreover, precisely *because* of this parallel relationship existing between the soul and the physical world, Plato cannot offer a more thorough or in-depth causal explanation, as one, I suggest, does not exist that is easily comprehensible. Plato's consistent lack of any such explanation attests to his awareness of this fact. Where there is a self-moving physical body, there is also a soul. This much Plato will say. He also chooses to affirm this, stating that by definition 'soul' is self-moving motion[526] or it is that which makes something 'alive.'[527] Thus, soul for Plato encapsulates and articulates the very self-movingness of the physical world, which he has argued in both the *Timaeus* and *Laws X* cannot belong to the physical world itself. The physical world, on the other hand, and its contents, inasmuch as they may (i.e. the Cosmos as a whole, the heavenly bodies, and living beings) or may not (i.e. individual inanimate objects in the world) be self-moving, attest to the hypothesized existence of soul, but more than this, especially with respect to the 'parallel' relationship which exists between soul and the Cosmos, Plato does not and cannot say.

2.3.4 The identity of soul with primal becoming and universal causality

In the second passage cited at §2.1/2, *Lg.* 896a5-b1, Plato seeks to complete his demonstration that not only is soul the universal cause of all change and motion, it is also to be identified with the primal becoming and movement of "all that is, has been, or shall be, and of their contraries." To achieve this he first eliminates the accessory causes or "motion induced by something else" as possible contenders for this position (passages iii and iv at §2.1/2), as he had done previously in the *Timaeus* when trying to discern the primary causal powers from the secondary causes. In an even stronger move, however, Plato drops the entire meta-

525. Saunders, *The Laws*, 431.

526. *Lg.* 895e10-896a5; passage (i) from §2.1/2.

527. *Lg.* 895c7-10. A body can be dead or alive. For this reason Plato hypothesizes that self-movingness is, in some way, a separate entity apart from the body, positing thereby the idea of 'soul.'

phor of generation from *Laws X*, as it is no longer needed and with that, and for the same reason, he also drops its primary causal powers, the Demiurge and Ananke. Plato's earlier purview from the *Timaeus*, therefore, has been completely dropped from *Laws X*, replaced now by a wholly post-generation vantage. As a consequence, the workings of generation and its key players are now deleted from the discussion currently taking place. This must be borne in mind when considering causality in *Laws X*. This is not to say, however, that Plato's metaphysics do not remain contiguous with the *Timaeus*, for they do, since by the end of this earlier work, Plato's vantage, too, had become post-generative, with all discussion of generation and its key causal players having similarly been dropped.

As a consequence of this change of focus, however, which begins in the *Timaeus* and culminates in *Laws X*, two related problems arise, namely, first, discerning the degree to which 'soul' in *Laws X* is distinguishable from the primal powers as they were articulated earlier in the *Timaeus* and second, discerning the degree to which 'soul' in *Laws X* can be identified with the generated 'soul.' The reasons for these problems arising are fairly obvious. With regard to the first problem, in *Laws X* Plato clearly wants to give 'soul' causal powers not earlier specified or ascribed to it in the *Timaeus*. As far as the second problem is concerned, with the metaphor of generation having been completely dropped from *Laws X*, the ontology and epistemology of 'soul' are more imprecise, as there is now an amassing of ideas previously articulated separately in Plato's earlier metaphysics.

On closer inspection, however, these problems resolve. Accordingly, in *Laws X*, passage (iv) above (*Lg.* 896b10-c3), it is not the primary causes which are being distinguished from the secondary causes, but rather, 'soul' as a *prior* cause over that which it governs, namely, body, a position which Plato has already affirmed in the *Timaeus*.[528] Thus, the earlier depiction of 'soul' in the metaphor from the *Timaeus* as having been generated by the Demiurge, which latter forms the first of two primary causes, the other being Ananke, is not contradicted by this later passage in *Laws X*. 'Soul' as understood in this passage is not a primary cause—it did not generate the Cosmos—rather, after its own generation (discussed in the *Timaeus*, but not in *Laws X*), soul provides the noetic movement for the generated universe. As mentioned, this change to a post-generative focus is contiguous between these two works, where by the latter half of the *Timaeus* all mention of the primary causes has ceased, with the remaining description having ceded to an explanation of certain aspects of the already generated world and its accessory causes, as well as to the workings of the human 'soul.' This change in causal vantage is prompted, I suggest, by a desire on Plato's part to give greater emphasis to the efficacious role of 'soul' in the generated world, combined with a development of his earlier view that 'nous' un-

528. See *Ti.* 34b10-35a1.

embodied ('rational motion' in *Laws X*) is to be identified with the Demiurge or Ananke (but the latter in a wholly negative sense as a power lacking in 'nous'), but when made particular and immanent, with 'soul.' Plato would have reasoned that if soul can be made particular and immanent (in the world), then it is begotten from that which is universal and transcendent. Accordingly, the Demiurge and Ananke become transformed in *Laws X*, represented metaphysically in this work as the Beneficent and Non-beneficent Souls, respectively.[529]

What *Laws X* does not do is define the boundaries of soul *within* the Cosmos, as seemingly, there are no boundaries—as a cause, soul is all pervasive and all powerful, always operating at a 'parallel,' psychic level with the universe itself. Moreover, being all-powerful, it is limitless with regard to its impact on change and motion. As such, there is no immediate requirement to understand soul as having been generated in *Laws X*, since there is nothing outside of soul or the universe, which is being considered. With respect to this, however, a distinction to which I wish to draw attention is Plato's use of the phrase 'universal cause' in *Laws X* (αἰτία ἅπασιν in passage (ii) above) as distinct from the idea of 'primary cause' in the *Timaeus*. We know from the *Timaeus* that the ἅπασιν or 'all' has already been generated and is in place. We also know from the *Timaeus* that the Demiurge was the primal cause of the generated world. Thus, successful in its task of offsetting the powers of the other primary force, Ananke, the Demiurge, I suggest, was depicted as *pre*-universal in its causal powers and not universal. According to the *Timaeus*, moreover, 'soul,' in the form of the World Soul, assisted by individual souls, was generated so as to be the primary causal power *within* the generated universe and therefore, to this extent, acts as the 'universal cause.' In accordance with this understanding, 'soul' in *Laws X*, therefore, is not pre-universal as a cause, nor is it generative, having neither the scope nor the power of the Demiurge. Thus, just as 'soul' was understood in the *Timaeus* as the primary causal power *within* the generated universe, so, too, 'soul' is aptly described in *Laws X* as the 'universal cause.'

As argued above, soul has 'parallel' causative powers in relation to the world it impacts, informing the Cosmos rationally while existing separately with regards to its ontology, having being derived (begotten) by the Demiurge from 'nous' and existing entirely at a noetic level. The Demiurge, on the other hand, is *necessarily* metaphysically distinct from both 'soul' and the Cosmos, being prior to both and not identifiable with either the Cosmos or its ongoing movement (physical or noetic). Its role, therefore, with regard to Plato's cosmology begins and ends with the generation of the Cosmos.

Laws X, however, makes no mention of this earlier process or to its players, e.g. the Demiurge and *Ananke*, beginning with the universe as already fully alive and informed by soul. This can be summarized as follows:

---

529. See passages (ix) *Lg.* 896e4-6 and (x) *Lg.* 896e8-897b5 above.

The Timaeus	***Laws X***
There is a chaos, out of which the Cosmos is generated.	The purview of *Laws X* is entirely within the scope of an already existing Cosmos.
The Demiurge is a *pre*-universal cause and so is *Ananke.*	The Demiurge is not mentioned, nor is the generation of the Cosmos or any of its other players, e.g. Ananke.
Soul is generated, but not out of chaos; rather, it is formed from 'nous.' Soul, moreover, being formed by the Demiurge shares in its divinity and to this extent is also eternal. It remains, however, metaphysically distinct from the Demiurge by virtue of its ontology.	There is no mention of the soul being generated. Rather, it is cited as the universal cause of all that is, has been, or will be, with no mention being made of its history. The soul is divine and therefore, it is also eternal.
The soul is the source of all causal power (rational movement) within the physical universe.	The soul is the source of all causal power (rational movement) within the physical universe.
The soul exists in a parallel causal relationship to the world.	The soul exists in a parallel causal relationship to the world.
Soul is identified with the movement of primal becoming.	Soul is identified with the movement of primal becoming.

As this chart shows, the scope of the subject matter is the major factor differentiating these works with regard to soul. To wit, whereas the *Timaeus* takes 'being,' 'space,' and 'becoming' into consideration, *Laws X* takes only the latter two, 'space' and 'becoming,' and not 'being.'

2.3.5 The soul, begotten and not generated

In *Laws X,* Plato drops all imagery of the soul as having been generated. This move, I suggest, is not at odds with the mythic representation of the soul's generation in the *Timaeus*, where its description is written in entirely different terms from that of the rest of generation, since, rather than being described as having been generated from the physical chaos as was the rest of the Cosmos, 'soul' is described as being generated from mixtures of 'difference,' 'sameness,' and 'existence,' which entities are purely noetic and metaphysical in kind and not physical. Thus, whereas the simple imagery of generation is appropriate for the physical world, it remains awkward for 'soul,' which is not an object *per se*. Plato's intention, therefore, in using this imagery must be seen from the perspective of the act of generation itself. Clearly its purpose was to show that soul, like the rest of the generated universe, was *designed* by the Demiurge and for a purpose, but nothing more, I suggest, was intended, since, unlike the 'stuff' of the Cosmos, soul was not made up out of various arrangements of triangles. Rather, and importantly, it was made up of noetic material, metaphysical in nature, taken or begotten from nous, which latter was not subject to generation.[530]

That 'soul,' therefore, was begotten and not generated, while at the same time, along with the generated Cosmos, was designed, is the position, I suggest, which is ultimately argued for by Plato in both the *Timaeus* and *Laws X*. To wit, first, in the *Timaeus* the World Soul is described as formed by the Demiurge *before* the body of the world, in other words, before the generation of the world and therefore carries no connotation of physical generation. Second, the Demiurge made this soul the "dominating and controlling partner" in its relation to the generated Cosmos. This carries with it the notion of authority, passed on by the Demiurge, but not the authority of brute force of which the Demiurge has no part, but of intelligence or 'nous.' Third, soul, as generated, is described as being composed, in part, of indivisible, eternally unchanging 'Existence,' a concept closely associated with nous. Although in the *Timaeus* eternally *unchanging* existence is directly descriptive only of 'the unchanging Form,'[531] to the degree that both the 'unchanging Form' and the Demiurge participate in nous, this description is applicable to both the Demiurge[532] as the direct agent and source of 'nous,' and 'the unchanging Form' as the object of thought, its indirect agent. Thus, although the Demiurge is described as having used the "highest and completely perfect of intelligent things" as the model after which it[533] generated the universe, and which model itself is later described as having unchanging and

530. For the generation of the World Soul see *Ti.* 34b10-37c5; for the human soul see *Ti.* 41d4-42e4.

531. *Ti.* 51e6-52a4.

532. See *Ti.* 30b4c1.

533. Described by Plato as the 'Father of the Universe' (*Ti.* 37c7) and hence, 'begetter.'

uncreated existence, this perfectly intelligent Form or 'being' (which according to Plato is inclusive of all the forms or ideas after which things have their intelligent pattern) remains the object of the Demiurge's own reflection upon itself or 'nous.'[534]

Soul is also now described as composed of divisible, *changing* 'Existence.' This had only been descriptive of 'Becoming,' which existed before the generation of the Cosmos and was not itself generated. This element was presumably incorporated by Plato to capture the ability of soul to remain and operate within the world of becoming as it was being organized, i.e. as the Cosmos was being generated. The third kind of 'Existence,' intermediary between the other two, of which the soul is composed, simply captures, I suggest, the soul's operative role or otherwise put, the compromised position of 'nous' within the generated physical sphere. Thus, Plato's reference to the three kinds of 'Existence' give voice to the complexity of the metaphysics of soul as it was designed by the Demiurge to operate within the generated Cosmos. Plato's reference, on the other hand, to the three kinds of 'Sameness' and three kinds of 'Difference' refer directly, I further suggest, to the noetic elements of nous (the ability to discern difference and sameness) from which the Demiurge chose to design the soul, listing them according to the three kinds of Existence, which soul would require to become fully functional. Accordingly, these noetic elements in turn align and correspond to the complex metaphysics (of existence), which soul as a designed noetic object has been given by the Demiurge in order to exist before, during, and after the generation of the Cosmos, either in it or in whatever specific object it inhabits (i.e. a living being). What is clear is that soul is comprised of noetic elements and is given a complex existence, but neither of which, the elements out of which it is composed or its existence, are themselves objects subject to generation. Plato's description, therefore, of the generation of soul is a metaphor giving voice to the above and in so doing, imaginative substance to his 'likely' account. Thus, although in the *Timaeus*, Plato goes into considerable detail regarding the composition and mathematical structure of the soul, having done so similarly for the physical objects within the generated Cosmos, he does this, I

534. Although it has been argued that the Demiurge understood as 'nous' is simply metaphoric language within strictly mythic material for intelligence or reason within the Cosmos, a position which Cornford maintains, this view does not take into consideration that the Demiurge is also the agent of nous and as a causal agent does not have strict identity with it. Furthermore, this view omits the fact that the intelligent being, or that aspect of 'nous' comprising the Forms and upon which the Demiurge reflects to bring about generation, is never described as being inclusive of everything on which the Demiurge *could* reflect and further, omits all discussion of what the Demiurge itself is comprised. Although we do know that the Demiurge participates in 'nous,' it could be that anything else of which the Demiurge is comprised (and *how* it is so comprised) is beyond the intelligence of man to comprehend, which position concerning the limitation of the human intellect is well documented by Plato.

suggest, to preserve the imagery and vitality of this likely account, and also perhaps to capture the general tenor of the soul's nature as complex and designed, but not to reflect in any way actual knowledge of the soul's structure.

2.3.6 God (the Demiurge), not begotten

Plato's understanding of the generation of the soul in the *Timaeus*, therefore, is congruent with his later understanding of soul as eternal and ungenerated in *Laws X*, the difference lying primarily in Plato's use of metaphor within his cosmology, but which is absent from the later work. Where greater clarification is still needed, however, is with regard to Plato's understanding of the Beneficent and Non-beneficent Souls in *Laws X* as they might relate to the Demiurge and Ananke of the *Timaeus*. There are a number of possibilities to consider regarding any purported relationship which might exist between these key causal players: (first) that the Demiurge and Ananke were purely mythic entities and the only causal force which exists for Plato is 'soul,' of which the Beneficent and Non-beneficent Souls are specific players, whose ends have moral significance; (second) that the Demiurge and Ananke remain as causal forces for Plato at the time of the generation of the Cosmos, but drop out after generation is completed, leaving only soul as agent(s) in a universe which has long had existence; (third) that by the time *Laws X* had been written, Plato had concluded that the Demiurge and Ananke were in fact, if not mythic, to be identified with soul, since upon reflection they seem to share identical attributes of 'nous' when considered in the light of an already-generated universe; or (fourth) that Plato's understanding of the workings of the Demiurge and Ananke as forces has broadened to include their ongoing causal impact at the level of 'nous' in the generated world, but only at the level of 'nous' (or lack of it) and thus, only in part (not the generative sphere), as the Beneficent and Non-beneficent Souls, which, moreover, exist separate from the notion of the World Soul. It is with respect to this latter proposal that I will offer support, the challenge remaining, however, that if this is correct, is there some degree to which the Demiurge and Ananke, extending into the world as Beneficent and Non-beneficent Souls, like all souls, are begotten or conversely, are these Souls in *Laws X*, like the Demiurge and Ananke, begetters and not begotten?

With regard to the first proposal, I have argued throughout this book that the Demiurge and Ananke were not mythic entities for Plato, but were understood by him as the primary metaphysical causal forces behind the generation of the Cosmos. Although Plato does use mythic material in the *Timaeus*, it is not with regard to either of these key figures in his cosmology. With regard to the second proposal, that the Demiurge and Ananke are causal forces during generation, but play no further role after that and therefore are in no way related to the Beneficent and Non-Beneficent Souls respectively of *Laws X*, I offer the following objections. First, although I would agree that there is not a direct one-to-one correlation between these two sets of causal players, on a closer look the similarities are too great to deny a link at least (and possibly, at most) at the noetic level or that of 'nous.' My claim, however, is not that the Demiurge or Ananke have now become elevated Souls for Plato, but rather, that the Beneficent and Non-beneficent Souls give voice to Plato's recognition and acknowledgment of the continuing presence of the Demiurge and Ananke in the generated world at the level of nous (or its lack), distinct from those activities, moreover, originally given to the World Soul. A number of points must be kept in mind, however. Plato's presentation and understanding of soul has changed considerably from the *Timaeus*. Passage §2.1/2 (vi) quoted above from *Laws X* shows the type of movement of which soul is now capable. The *Timaeus*, on the other hand, does not specify the soul's movements other than those associated with intelligence and reason.[535] Its three-tiered structure of soul carried over from *The Republic*, moreover, is no longer mentioned in *Laws X*. Memories, moods, and wishes are listed alongside calculations and true judgments in this latter work. Thus, soul appears now to be viewed as a complex noetic whole, rather than as a neat supra-physical structure. Further, whereas the concept of 'will' is not viewed as a characteristic of soul in the *Timaeus*, but is attributed to the Demiurge and Ananke[536] as Unmoved Movers, in *Laws X* the idea of 'will' is now acknowledged to be attributable to the notion of soul as a 'self-moved mover.' It is with particular attention to this last change that I note Plato's possible inclusion of the primary causal powers into the world of becoming as Beneficent and Non-beneficent Souls, albeit, whilst maintaining the strictly parallel relationship of the soul's movement to the world's physical movement and the existence of the Demiurge and Ananke as the primal powers, from which, I argue, the Beneficent and Non-beneficent Souls have emerged. By the time Plato wrote *Laws X*, the particular work of a World Soul, i.e. to give rationality to physical movement so as to direct it towards an end, appears to be assumed and therefore is not highlighted, nor does this particular role appear to have become subsumed under the larger concept of the Beneficent Soul, which according to Plato has quite broad powers, whereas the World Soul was specifically begot (generated) at the time of the generation of the World for the above-noted purpose. Thus, my objection to proposal two is that Plato, in presenting his notion of soul not simply as a supra-physical structure closely aligned to the body in or to which it is attached,

535. *Ti.* 34a1-5.

536. See *Ti.* 41a7-b6. Although the World Soul is described in the *Timaeus* by Plato as having been generated so as to be the "dominating and controlling partner" over the body of the world to which it is attached, he soon specifies the activity that the World Soul has, which is to provide the Cosmos with "a divine source of unending and rational life for all time," a much more static role than that given to the Beneficent and Non-beneficent Souls of *Laws X*, where most certainly the concept of an active 'will' is now assumed and so, too, non-rationality (i.e. activity of the Non-beneficent Soul). See *Lg.* 896e4-6.

thereby enabling intelligent movement, as it was in the *Timaeus*, but now cast as whole and complex psychic entities capable of self-movement, especially as regards the existence of the Beneficent and Non-beneficent Souls, imitates too closely those movements of which the Demiurge and Ananke would be capable in a post-generative world, to not in fact have some relation one to the other. It would appear, moreover, that although the specific concept of a World Soul has dropped from Plato's metaphysical pallet, the role of 'World Soul,' although clearly not its inception or generation, could easily be subsumed under the broader concept of 'soul(s)' now in place in *Laws X*.[537]

This brings me to the third proposal, which I also reject but with qualification, namely, that the concept of soul in *Laws X* is now identifiable with what were formerly described by Plato as the Demiurge and Ananke in the *Timaeus*. The reasons why this position is untenable are the following. First, *Laws X* is not a cosmology and deals in no way with the generation of the universe. Its vantage, therefore, lies entirely within the scope of an already generated Cosmos of which the viewer has no perspective of its earliest beginnings. Hence, its understanding of the soul is at all times post-generative and from this perspective 'soul' is the only causal power. This vantage, however, does not automatically subsume the earlier primal causes, nor does it assume that they did not exist or deny a beginning to the Cosmos. Second, there is a qualitative difference between the ideas of 'universal cause' as it appears in *Laws X* and that of 'cause of the universe,' which is referred to in the *Timaeus*. The first idea of 'universal cause' leaves the question 'of what' open, although we are told that with reference to the soul the answer is 'movement.' The same question, however, is not raised by the second idea, 'cause of the universe,' since, as stated, it is clearly not movement, but 'the universe' itself wherein movement takes place and which includes soul as another element or causal factor within it, which is being considered. Although I have argued that soul was actually begotten by the Demiurge and not generated, it was nonetheless described by Plato as having been generated, intimating the idea that it was *designed* by the Demiurge as part of the universe. Thus, soul, which according to the *Timaeus* is clearly not self-generated, cannot possibly be the cause of what it is claimed to be a part. Accordingly, the idea of a universal cause of all motion, which is the role of soul in both *Laws X* and the *Timaeus*, is not the same thing as a primal cause, which is the role given to the Demiurge (and Ananke indirectly) in the *Timaeus* as cause

537. See §2.1/2 (viii), *Lg.* 896d10-e2. In the case of the World Soul, although its ongoing rational powers appear to be subsumed under those of the Beneficent Soul (clearly not those of Non-beneficent Soul), its inception remains the work of the Demiurge or 'Father of the Universe.' In other words, it did not beget (generate) itself or, as a discrete entity, has it always been in existence. Plato has not worked out all of the details relating to this change of viewpoint, as sometimes can be the case in his writings, leaving awkward loose ends.

of the universe. With regard to the above-noted qualification in respect of this third proposal, there are nonetheless grounds for claiming that the Beneficent and Non-beneficent Souls are in some way related to the primal causes, the Demiurge and Ananke. This relationship, however, is not one, I suggest, of strict identity, nor is it entirely congruent, but remains for Plato a loose or free adaptation in which he is able to incorporate new ideas, while abandoning those which are no longer useful. This, therefore, suggests a fourth and final proposal.

In this proposal the Demiurge and Ananke are understood as being present (to the degree and manner in which it is possible) in the generated world, manifest now as the Beneficent and Non-beneficent Souls, extending from their primal source into the Cosmos. This idea of the Demiurge and Ananke as emergent souls in the Cosmos presupposes not only the presence of 'nous' in the world (or its absence, as in the case of Ananke) from which 'soul' has consistently been understood to derive its rationality (or irrationality), but also the presence of active 'primal wills' or powers within the universe, who, in the form of 'Souls' or self-moved movers, desire and are able to bring about certain ends within it, either of order or disorder.

2.3.6.1 The power of 'will' and the achieving of ends

In the *Timaeus*, the idea of 'will' is understood as something that is possessed by both the Demiurge and Ananke, as well as by the generated soul. As an unmoved mover, the Demiurge desires and sets out to achieve order or, as in the case of Ananke, maintain disorder. The idea of 'will,' therefore, is based on the notion of desire and the ability or power of the agent of this desire to achieve its object. Timaeus describes the 'will' or ἐθέλοντος of the 'Father of the Universe' (the Demiurge) as being a "stronger and more sovereign bond than that which existed before its generation" (that latter being the 'will' of Ananke which previously held the universe in disorder).[538] The human soul is also described in the *Timaeus* as possessing a 'will' or of desiring good ends and of attempting to bring these about, although usually in the somewhat weaker guise of 'not desiring' bad ends.[539] Thus, having been generated ('begotten') by the Demiurge and

538. *Ti.* 41a7-b6.

539. See *Ti.* 73a7 (ἀνυπήκοον), 'unwilling' as in 'disobedient'; see also *Ti.* 87b3 (ἀκουσιώτατα), 'involuntary.' The idea of the human soul possessing a 'will,' although present, is a fairly weak notion in the *Timaeus* and almost invariably it is the negative idea of 'not willing' that is being invoked. The reason for this is that for Plato the soul can only desire what is good ('evil' for man is ἄκοντι [*Ti.* 86e2] or something which is never achieved with human effort or toil, in other words, is never 'willed'). Moreover, because the soul's understanding of what is good for it is dependent upon its different parts exercising their corresponding motions proportionately, the soul sometimes mistakenly wills or desires what is evil or is not good for it. Accordingly, whenever these motions, due to unfavorable conditions, are imbalanced, they suggest to the soul equally imbalanced ends from which

subject to the principle of 'like from like,' the human 'will' shares infinitesimally with the divine will, limited by its own finiteness and human condition, but like the Demiurge, always desiring good and never its own destruction. The human soul's capacity, however, to know what is good for itself is described by Plato as being greatly limited due to both its incarnate state and its later sloth, resulting in ignorance and error and accordingly, requiring careful tending. The idea of 'will' is not developed further in the *Timaeus.* It reappears, however, in *Laws X,* where the idea of 'will' or the desire and power to bring about certain ends, good or evil, depending on the agent, is again introduced, but this time brokered by beneficent and non-beneficent souls interacting with an already generated Cosmos. Again, there is a call to careful vigilance, but this time not towards one's own soul, but with respect to the outside influence of other souls. Thus, human souls continue to desire good ends, but depending on their alignment with wisdom or folly, either achieve these or fail. Although 'soul' is referred to in *Laws X,* and not specifically 'will,' this aspect or function of the human soul is most certainly meant. Later in *Laws X* Plato will point to the source of this wisdom or folly, where order and disorder have now become immanently infused into daily life by the assistance of powerful souls robustly or dangerously present.

2.3.6.2 Primal Souls (causal), not begotten

The idea of primal souls existing within the Cosmos itself as 'self-moved movers' does not contradict or duplicate the idea of a World Soul also being present, nor does the concept of 'World Soul' in any way require modification, although as I have argued earlier, its existence simply appears to be assumed in *Laws X* and not discussed. The World Soul, as understood from the *Timaeus*, is the matrix of rationality that saturates the world with its countless natural ends. This idea is still extant in *Laws X* and does not extend to the activities ascribed to the Soul.[540] Likewise, the existence of individual human souls is also assumed in *Laws X*[541] and is mentioned, but only in passing, in Plato's description of the soul's ability to company with wisdom or folly. This ability of the human soul to

---

to choose its good, but falsely. Thus, the soul can inadvertently choose (will) an end which it does not desire, which is a corruption of the idea of 'will' understood as desire and the ability to achieve its object or end. The positive notion of 'willing,' however, does enter here indirectly with respect to the human soul, namely, in Plato's insistence that attention be given to the proper proportioning of the soul's various motions or in other words, by introducing the idea of *choosing* to keep the soul's motions in their proper balance by proper attendance, regardless of any unpropitious circumstances (*Ti.* 90c6-d7).

540. The 'Benevolent Soul' is to be understood as being synonymous with the 'Beneficent Soul.'

541. See *Lg.* 903d3-5.

sometimes choose wrongly was noted in the *Timaeus*, but with the soul's integrity stalwartly defended by Plato against ever knowingly choosing evil. The source of this evil, however innocently chosen, is left open. Imbalance, sloth, and overindulgence are descriptions of disorder, but do not name the cause. Equally, in *Laws X*, the source of the wisdom or folly that befall the human soul again raises questions. These questions and their answers, however, are anticipated by Plato as he begins to focus on and center his discussion upon the idea of soul as primal cause, presented now in the form of two distinct causal powers, the Beneficent and Non-beneficent Souls.

The provenance of the Beneficent and Non-beneficent Souls, or their ontological histories, although not explicitly stated, can be to some extent extracted and further hypothesized. With respect to Plato's metaphysics, their ontologies are, in a way, eschatological, not having existed in the *Timaeus* or the earliest stages of generation, appearing for the first time in the fully functional world of *Laws X*. An essential difference, therefore, between these Souls as distinct from the World Soul and the human soul, is that whereas it is congruent to understand these latter as having been begotten (generated) by the Demiurge and as existing in both the *Timaeus* and *Laws X*, their ontologies intact in both works and requiring no emendation, this is not the case for the Beneficent or Non-beneficent Souls, which did not apparently even exist in the *Timaeus*. Is their existence due to the earlier 'likely tale' having turned fanciful over time or forgetful or alternatively, found to be fanciful and thus, forgotten, so that *now* the truth of causality can be told? Such conclusions, I would caution, require specious partiality with regard to Plato's metaphysics, abandoning one part or another. The fourth proposal noted above, however, approaches Plato's metaphysics with an inclusive view of causality in mind and for this reason, I suggest, is preferable, requiring little, if any, amendment to Plato's causal theory. According to this theory, the primal (causal) souls introduced in *Laws X*, rather than being designed by the Demiurge (who would never design a non-beneficent soul, being unlike itself) or becoming *de facto* a replacement or refinement of the earlier notion of generative primal powers, are the psychic counterparts of the Demiurge and Ananke, noetic and anoetic respectively, as they manifest themselves in the generated world or in other words, as they extend themselves, not as 'unmoved movers' outside the universe, but as 'self-moved movers' within the universe after its completion.[542]

Although the idea of the primal powers extending their influence into the daily life of the generated universe was a possibility in the *Timaeus*, Plato would not have and did not develop this idea then, as it required a contemporary timeframe, which, being beyond the requirements of his cosmology, did not interest him. By the time Plato wrote *Laws X*, however, matters had changed. Contemporary life and its exigencies pressed down upon the philosopher. Moral vacuity

542. See *Lg.* 894e4-895a3.

in the form of false belief, agnosticism, and atheism demanded redress. It is within this redress and its appeal to reason that Plato chose to identify what he believed to be the sources of wisdom and folly within the world. This could not be the World Soul (the source of rationality within the world), nor could it be human souls (who sometimes mistook folly for wisdom and never knowingly chose 'evil'). Accordingly, Plato was moved to introduce new players into his post-generative causal scheme, the Beneficent and Non-beneficent Souls, as the causes, respectively, of good and evil within the Cosmos, while not concerning himself with how their causal agency tallied up with his earlier cosmology. The physical world of the *Timaeus* was far removed from the religious concerns of *Laws X*. Thus, here, for the moment at hand, Plato delved exclusively into the world of soul, where only moral truths mattered and their impact on the human soul.[543]

With regard to gaining an understanding, however, of what appear to be irresolvable differences in causal agency between these two works, this fourth proposal again provides clarity. According to this view, the Beneficent and Non-beneficent Souls are not only an extension of the key causal players from the *Timaeus* into the generated world, but it is also an appropriation of their powers, being an amalgamation of the immanence of soul within the universe with the powers of the causal forces outwith it. By this extension and amalgamation, moreover, Plato, in developing his metaphysical theory, is able to introduce his moral theory. In other words, primal metaphysical movement in the *Timaeus* becomes the source of wisdom and folly in *Laws X*, whereby the association of human souls with the Beneficent and Non-Beneficent Souls through works and conduct becomes moral choice. In this way continuity is achieved, but without destroying the integrity of either explanation, the cosmological or the religious.[544]

The idea of 'self-movingness,' both in the *Timaeus* and *Laws X*, appears as a feature of anything possessed of a soul, wherein both 'nous' and 'will' reside. The principle of being a 'mover,' however, unless it is an accessory cause or inanimate object,[545] requires something else. The primal powers in the *Timaeus*, the Demiurge and Ananke, did not possess a 'soul' or a body, yet they were movers. The Beneficent Soul and Non-beneficent Soul from *Laws X*, although souls, are also movers. What these 'movers' in fact share is the desire and ability to achieve ends, good or evil, independently of inhabiting a body themselves.

543. See the speech of the Athenian at *Lg.* 887b1-888d5: "the greatest of these . . . is that of thinking rightly about the gods and so living well or, the reverse," *Lg.* 888b2-4 in Taylor, 278.

544. By the end of *Laws X*, Plato refers to these primal Souls as 'gods,' omitting all further references to these gods as 'Souls.'

545. In this case the accessory cause or object, e.g. a hammer, is moved and in turn, moves something else, e.g. a nail. It has, however, no intrinsic power itself to move anything without first being moved.

Accordingly, the essential difference between these powers, I suggest, is one of aspect. When these primal forces, ultimately of order and disorder, are viewed outwith the organized Cosmos and understood in terms of generation, they are known as the Demiurge and Ananke. These same powers, however, manifest within the finished Cosmos and affecting its moral agency, become Beneficent and Non-beneficent Souls.

Plato's description of 'soul' in *Laws X*, quoted at §2.1/2(x) above (*Lg.* 896e8-897b5), is a general description, capturing the essence of soul as it manifests itself prodigiously and diversely within the Cosmos, whether (according to this fourth proposal) emanating directly from the Demiurge in the form of the Beneficent Soul, or being generated (begotten) as in the case of the human soul (where it will company with Wisdom or Folly, i.e. the Beneficent or Non-beneficent Souls), or conversely, emanating from Ananke (the Non-beneficent Soul), or again, simply manifesting itself as the World Soul. Although, there is a metaphorical or anthropomorphic component to the above-noted passage, even a generic quality, I would suggest that this is due to the generality of its appeal or its breadth, and not for any other reason. Thus, although admitting to all of the following *kinds* of souls, it is not the Beneficent Soul or the Non-beneficent, the World Soul or individual souls, that Plato is discussing here, rather, it is a passage attempting to capture the essence of 'soul' however it might manifest itself. The view that the Beneficent and Non-beneficent Souls represent for Plato the Demiurge and Ananke as they exist emergent in the generated Cosmos argues against the possibility that these souls are begotten (generated) by the Demiurge, as were, for example, the World Soul and individual souls in the *Timaeus*, since not only would this be impossible (the Demiurge could not generate anything unlike itself, i.e. a soul that was a source of folly), this is also prevented by the fact of their ontology, whereby 'soul' is described in similar terms to the primal movers, namely, as "the first cause of the birth and destruction of all things."[546] Although Plato describes 'soul' near the beginning of *Laws X* as being "one of the first creations before all physical things," this is not at odds with Plato's understanding of primal Souls, since the World Soul and individual souls are assumed to have been generated (begotten).[547] It is only later in *Laws X*, as Plato develops his idea of soul as being prior and as "either the best kind of soul or the other sort,"[548] that he introduces the notion of the Beneficent and Non-beneficent Souls. These Souls are not described as being either born or generated by Plato, since by his own account this kind of 'soul' is now defined as 'self-generating

546. See *Lg.* 891e4-9. In this passage the prior division between 'soul' (as now described) and the gods breaks down, preparing by the end of this work for a certain identity between them.

547. *Lg.* 892a2-b1.

548. *Lg.* 898c1-5.

motion.'[549] Neither the World Soul nor the human soul was described in the *Timaeus* as 'self-generating motion,' although they were recognized as the source of motion in living things. Finally, just before his discussion of 'the god' or 'gods' in *Laws X*, in what becomes a very strong move for Plato given his understanding of the generated (begotten) soul from the *Timaeus*, he soon identifies the Beneficent and Non-beneficent Souls or these 'self-moving movers' as gods.[550] In this way, Plato leaves broad rein for both the Beneficent Soul and the Non-beneficent Soul, now described as gods, to assume complete identity with the Demiurge and Ananke of the *Timaeus*, where, like their predecessors, these Primal Souls stretch back into infinity, ungenerated and unbegotten.

---

549. *Lg*. 895e10-896a2.

550. *Lg*. 899b3-10. See also *Lg*. 903e3-5, where, in Plato's discussion of the gods, he describes their work in exactly the same terms as he did earlier the Soul, namely, in their controlling the universe. Nowhere in *Laws X*, however, are these gods ever described in terms of having been generated, as were the heavenly bodies (understood as gods) in the *Timaeus*.

Chapter 3

3 *LG.* 897B7-D1 (omitting 897D2-898A2) *LG.* 898A3-C5

3.1/2 Original Greek/Translation

(i) *Lg.* 897b7-c2

ΑΘ. Πότερον οὖν δὴ ψυχῆς γένος ἐγκρατὲς οὐρανοῦ καὶ γῆς καὶ πάσης τῆς περιόδου γεγονέναι φῶμεν; τὸ φρόνιμον καὶ ἀρετῆς πλῆρες, ἢ τὸ μηδέτερα κεκτημένον; βούλεσθε οὖν πρὸς ταῦτα ὧδε ἀποκρινώμεθα;

Then which manner of soul, must we say, has control of heaven and earth and their whole circuit? That which is prudent and replete with goodness, or that which has neither virtue? Shall we, if you please, give the question this answer?

(ii) *Lg.* 897c4-9

ΑΘ. Εἰ μέν, ὦ θαυμάσιε, φῶμεν, ἡ σύμπασα οὐρανοῦ ὁδὸς ἅμα καὶ φορὰ καὶ τῶν ἐν αὐτῷ ὄντων ἁπάντων νοῦ κινήσει καὶ περιφορᾷ καὶ λογισμοῖς ὁμοίαν φύσιν ἔχει καὶ συγγενῶς ἔρχεται, δῆλον ὡς τὴν ἀρίστην ψυχὴν φατέον ἐπιμελεῖσθαι τοῦ κόσμου παντὸς καὶ ἄγειν αὐτὸν τὴν τοιαύτην ὁδὸν ἐκείνην.

Why, man, if the whole path and movement of heaven and all its contents are of like nature with the motion, revolution, and calculations of wisdom, and proceed after that kind, plainly we must say it is the supremely good soul that takes forethought for the universe and guides it along that path.

(iii) *Lg.* 897d1

ΑΘ. Εἰ δὲ μανικῶς τε καὶ ἀτάκτως ἔρχεται, τὴν κακήν.

But the evil [soul], if the procedure is distraught and without order.

(iv) *Lg.* 898a3-6

ΑΘ. Τούτοιν δὴ τοῖν κινήσεοιν τὴν ἐν ἑνὶ φερομένην ἀεὶ περί γέ τι μέσον ἀνάγκη κινεῖσθαι, τῶν ἐντόρνων οὖσαν μίμημά τι κύκλων, εἶναί τε αὐτὴν τῇ τοῦ νοῦ περιόδῳ πάντως ὡς δυνατὸν οἰκειοτάτην τε καὶ ὁμοίαν.

Of these two movements, that confined to one place must in every case be performed about a center, after the fashion of a well-turned cartwheel, and it is this which must surely have the closest affinity and resemblance that may be to the revolution of intelligence.

(v) *Lg.* 898a8-b3

ΑΘ. Τὸ κατὰ ταὐτὰ δήπου καὶ ὡσαύτως καὶ ἐν τῷ αὐτῷ καὶ περὶ τὰ αὐτὰ καὶ πρὸς τὰ αὐτὰ καὶ ἀνὰ λόγον καὶ τάξιν μίαν ἄμφω κινεῖσθαι λέγοντες, νοῦν τήν τε ἐν ἑνὶ φερομένην κίνησιν, [σφαίρας ἐντόρνου ἀπεικασμένα φοραῖς,] οὐκ ἄν ποτε φανεῖμεν φαῦλοι δημιουργοὶ λόγῳ καλῶν εἰκόνων.

Why, of course, that if we say that intelligence and movement performed in one place are both like the revolutions of a well-made globe, in moving regularly and uniformly in one compass about one center, and in one sense, according to one single law and plan, we need have no fear of proving unskilled artists in imagery.

(vi) *Lg.* 898b5-8

ΑΘ. Οὐκοῦν αὖ ἥ γε μηδέποτε ὡσαύτως μηδὲ κατὰ τὰ αὐτὰ μηδὲ ἐν ταὐτῷ μηδὲ περὶ ταὐτὰ μηδὲ πρὸς ταὐτὰ μηδ' ἐν ἑνὶ φερομένη μηδ' ἐν κόσμῳ μηδ' ἐν τάξει μηδὲ ἔν τινι λόγῳ κίνησις ἀνοίας ἂν ἁπάσης εἴη συγγενής;

And again, motion which is never regular or uniform, never in the same compass, or about the same center, or in one place, motion which has no order, plan, or law, will have kinship with folly of every kind.

(vii) *Lg.* 898c1-5

ΑΘ. Νῦν δὴ χαλεπὸν οὐδὲν ἔτι διαρρήδην εἰπεῖν ὡς, ἐπειδὴ ψυχὴ μέν ἐστιν ἡ περιάγουσα ἡμῖν πάντα, τὴν δὲ οὐρανοῦ περιφορὰν ἐξ ἀνάγκης περιάγειν φατέον ἐπιμελουμένην καὶ κοσμοῦσαν ἤτοι τὴν ἀρίστην ψυχὴν ἢ τὴν ἐναντίαν—

Now there can be no further obstacle to positive assertion, since we have found that it is soul which conducts the revolutions of all things, and are also bound to say that the soul by which the circle of the heavens is turned about with all foresight and order is either the supremely good, or its contrary—

3.3 Commentary and Notes

3.3.1 The nature of soul

These seven passages together raise and answer the question of the nature of soul, the existence of which Plato has just previously argued in *Laws X* as that whose activities are behind the various movements of the Cosmos. It is for the most part, however, an idea of soul quite removed from that of the *Timaeus*, finding few similarities with either the World Soul or the human soul. The key question Plato now asks is what manner of soul, that is, what *kind* of soul controls the universe, that which exhibits prudence and goodness or which has neither virtue? Thus, just as the Demiurge was the ultimate source of all ordered motion in the *Timaeus*, in *Laws X* Plato posits a soul, or possibly more than one, i.e. souls, within the generated universe as being responsible for all ordered motion. Similarly, just as Ananke was the source of all disordered motion before the generation of the Cosmos, Plato now assigns in *Laws X* a soul or souls as responsible for all disordered motion within the Cosmos. The 'likely tale' and its key players upon which Plato relied so heavily in order to explicate his cosmology in the *Timaeus* are now left within a distant past as no longer needed or referred to. Plato, however, once again requires to explain the causality of the Cosmos, which he does, but with new players and in slightly different terms. In *Laws X* Plato's major concern is not with the Cosmos *per se*, but with the soul of man in a Cosmos long generated and specifically, with man's diminishing belief in gods long held to be active in the world. Something has gone wrong within the soul of man, he contends, not to believe in the gods. Thus, it is now Plato's passionate concern to search out and name the source of this disbelief and in doing so, to appeal both to man's reason, which for Plato is all good and seeks only the soul's good, and to man's 'will,' but which is weakened and mistaken by the lower (but integrated) powers of soul, warning him ('man') of a non-benevolent source, its evil intent, and to stay clear.

There is a gap, however, between these two works, which Plato does not clearly articulate. Somewhere between his understanding of a generated soul and a soul which generates, or between the *Timaeus* and *Laws X*, is Plato's realization that there are more than mere vestigial remains of Ananke in the Cosmos, there is also a source of disorder or evil within the Cosmos, a powerful one, which is able to attach itself to and influence the soul of man if he ('man') is not careful. This raised a problem, however. If Ananke is not a soul, which at no time is Plato's claim, but if soul is the only source of non-physical or psychic movement within the Cosmos, which in both the *Timaeus* and *Laws X* is Plato's position, then the evil power now referred to is a soul. There is no other option. However, this soul cannot possibly be made by the Demiurge, who cannot create (or beget) anything evil or unlike itself. Therefore, this soul and equally the benevolent soul, who are never described in *Laws X* as generated, must each extend into that ancient past which Plato has left behind with the *Timaeus*.

In the *Timaeus* the question of the kind of soul would never have been raised, except perhaps to distinguish between the World Soul and the human soul as regards their function and separate ontologies. The soul that was generated, precisely *because* it was generated by an altogether good causal power or Demiurge, was entirely good, be it the World Soul or the human soul. Although Plato allowed in the *Timaeus* for vestiges of Ananke to remain within the Cosmos as residual effects of this disordering cause,[551] the principle of evil as an active power within the generated Cosmos had not been developed at the time of the writing of this earlier work. Rather, the 'likely tale' explains the remaining disorder within the generated Cosmos as a result of the disordering power, Ananke, having not been wholly subdued and thus, as primarily residual. It is possible that Plato did not believe that such a separate power existed within the world prior to his writing of *Laws X*, believing earlier only in the vestigial remnants of an ancient disordered past and equally disordering power, but not for the power itself to have entered into the worldly domain. The World Soul was no doubt entirely good, having been generated by the Demiurge, and as explained in the last chapter, can still be presumed to be extant in the generated

551. See my discussion of the phrase "as far as he could" in Part I, §2.3.4.1–5; see also §3.3.1 of Part I concerning the idea of Ananke being "persuaded" and my earlier discussion at §1.3.20 on the "persuasion" being only "for the most part."

world in *Laws X*, although not specifically referred to, being the underlying rationality in the natural world and never straying from this role. Moreover, although the World Soul is described in the *Timaeus* as the cause of motion in the world, it is never described as a primal causal power or an independent power capable of anything other than natural ordering. It does nothing but execute the rationality of the Demiurge within the natural world and is given precisely this role of bringing about ends within the world and nothing else. The human soul, moreover, is not influenced by the World Soul, being an entirely separate noetic entity and having its own realm of ordering. Plato's introduction in *Laws X*, therefore, of causal souls empowered to interact with human souls, thereby influencing them towards good or evil ends, is an entirely new idea for Plato, not found in the *Timaeus* or felt even necessary to develop until now.

The passages cited above from *Laws X* include powers of soul, which in the *Timaeus* lie entirely within the scope of the World Soul, with one exception. Although Plato would ascribe goodness to the World Soul, never would he ascribe to it the prudence or foresight of discerning the good ends that are to be carried out. This power belongs solely to the Demiurge, who, as the source of all goodness discerns what is good and disseminates the power to carry out these ends to others, in this case, the World Soul. I would suggest, therefore, that in *Laws X* an amalgamation of players and roles has taken place. Thus, the role of the World Soul, rather than being articulated, is now subsumed under the powers of the Beneficent Soul(s), just as the Demiurge is drawn upon as its unspoken primal source outwith the Cosmos. Accordingly, the metaphysical realm in *Laws X* becomes less populated as its characters become more complicated. A similar amalgamation also takes place between the vestigial remains of Ananke under the auspices of the Non-beneficent Soul, as indeed the Non-beneficent Soul echoes similar refrains of disorder as did its ancient primogenitor outwith the Cosmos, Ananke. The human soul in *Laws X* is also similarly affected, no longer understood as tripartite, but whole and complexly so, free of easy divisions. The entire mood of *Laws X* with regard to its metaphysics, therefore, carries an undertone of worldly sophistication and a slightly modern touch. The divides of category and characterization permissible of a perspective speculating from beyond the generated world have been removed, giving voice to a distinctively Platonic trait found in both the *Timaeus* and *Laws X*, that all elements be pertinent to the story at hand and to ensure this, when necessary, telling the story anew.

3.3.2 The hypothetical non-virtuous soul

In this retelling, therefore, there is also another element which is introduced, but at first only hinted at and whose immediate presence, at least with regard to the ordering of the Cosmos, is flatly denied and that is the presence of the hypothetical non-virtuous soul. Plato asks in the first passage cited above whether a soul without prudence or goodness could possibly control the heavens and their circuit. The answer is, of course, no, where he explains at passage (ii) that the motions in question proceed of like nature and kind requiring forethought and the wisdom of calculation, and therefore require a supremely good soul and no other kind. We are reminded here of the *Timaeus* in which, until the Demiurge had set the world in order according to number and calculation,[552] there had been no such order. In *Laws X* we are now told similarly, but not in detail and this time with the supremely good soul held responsible for this order. Denied as ever being a source of order in the prior two passages, but which is nonetheless fully articulated in passage (iii), is the notion of an evil soul existent in the world as the source of disorder or any procedure which is distraught (imbalanced) and without order. Thus, rather than settling with the answer to the source of order in the world, Plato chooses to introduce a power capable of the very opposite causal effect. Moreover, he further prepares the reader for the possible existence of such a soul as he compares these opposite powers. Accordingly, in passages (iv) and (v) the motions of the supremely good soul, as distinct from the non-virtuous soul, are described as being akin to that of a well-turned cartwheel and thereby having the closer affinity to the revolutions of intelligence (than that of the so far hypothetical non-virtuous soul, which presumably would only be akin to something without form and quite useless). In the framing of the motions of the universe at least, posits the Athenian, we have no fear of an unskilled artist at work, which ordered motions are perceived as the handiwork of a supremely good soul. According to *Laws X*, therefore, the foresight of the Demiurge in its ordering of the world as understood from the *Timaeus*, along with its work carried out by the World Soul, both remain intact, their respective powers (that of the Demiurge and the World Soul) having in no way been breached, but which are now understood, rather, from *within* the world as belonging to, and thus acquiesced by (when compared), the composite notion of a supremely good soul.

3.3.3 The Non-beneficent Soul, the hypothetical becomes real

It is within the next two passages, (vi) and (vii), where Plato moves to a more serious consideration of the irregular motions previously mentioned and their source, making way for his pending introduction of the Non-beneficent Soul within his metaphysical scheme and Cosmos, in addition to the supremely good or Beneficent Soul. Plato does not allow for the revolutions of the heavens to be influenced in any way by disorder. However, in passage (vi) he does admit to the possibility of motion which is never regular or uniform existing separately from the ordered motion of the Cosmos and the ordered revolutions of the heavens, akin, he suggests, to every kind of folly in the world, with foolishness in the moral order reflecting similar confusion and lawlessness as disorder in the

552. *Ti.* 53a7-b6.

physical realm. In passage (vii), however, Plato reaffirms that it is "soul which conducts the revolutions of all things" and as a consequence, what is ordered (or disordered) in the world must be ordered by either the "supremely good soul" or "its contrary." Having already admitted to the possibility of disordered motion in the world, it takes very few further steps for Plato to also introduce the existence of this "contrary" power. Thus, soon to be introduced as the disordering or Non-beneficent Soul, this metaphysical power is no longer hypothetical, but emerges fully existent and powerful, perhaps, Plato suggests, even more powerful than the Beneficent Soul.

Chapter 4

4 *LG.* 898D9-E3
LG. 899B3-10
LG. 905D8-E3
LG. 906A2-C6

4.1/2 Original Greek/Translation

(i) *Lg.* 898d9-e3

ΑΘ. Ἡλίου πᾶς ἄνθρωπος σῶμα μὲν ὁρᾷ, ψυχὴν δὲ οὐδείς· οὐδὲ γὰρ ἄλλου σώματος οὐδενὸς οὔτε ζῶντος οὔτε ἀποθνῄσκοντος τῶν ζῴων, ἀλλὰ ἐλπὶς πολλὴ τὸ παράπαν τὸ γένος ἡμῖν τοῦτο ἀναίσθητον πάσαις ταῖς τοῦ σώματος αἰσθήσεσι περιπεφυκέναι, νοητὸν δ' εἶναι. νῷ μόνῳ δὴ καὶ διανοήματι λάβωμεν αὐτοῦ πέρι τὸ τοιόνδε.

The sun, whose body can be seen by any man, but his soul by no man, any more than that of any other creature's body is to be seen, during life or at the time of death. We have every reason to believe that it enfolds us in a fashion utterly imperceptible to all bodily senses, and is only to be discerned by the understanding. So here is a relevant consideration, which we must apprehend by an act of pure understanding and thought.

(ii) *Lg.* 899b3-10

ΑΘ. Ἄστρων δὴ πέρι πάντων καὶ σελήνης, ἐνιαυτῶν τε καὶ μηνῶν καὶ πασῶν ὡρῶν πέρι, τίνα ἄλλον λόγον ἐροῦμεν ἢ τὸν αὐτὸν τοῦτον, ὡς ἐπειδὴ ψυχὴ μὲν ἢ ψυχαὶ πάντων τούτων αἴτιαι ἐφάνησαν, ἀγαθαὶ δὲ πᾶσαν ἀρετήν, θεοὺς αὐτὰς εἶναι φήσομεν, εἴτε ἐν σώμασιν ἐνοῦσαι, ζῷα ὄντα, κοσμοῦσιν πάντα οὐρανόν, εἴτε ὅπῃ τε καὶ ὅπως; ἔσθ' ὅστις ταῦτα ὁμολογῶν ὑπομενεῖ[553] μὴ "θεῶν εἶναι πλήρη πάντα" ;

Of all the planets, of the moon, of years and months and all seasons, what other story shall we have to tell than just this same, that since soul, or souls, and those souls good with perfect goodness, have proved to be the causes of all, these souls we hold to be gods, whether they direct the universe by inhabiting bodies, like animated beings, or whatever the manner of their action? Will any man who shares this belief bear to hear it said that all things are not 'full of gods'?

(iii) *Lg.* 905d8-e3

ΑΘ. Φέρε δὴ πρὸς θεῶν αὐτῶν, τίνα τρόπον παραιτητοὶ γίγνοιντ' ἂν ἡμῖν, εἰ γίγνοιντο αὖ; καὶ τίνες ἢ ποῖοί τινες ὄντες; ἄρχοντας μὲν ἀναγκαῖόν που γίγνεσθαι τούς γε διοικήσοντας τὸν ἅπαντα ἐντελεχῶς οὐρανόν.

Why, then, I ask you, in the name of these same gods, what can be the mode of the perversion, if indeed they are to be perverted? And what or what manner of beings must they be themselves? Governors, to be sure, they must be supposed to be, if they are to have effective control of the whole universe.

---

553. England explains (482) in his footnote b 8 f. that "the MSS. probably had ὁμολογῶν ὑπομενεῖ, and not ὁμολογεῖ, divined what all subsequent editors . . . hold to be the right reading, i.e. ἔσθ' ὅστις ταῦτα ὁμολογῶν ὑπομενεῖ, which I have retained.

(iv) *Lg.* 906a2-c6

[ΑΘ.] ( ). ἐπειδὴ γὰρ συγκεχωρήκαμεν ἡμῖν αὐτοῖς εἶναι μὲν τὸν οὐρανὸν πολλῶν μεστὸν ἀγαθῶν, εἶναι δὲ καὶ τῶν ἐναντίων, πλειόνων δὲ τῶν μή, μάχη δή, φαμέν, ἀθάνατός ἐσθ' ἡ τοιαύτη καὶ φυλακῆς θαυμαστῆς δεομένη, σύμμαχοι δὲ ἡμῖν θεοί τε ἅμα καὶ δαίμονες, ἡμεῖς δ' αὖ κτῆμα θεῶν καὶ δαιμόνων· φθείρει δὲ ἡμᾶς ἀδικία καὶ ὕβρις μετὰ ἀφροσύνης, σῴζει δὲ δικαιοσύνη καὶ σωφροσύνη μετὰ φρονήσεως, ἐν ταῖς τῶν θεῶν ἐμψύχοις οἰκοῦσαι δυνάμεσιν, βραχὺ δέ τι καὶ τῇδε ἄν τις τῶν τοιούτων ἐνοικοῦν ἡμῖν σαφὲς ἴδοι. ψυχαὶ δέ τινες ἐπὶ γῆς οἰκοῦσαι καὶ ἄδικον λῆμμα κεκτημέναι δῆλον ὅτι θηριώδεις, πρὸς τὰς τῶν φυλάκων ψυχὰς ἄρα κυνῶν ἢ τὰς τῶν νομέων ἢ πρὸς τὰς τῶν παντάπασιν ἀκροτάτων δεσποτῶν προσπίπτουσαι, πείθουσιν θωπείαις λόγων καὶ ἐν εὐκταίαις τισὶν ἐπῳδαῖς, ὡς αἱ φῆμαί φασιν αἱ τῶν κακῶν, ἐξεῖναι πλεονεκτοῦσιν σφίσιν ἐν ἀνθρώποις πάσχειν μηδὲν χαλεπόν· φαμὲν δ' εἶναί που τὸ νῦν ὀνομαζόμενον ἁμάρτημα, τὴν πλεονεξίαν, ἐν μὲν σαρκίνοις σώμασι νόσημα καλούμενον, ἐν δὲ ὥραις ἐτῶν καὶ ἐνιαυτοῖς λοιμόν, ἐν δὲ πόλεσι καὶ πολιτείαις τοῦτο αὐτό, ῥήματι μετεσχηματισμένον, ἀδικίαν.

For since, as we have agreed among ourselves, the world[554] is full of good things, but no less full of their contraries, and those that are amiss are the more numerous, the fight we have in mind is, we maintain, undying and calls for a wondrous watchfulness; gods and spirits are our allies in the warfare and we, moreover, the property of these gods and spirits. Wrong, arrogance, and folly are our undoing, righteousness, temperance, and wisdom our salvation, and these have their home in the living might of the gods, though some faint trace of them

---

554. I would suggest that 'world' here does not mean 'earth,' but carries the broader meaning of 'universe,' particularly the 'heavens' as in the 'seat' or 'dwelling place' of the gods above the firmament, wherein the gods ("souls") which bring about good and evil, are pictured to exist. Thus, in using the word 'οὐρανός' Plato is emphasizing a particular aspect of the world, i.e. its heavenly component.

is also plainly to be seen dwelling here within ourselves. Yet it should seem there are souls inhabiting our earth in possession of unrighteous spoil—bestial souls, these, beyond a doubt—who grovel before the souls of our guardians—watch-dogs, shepherds, supreme masters of all, alike—and would fain persuade them by fawning speeches and witcheries of supplication—such is the tale told by the wicked—that it is lawful for them to encroach upon mankind without grievous consequence. But our contention, I take it, is that this vice I have just named of encroachment when found in bodies of flesh and blood is what is called *disease*, when found in seasons and whole years, *pestilence*, while in societies and politics it shows itself once more under the changed designation of *iniquity*.

4.3 Commentary and Notes

4.3.1 A theory of disorder

The above four passages together summarize succinctly Plato's understanding or 'theory' of disorder: its epistemological origins, its metaphysical relationship with the generated world, and lastly, its effect upon the generated world. In response to the latter is also Plato's call to constant vigilance against this power(s) of disorder, understood now from a different perspective, thus, not as a brute disorganizing force outwith the universe, but as a Non-beneficent Soul(s) within the Cosmos fully capable of engaging the world in its ἁμάρτημα or erroneous ways.

4.3.2 An epistemology

In this first passage the Athenian explains that the soul is "utterly imperceptible" to all bodily senses. As such, 'soul' can only be discerned by the understanding, and further, *apprehended* by an act of pure understanding and thought. In the *Timaeus*, through similar acts of understanding (reasoned discourse setting out premises leading to a conclusion, which in turn may become a further premise), Plato had discerned both the existence of a World Soul, as well as human souls, hypothesizing in each case their generation and existence. A full mapping of 'soul,' however, was not Plato's principal concern in this work. Rather, his main project here was a description of the Cosmos as a whole, from its very early beginnings down to the origins of man, which included the postulation of 'soul,' but little beyond that. To achieve this Plato thoroughly ensconced his cosmology in the imagery of story-telling, importantly 'likely' storytelling, theorizing not only the existence of souls, but also a maker of souls and of gods, the Demiurge, and postulating a power which sustained original chaos and constantly sought its return, Ananke. Thus, the work of apprehending 'soul' or of acquiring a *deeper* knowledge of it in the light of "pure" reasoning was inhibited by the necessity and prerequisite of telling the larger story of the Cosmos through myth-making and the telling of a 'likely tale.' Accordingly, it was not possible to instigate an inquiry into a purer understanding of 'soul' without compromising these venerable, albeit inexact, modes of description. This imagery made possible the working out of a vastly larger subject, namely, a cosmology of the whole universe, in which both its physics and its metaphysics could be broadly depicted. If Plato wanted a more exact understanding or comprehension of 'soul,' this would have to wait. It is also possible that at the time of writing the *Timaeus* Plato did not feel a need to develop his theory of soul to the degree that he does in *Laws X*. Nonetheless, the *Timaeus*, through the hypothesizing of both souls and the metaphysical players which preceded them—the Demiurge and Ananke—paved the way for his more formal investigation of 'soul' in *Laws X*. Now, not only would Plato's understanding of the soul have a more thoroughgoing rational base, but importantly, *because* of this, it would also become more inclusive.

Plato's claim in *Laws X* is that the soul cannot be discerned or known to exist through the senses and to this degree he is in accord with his supposed opponents (non-believers or those who believe erroneously). On this basis, however, Plato does not then deny the existence of 'soul' altogether (which 'souls' he will later identify with and refer to as 'gods'), as do certain "young men of Athens," whom he points out and so bitterly opposes. Rather, Plato makes the further claim that 'soul' can indeed be apprehended, but *only* through pure understanding. This proposal goes beyond what he was able to say in the *Timaeus*. In this earlier work all his metaphysical players were hypothesized and the rationale that was brought to bear bore a certain congruence with "likeliness" as it related to the world, thus, was a component of rational imagery, but was not an act of pure reasoning *per se*. The rational imagery that Plato used, for example, was of the sort: all things that have discernible ends have a maker; the Cosmos is comprised almost wholly of discernible ends; therefore, the Cosmos also had a maker. The same sort of reasoned imagery was also behind Plato's postulation of the figure of Ananke, i.e. since all intelligent things are designed and being designed, have a beginning, then the world had a beginning. If the world had a beginning, then what existed beforehand could not have been designed and what was not designed was disordered. Since there is a power behind all movement, there is also a power behind disordered movement, namely, Ananke. Thus, for his imagery of disorder outwith the world, Plato has alluded to, as its basis, the imagery of power within the world. All things that move in the world, whether the end of this movement is organized or random, move not of their own accord, but through the causal force of some power.

What Plato does in *Laws X* is to add to this vision of movement in the world a rational basis or, as he writes, an apprehension of it through "pure understanding." The end result of this new investigation becomes his acknowledgment of 'soul(s),' both as primal powers and as individual movers, within the Cosmos. Moreover, because Plato is intending to use the argument which follows from this as his basis for postulating the existence of metaphysical powers which are active *within* the universe, in order to avoid circular reasoning[555] it becomes necessary for him to abandon the notion of "likeliness" or as I have named it, 'reasoned imagery,' in favor of reasoning alone. He therefore now proposes and deduces premises, which earlier had been merely aspects of imagery, within the context of argumentation. In this way he lays the foundation for a "pure understanding" with regard to 'soul,' which ultimately also becomes the basis of his argument for the existence of the 'gods.' Accordingly, Plato offers a proof for the existence of the gods, first, by proving the priority of soul, second, by defining soul in terms of being a 'self-moved mover,' and finally, by equating our understanding of the gods with this type of movement. After satisfying himself that he has achieved a satisfactory proof, he then drops all further references to 'soul,' using 'god' instead, thereby equating the meaning of these two terms by apposition, simply exchanging the word 'god' where earlier he would have used the word 'soul.'

4.3.3 Arguing for the divine and primal souls – an ontology

In the second passage quoted above Plato pleads passionately for confidence in the belief that the gods must exist, providing various arguments for the basis of this confidence. There are four minor arguments in *Laws X*, which Plato first puts forward as proofs for the existence and goodness of the gods before embarking on his main argument, the priority of soul, which will take up the rest of *Laws X*. These initial arguments are at best weak and at worst false, and include the following: (1) the cosmological: observation of the earth, the sun, and the planets, and the ordering and beauty of the seasons, the implication being that design and intelligence are behind this (*Lg.* 886a2-6); (2) the belief that all humankind, Greeks and non-Greeks alike, believe in the gods, from which Plato derives an argument from numbers: not everyone can be wrong (ibid.); (3) an argument from faith: non-believers lack faith (*Lg.* 887c5-888a7); and finally, (4) an argument from tradition: "the spectacle of parents addressing their gods[;] . . . they have heard the universal prostrations and devotions of mankind" (ibid.).[556] In this way Plato leads up to what he considers to be the most persuasive argument in support of the "old traditional belief in the gods" (*Lg.* 890d1-8), namely, (5) the primacy of the soul as "elder-born than all bodies and prime source of all

555. Whereas in the *Timaeus* Plato used the imagery of force in the world to postulate by analogy force outwith the world, he now needs a stronger basis for his argument, since in *Laws X* he is trying to postulate the existence of metaphysical forces *within* the world as well. Thus, it does not suffice as an argument to move by analogy from physical force in the world to metaphysical force also within the world—something more is needed.

556. This fourth argument is possibly a fallacy of the type: A is a reliable authority concerning P; A asserts P; therefore P, also known in informal logic as the *ad Verecundiam* fallacy.

their changes and transformations" (*Lg.* 892a2-b2). In this final and main argument he proposes that all which is akin to soul must therefore be prior to all which is proper to bodies (ibid.). He supports his position by first equating the word 'soul' with the definition of "being alive" or "self-movement" (*Lg.* 895e10-896a4). Thus, by identifying 'soul' with being alive, 'soul' is then argued to be the source of all movement (*Lg.* 896a5-b1) and hence, prior. Moreover, the soul that has control over the heavens and earth and their whole circuit is the supremely good soul (*Lg.* 897b7-c9). Hence, this soul and the other soul(s) directing the universe, either by inhabiting bodies like animated beings, by acquiring their own body, or as entirely immaterial beings, guiding the heavens by some wonderful power, give evidence that all things are "full of gods" (*Lg.* 898e8-899b10).

Plato's principal argument for the existence of the gods in *Laws X* ends by equating god with soul after first defining 'soul' as being alive or a 'self-moved mover.' In the *Timaeus*, however, Plato did not equate the primal powers, the Demiurge or Ananke, with soul, although such an inference would not have been unreasonable given that the human soul, although formed by the Demiurge, was the one immortal, divine part that he bestowed to human beings, and thus had a divine origin (*Ti.* 69c3-6). Instead, in this work, God was defined as the "best of causes" (*Ti.* 29a5-6) and "the best of intelligible and eternal things" (*Ti.* 37a1-2), by whose agency soul ultimately came into existence. By the time Plato wrote *Laws X*, however, the idea of 'God' as the *source* of the "one immortal divine part" (or 'soul') and as the "best of intelligible and eternal things" had met a further development since his writing of the *Timaeus*. God is now understood by Plato to be an active power *within* the Cosmos as well[557] and thus is no longer limited to the role of Demiurge or in solely having generated the universe. A change of perspective also takes place in *Laws X*, from that of timeless eternity to timely events, with the consequence that in this retelling the 'unmoved mover' of generation ultimately gives way to the 'self-moved mover' of regeneration.[558] Accordingly, God is now identified by Plato within the world as "the supremely good soul," since it is only the self-moved (by definition) which is able to move and direct the whole universe from within, becoming in this sense both the Cause within the universe and the "universal cause."

This is not to say that the World Soul of the *Timaeus* has been removed from *Laws X*. Rather, I suggest, it has been surpassed in importance by a dynamic new notion of the divine Cause as 'soul,' not earlier envisaged or indeed required in Plato's nascent cosmology. The World Soul would still, therefore, be understood by Plato as having been constructed (understood differently, how-

---

557. Plato never denies, however, the existence or role of the Demiurge as an 'unmoved mover' in the generation of the Cosmos.

558. See my discussion at §2.3.2.1.3 on 'Physical movement and the question of dualism.'

ever, from the physical generation of things within the Cosmos) by the Demiurge and "appointed" the particulars of its motion (i.e. the direction and speed of its circular strips)[559] when the Cosmos was first made (at the very least, not denied), thereby accounting for its rational basis. By the time of *Laws X*, however, the world that is being addressed has long been operational as an intelligible entity, its reasonableness not in question. What Plato attempts to account for here is the unreasonable behavior of man, whose noetic powers are governed not by the World Soul, but each by his own individual soul and the soul's will. The World Soul, therefore, is allowed to slip away unaddressed into obscurity. There are power(s), however, according to Plato, which are capable of interacting with and thus influencing the human soul, namely, the primal powers themselves (formerly understood as the Demiurge and Ananke in the *Timaeus*), but transformed into supremely good and bad souls post-generatively, bringing to bear the principle of 'like with like' as regards their influence. In this way Plato is able to account for the *freedom* of human souls to make rational and irrational choices (the irrational choices mistakenly thought to be rational or good with respect to the individual).[560] It is this noetic 'freedom,' moreover, existing within the world through the human soul, which is subsequently focused upon.

559. See *Ti.* 36c2-d7. This physical description forms part of Plato's mythic imagery. The nature of soul at all times remains entirely psychical and its movement, noetic. See §3.3.1 of Part II. Moreover, since 'soul' is not a physical object, there is no question of it being "made to move." See my later discussion in Part II at § 4.3.6, 'Primal powers and primal souls.'

560. The determinism that Ritter understands to be present in Plato's philosophy is not supported by the *Timaeus* or *Laws X*, except when it is subject to an interpretation leaving out the fuller picture (see Ritter, *The Essence of Plato's Philosophy*, 381 ff.; see also 389–90). He writes: "With the purpose of establishing the perfect happiness of the whole, the fundamental laws of all Being and Becoming are *so determined* [italics my own] that the infinite, individual motions and changes of each thing follow of themselves from them, and that all the consequences ensuing from these motions result from meeting other changing objects. Even freedom of choice, which the human soul retains, cannot interfere with the fixed, divine order of things, cannot annul divine plans. Every decision and action, which transforms the soul, in that they move it, have predetermined consequences for it. At the most, there remains for the demiurge, who beholds human activity which takes place within fixed bounds, the task of moving the individual soul from place to place, as a player moves a pawn in chess in accordance with rules which he himself has laid down." This analogy of a "divine Chess Player" is also picked up by T. M. Robinson (*Plato's Psychology*, Toronto: University of Toronto Press, 1970, 153) when trying to understand Plato's notion of the god, understood as the Demiurge's counterpart, in *Laws X* (*Lg.* 902e4-5). I would suggest that the determinism which is argued by Ritter to be in place in Plato's metaphysics, depicted through this 'divine chess player' metaphor, is not what Plato had in mind in discussing either the "superintendence of the gods" or a god whose work is "not inferior to human workmen." Rather, Plato's purpose in drawing up his cosmology as being the work of a divine craftsman was to show the noetic nature of the universe, which he could demonstrate, but only for the most part, as well as to depict its causal foundation. The idea of a 'will' or the active participation by a generated

With this further idea of 'freedom' in mind, therefore, in the form of participation in or departure from reason ('rational freedom'), not only is the idea of God as 'Soul' introduced in *Laws X*, but so, too, is the idea of Ananke as 'Soul,' whereby Plato acknowledges the existence of and introduces the notion of the 'non-beneficent' or 'evil' soul(s). As with the Demiurge, Ananke is primarily understood in the *Timaeus* as a causal force outwith the universe,[561] with only vestiges of its power for disorder remaining within the generated Cosmos. In *Laws X*, however, Ananke re-emerges as a self-moving cause or 'Soul,' whose power and effects, like those of the Demiurge (also re-emergent as 'Soul'), are now felt dynamically within the Cosmos. In the *Timaeus*, moreover, as previously explained, although Ananke is presented as a co-generator of the Cosmos along with the Demiurge, it is nonetheless considered to be a lesser power, since it was ultimately subdued (for the most part) by the Demiurge upon order taking precedence over disorder. In *Laws X*, however, the disordering power of Ananke is set up in direct contrast to the supremely good soul, now also cast as a soul, but in keeping with its nature, one which brings about results "without forethought" (*Lg.* 897c4-9) and which are "contrary to beneficence" (*Lg.* 896e4-6). Thus, Ananke, as a metaphysical power, is ultimately also understood by Plato to be active within the world and is accordingly transformed into a 'Non-beneficent Soul' which "effects evil and disorder" (*Lg.* 897d1), challenging its earlier position as a lesser power.

---

being in that noetic structure is a different matter. If a 'divine chess player' analogy is to be employed, then it would have to be a version of the game where the 'soul stuff' (whatever that would entail) of the divine chess player was given to the chessmen by the divine chess player, so that they could discover the rules and follow them or mess up the game (by ignorance or design) if they so chose. That Plato did not and *could* not highlight this aspect of the 'divine game' in the *Timaeus* is clear—without a greater development of the idea of 'soul' he did not have the means to do this. However, by the time Plato wrote *Laws X*, the notion of 'soul' was more fully developed and the divine chess game could now be taken over by the divinely wrought chessmen themselves. I would also suggest in this divine chess game that for Plato there is not one, but *two* chess players, the Demiurge and Ananke, or by the time he wrote *Laws X*, the Beneficent Soul and Non-beneficent Soul, who avail themselves noetically (but *how* Plato can only venture) to the chessmen.

561. The primary power from the *Timaeus* has also been understood as the 'Unmoved Mover.' This concept, I suggest, has been notoriously misunderstood, where attempts are made to trace all subsequent physical movement within the Cosmos back to the first physical movement initiated by a metaphysical God or Demiurge. What results is that either the pure dualism inherent in Plato's cosmology is compromised or alternatively, in opting for a mythic interpretation of the God and for discrete metaphysical powers generally, it is denied. Against this, however, is the fact that nowhere in the *Timaeus* does Plato offer support for such compromise and in *Laws X* he vehemently argues against the non-existence of the gods. If Plato's God is to be accurately described as an 'Unmoved Mover,' this movement must be strictly understood solely in the sense of noetic movement.

4.3.4 The metaphysics of disorder

In the third passage quoted above, Plato is disarming the last of the arguments which had been put to him concerning the nature of the gods, namely, that the gods can be perverted. To achieve this he first offers an argument for why the gods are good. Their goodness, he explains, pertains to them by definition of what is good or beneficent, which is always, according to Plato, some activity that brings about an ordered or rationally discernible end, which end is understood as a particular good with respect to that activity. In the case of the gods, their 'goodness' (virtuous activity) takes the form of watching over the entire universe, which is both their principal virtue and principal activity. It is within their overall care of the universe, moreover, that all particular acts of caring lie, with the latter constituting the former. The gods, therefore, Plato argues, must give greater attention to small things than to great,[562]since it is only through instances of individual care, such as caring for mankind, of which the totality of the gods' care is comprised.[563] In other words, you cannot care for the whole without first caring for its individual parts. Accordingly, in order to care for the generated universe, the gods must first care for its many ordered parts, of which mankind forms one.

Thus far Plato has argued that the gods are completely good and as an expression of their perfect virtue, their special job is to watch over the universe. However, when he begins to name the particular virtues which qualify the gods as being good, e.g. moderation, possession of reason, courage, i.e. naming their 'admirable' qualities, thereby distinguishing these from those which are their opposite or vices (i.e. those qualities which are 'disgraceful'), Plato, by default, is forced to introduce into his post-generative metaphysics the source of these vices or this evil and posit its existence, since not only are these disgraceful traits the proposed opposites of admirable qualities, they are clearly present in man and must have a source. He denies that any of these negative qualities could ever belong to the gods or yet other bad qualities, such as neglect and idleness or riotous living (or indeed any of the 'disgraceful' qualities), since the gods, it is argued, are defined by their good qualities and could not love (and therefore possess) what is opposite in nature to their being. Mankind, moreover, being the handiwork of the gods,[564] although displaying at times these bad qualities and thus, participating in evil, could never itself be the source of that evil. As a con-

562. *Lg.* 900c8-d3.

563. *Lg.* 902e4-903a3.

564. Whereas in the *Timaeus* man is understood to be generated by the Demiurge (God), who is a noetic, organizing power external to the world, in *Laws X* the primal causal power is understood to be active within the universe as a Self-moved Mover in the form of a Beneficent Soul, since according to Plato's understanding it is 'soul' alone which actively participates *within* the generated universe (Cosmos) as a noetic causal force.

sequence, Plato must look elsewhere for a source involving neither man nor gods.

Within his argument supporting the position that the gods care for mankind, Plato has also acknowledged as premises (1) that the gods know, see, and hear everything, and that nothing within the range of our senses or intelligence escapes them;[565] (2) that they can do anything within the power of mortals and immortals;[566] and finally, his earlier agreed-upon premise, namely, (3) that the gods are supremely good.[567] In arguing now for what is effectively the omnipresence, omniscience, omnipotence, and the goodness of the gods, Plato does not at the same time deny the existence, and hence, a source for, the aforementioned 'disgraceful' qualities, having already admitted that they exist at the very least in mankind. Thus, on account of his acknowledgment of the existence of these "disgraceful" qualities in mankind, he must also acknowledge a source for this evil. What had earlier been tacitly accepted by Plato, that evil exists in the world, is now searched out as to its source.

Mankind, as previously noted, on account of its generation by the Demiurge (now understood from the perspective of a Beneficent Soul) is exonerated from ever being a source of its own evil, except perhaps as an unwitting accomplice in seeking out what is wrongly thought to be good. The gods, too, are made exempt by the fact of their nature having been qualified as 'goodness' itself. Accordingly, in *Laws X* Plato must look not to a generated being, but to another causal power within his metaphysics[568] as the source of all instances of disorder or 'evil' within the world. As a consequence, this power for disorder, understood earlier outwith the Cosmos as Ananke, is subsequently encountered within the world as the Non-beneficent Soul.[569]

---

565. *Lg.* 901d2-6.

566. *Lg.* 901d8-9.

567. *Lg.* 901e1-2.

568. As in the *Timaeus*, in *Laws X* all secondary causal movement, which is inclusive of all physical movement, is dependent on a primary metaphysical source, be it for order or disorder, within the Cosmos, since, as the name implies, i.e. in being "secondary," secondary causes have no true causal power of their own. In the *Timaeus* these primary sources of movement were understood to be the Demiurge (and its use of 'soul') and Ananke respectively and in *Laws X*, the Beneficent and Non-beneficent Souls. In both works, moreover, in the *Timaeus* explicitly and in *Laws X* implicitly, the physical movement of the universe and its basic building blocks or triangular material are understood to have always existed. See *Lg.* 895a1-b7. This secondary movement, therefore, is not considered by Plato to be initiated by its primary source, but rather directed to some end through the workings of 'soul' or alternatively, not so directed (i.e. when Ananke is the source).

569. In the *Timaeus*, although Plato acknowledged that remnants of disorder remained in the world (Ananke was not completely subdued by the Demiurge) he did not acknowledge any further activity of either causal power within the Cosmos. In fact they drop out of the story completely once the Cosmos has been generated, with the World

4.3.5 Evil and the generated world

In this fourth and final passage for consideration Plato confronts what he views as the "contrary" to those good things, which, along with the good, exist in the world. He states that not only is the world as full of the things which are amiss as it is of the good, but that the "bad" are more numerous. In so stating, he has moved far from his earlier position in the *Timaeus*, removing all complacency that in presuming the good to be more prolific, they necessarily prevail. In the *Timaeus*, where the ordering of the universe and its generation are the chief concerns, Plato's position was that, of the two primal powers, the Demiurge was a greater power than Ananke, having for the most part subdued the original chaos when organizing and generating it into a living Cosmos, with only remnants of disorder remaining. Thus, from the perspective of the *Timaeus*, at the outset of generation, the world was exceedingly good. There is no question in this work of "bad things" being addressed, let alone of their being prevalent. Moreover, the possibility of there being anything "contrary" to the good was only hinted at within the idea of there being a "remnant" of disorder left in the world, since no focus was given at this time by Plato to "disordered" things or to their realization. Accordingly, the moral consequence of the world being fully operational over time and of being divisible into good and bad things or events (the ordered versus the disordered) was not a consideration in the *Timaeus* and was not addressed by Plato until the Laws, where in *Laws X* he confronts his opponents against their non-belief or wrong belief in the gods.

In *Laws X* Plato's chief concern and hence, his perspective have now both changed. In this work, although he considers the world to still be full of good things, he nonetheless now considers there to be more "bad things" in it, the non-believer being a prime example of wrong thinking and the efficacy of evil. Thus, no longer content with viewing the world from outwith the universe through the idealistic eyes of a believer, this very world and perspective being challenged, his main job in *Laws X* is to take to task the non-believer or those who have wrong ideas about the gods. By viewing it through their eyes and with their understanding, from *within* the world, a world, moreover, presumed long in existence, he sets out to prove them wrong. To this view, however, he now adds a carefully chiseled argument concerning the existence of 'soul' and more importantly, powerful and influential 'Souls.' I would suggest that Plato's position with regard to the primal powers has not essentially changed, but rather, his un-

---

Soul taking over as the metaphysical power active within the world. In *Laws X*, however, the primal causal powers are now acknowledged to be active within the world in the form of good (beneficent) and bad (non-beneficent) souls. This, I suggest, is a further development of Plato's metaphysics, which was necessitated by his acknowledgment of ongoing evil in the world (no longer merely remnants of disorder, but of fresh disorder, particularly where the human soul was concerned). See *Lg.* 904a6-905d6.

derstanding has enlarged to include how they interact with the world. Thus, what has changed is Plato's perspective and because of this he introduces into his metaphysics once again the primal powers, but this time transformed, as 'Souls' interacting with the world they had earlier generated.[570]

4.3.6 Primal powers and primal souls – the persistence of evil

From the outset of the debate in *Laws X* Plato refuses to allow into the argument the notion that evil stems from weakness of will, i.e. of not choosing the good when a man knows what it is. Thus, he has the Athenian state, "You think it is only because they cannot resist temptation and desire that they are attracted to the godless life [referring to the atheists]."[571] Slowly, the remaining dialogue of *Laws X*, almost to its end, forges an answer to the question which Clinias raises back, "what other reason could there be, Sir?" Initially the Athenian answers Clinias with the obvious reason for disbelief, namely, the philosophic influence of the pundits, particularly the modern ones, where the gods and divine beings are said to be just "earth and stones."[572] However, in explaining what he believes to be the truth concerning the gods (which gods are presented initially as 'souls'), namely, that they exist, they care for mankind, and cannot be bribed (or in Clinias's terms, that they exist, are good, and respect justice more than men do),[573] Plato's inquiry soon points to a deeper reason for the non or wrong belief in the gods. To wit, by retaining from his earlier works, i.e. in the *Meno*, the view that men choose wrongly mistakenly thinking something is good when in fact it is evil[574] and in *The Republic*, the view that no god or man would of his own will worsen himself in any way,[575] and by adding to these views a belief in good and evil souls that exist prior to the material world,[576] Plato provides an argument for *how* wrong belief in the gods takes place. The true cause, he sug-

570. See *Lg.* 899b3-10 and 899c6-d2.
571. *Lg.* 886a6-b3.
572. *Lg.* 886b10-e2.
573. *Lg.* 887b5-8.
574. Plato, *Meno*, trans. W. K. Guthrie, in *Plato: Collected Dialogues*, eds. Edith Hamilton and Huntington Cairns (Princeton, NJ: Princeton University Press, 1971), 353–84. In this much earlier work, the *Meno*, Plato states his position regarding whether a man chooses evil. Through lengthy dialectic the central character, Meno, admits to the view that some people wish for evil things. Moreover, while some individuals suppose such evils to be good, others will still desire them although aware that they are evil (*Men.* 77c). Through Socrates's character, however, Plato opposes this view, when Socrates concludes that the people who suppose evils to be good are mistaken. Further, the people who desire something evil knowing that it is evil are unhappy, since this, he proclaims, is all that unhappiness is: "desiring evil things and getting them." He concludes that since no one wants to be unhappy and unfortunate, "nobody desires what is evil" (*Men.* 78b).
575. Plato, *Republic*, Bk. II, 381c in *Plato: Collected Dialogues*, 629.
576. *Lg.* 896e4-6.

gests, behind wrong belief in the gods is neither a weak will, nor simply belief in a false opinion. It is more complex than this. Rather, wrong belief is the result of the human soul freely participating in the same motion as an evil or non-beneficent soul, whose particular motion in this case is 'false opinion,' the human soul mistakenly thinking it to be a true opinion or good motion.[577] Plato's essential argument for positing at least one evil soul, in addition to the good, is that although 'soul' is the cause of all things,[578] just as good qualities can only be ascribed to a soul's virtue, so, too, bad qualities can only be ascribed to a soul's vice. Since God (identified with the Beneficent Soul) is perfectly good and is the cause of everything that is good,[579] he cannot have precisely the sort of character that he detests.[580] Therefore, the source of vice must lie elsewhere, which source is ultimately identified by Plato as being that of a Non-beneficent Soul(s).

To show this, Plato has had to redraw his metaphysical scheme, depicting it from a different perspective, with the primal powers now intervening in human affairs, where earlier they remained as pure forces outwith the generated world.[581] Hence, with respect to their perceived form but not, I suggest, to their essential form[582] or their power,[583] within the generated world the infinite (the

---

577. *Lg.* 896e8-897b5.

578. *Lg.* 896d5-8.

579. See also *Rep.*, Bk. II, 380c in *Plato: Collected Dialogues*, 629.

580. See *Lg.* 900e6-901a9.

581. See *Lg.* 891e4-9. In this passage Plato refers to the soul as being prior, describing it as "the first cause of the birth and destruction of all things." Here also, which is very near to the beginning of the dialogue, he begins to interchange the word 'soul' with that of 'god,' finally exchanging the word 'god' for 'soul' towards the end of the dialogue. To wit, at the end of this passage when Plato refers to the doctrine which produces an impious soul, he states that 'soul' is asserted to have come later when it actually came first, explaining further that this is "the mistake these people have made about the real nature of the gods." Towards the end of the dialogue, after proving the existence and priority of soul, the word 'soul' is dropped, favoring the use of the word 'god' alone. Plato makes the identity between the actions of 'soul' and our understanding of what a 'god' is at *Lg.* 899a7-b10, after which Plato drops all references to 'soul(s)' and instead uses the word 'god(s).'

582. Plato's argument from the *Rep.*, Bk. II, 380d-381c, that God (and everything that belongs to god), being in every way the best possible state, would not change or alter himself and thus, being the fairest and best possible forever abides simply in his own form, does not exclude the idea of God being understood as a causal power outwith the Cosmos, while at the same time, being understood as a Soul within the Cosmos, since it is only the perspective which has changed, and not the essential form of God as the immaterial primal cause of everything that is good.

583. By positing 'soul' as "the cause of all things" in *Laws X* (see *Lg.* 896d5-8), Plato challenges the equally powerful primal powers, the Demiurge and Ananke, found in the *Timaeus*. These causal souls match the causal powers in the *Timaeus* in everything except their form and thus, their role, making a causal link between the two possible and, as I have argued, likely.

Demiurge and Ananke) now enter into the finite as Beneficent and Non-beneficent Souls[584] or as they are latterly referred to in *Laws X*, as "gods" within the world, the chief god (understood to be the 'Demiurge' in the *Timaeus*) also referred to as 'king' and 'supervisor of the universe.'[585] Accordingly, these powers are now understood as primal souls engaged within the world effecting ongoing change. Any changes which take place, moreover, and their respective ends, are determined in part by the choices and associations of the human soul as it participates in the various motions of these primal souls. Plato does not fully articulate his views on how the human soul interacts with the primal souls, leaving the reader to fill in or assume part of the process. He abjures, moreover, on the human mind ever knowing exactly how soul is able to move a body, or as he describes it, rational (or irrational) motion,[586] presenting instead three possibilities describing the relationship of the soul to the body.[587] What Plato does say concerning the primal soul is that "soul, by virtue of its motions, stirs into movement everything in the heavens and on earth and in the sea," having prefaced this by stating that there are at least two souls, a Beneficent Soul and one that has "the opposite capacity."[588]

Two problems arise, however, in identifying the primal souls of *Laws X* with the primal powers of the *Timaeus*. The first problem is that Plato refers twice in *Laws X*, and in close proximity (*Lg.* 892a2-b1 and 892c2-7), to the soul being the first ("elder-born") of generated things. If this is so, how can a claim be made that the 'self-moved movers' or primal souls which Plato now refers to in *Laws X* have any direct relation to the 'unmoved movers' or primal powers of the *Timaeus*? My proposal, both earlier and now, is that Plato does not consider that any souls, primary (which he ultimately differentiates as 'Beneficent' and 'Non-beneficent' Souls) or otherwise, were created or generated in the same sense that physical objects in the Cosmos were generated, i.e. from pre-existing triangles. Rather, 'souls,' whether primal or those which have been begotten by the Demiurge, as is the case of the World Soul and the human soul, are eternal and divine according to their proximity and thus identity with their particular primal source.[589] This ranges, I suggest, from complete or near complete identity

584. *Lg.* 896e4-6.

585. See *Lg.* 904a6 and 904a3-4 ("Supervisor of the Universe" is also more literally translated as "He who provides for the Universe." See Taylor, *Laws X*, 297.)

586. *Lg.* 897d3-e2.

587. *Lg.* 898e5-899a4.

588. See *Lg.* 896e8-897b5. The three possibilities Plato presents are: (a) the soul resides within the visible body, e.g. the sun; just as our (the human) soul takes us around from place to place; (b) acquires its own physical body; or (c) [soul] is entirely immaterial carrying the body along its path by virtue of possessing some other prodigious and wonderful (miraculous) powers.

589. There are several factors, which argue against the position that because 'soul' was generated (i.e. the World Soul and human souls), it is therefore not eternal. These

as is the case of the Beneficent Soul with the Demiurge and the Non-beneficent with Ananke (aspect being the only real difference), as well as the lesser divini-

---

can be listed as follows: (1) the description of the construction of the 'soul' is written entirely in metaphorical and mythic terms, which terms do not indicate a finite object (see *Ti.* 34b10-37c5 for the construction of the 'World Soul' and 41d4-42e4 for construction of the 'human soul'); (2) 'soul' was not generated in the same sense as the physical world was generated, namely, out of arrangements of triangles; thus, although Plato states at *Ti.* 41a7-8 that what is bonded together can be dissolved, clearly what he is referring to here are the triangles which make up the physical objects in the world and not to what soul is composed. Soul, rather, was constructed within Plato's myth largely along noetic and hence, metaphysical or psychic lines (i.e. out of different mixtures of 'Existence,' whether of what is eternally unchanging or of the material world, 'Sameness,' and 'Difference') and was done in such a way, I suggest, so as to accommodate the eternal in what was not eternal, namely, a physical body; (3) at no place in the *Timaeus* does Plato state that the 'soul' is finite or not eternal; to the contrary, he specifically states within the passage at *Ti.* 41d4-42e4 noted above, as well as at *Lg.* 904c6-905c4, that the human soul is subject to transmigration between bodies and while the body is referred to as entirely mortal, the soul is not so described, which, moreover, according to certain laws of destiny, if it controls the mortal creature with which it is bound for the best, will itself ultimately reside in its first and best form, namely, with its appointed star apart from a body, which stars in turn are previously described as being eternal and divine (*Ti.* 40b4-8); (4) in *Laws X* Plato argues unequivocally for the position that not only is 'soul' prior to matter, but that it is the most ancient thing there is (*Lg.* 896b2-3) and hence, is also master of matter (*Lg.* 896b10-c3), concluding that in truth 'soul' is a divinity (see *Lg.* 897b1-b4 and 899b3-10) and as a god is in possession of mortal creatures and the universe as a whole (*Lg.* 902b8-9), and as First Cause (*Lg.* 899c6-d2) is ultimately immortal itself, running the entire universe for ever (*Lg.* 905d8-e3); (5) similarly, at *Ti.* 36e2-5 Plato states that on its completion, the 'World Soul' provided "a divine source of unending and rational life for all time." For Plato, what is divine is eternal, whether placed within an instrument of time or outside of time; (6) the 'World Soul' is described by Plato as "the best creation of the best of intelligible and eternal things" (*Ti.* 36e5-37a2); however, see Cornford's discussion of this passage in *PC*, 94, fn. 2; (7) when Plato refers to the '*mortal* creatures' which the Demiurge is about to generate at *Ti.* 41b7, 'soul' is not described as one of these. Furthermore, at *Ti.* 41c6-d4, when the Demiurge announces to the Gods that

> in so far as there ought to be something in them [the mortal creatures] that can be named immortal, something called divine, to guide those of them ready to follow you and the right, I will begin by sowing the seed of it and then hand it on to you; it remains for you to weave mortal and immortal together and create living creatures. Bring them to birth, give them food and growth, and when they perish receive them again (Lee, 57),

'soul' is the only object which the Demiurge could be referring to as being immortal in relation to living creatures; and (8) at *Ti.* 41e4-42a1, the Demiurge announces that each soul will be sown in its appropriate instrument of time, implying that the soul itself resides outside of time. What is eternal, therefore, is the soul itself, but not the combination of body and soul (see *Lg.* 904a6-c4).

ties with the heavenly gods, to being a divinely wrought mixture as was the World Soul (not mentioned in *Laws X*), to an even less pure mixture as was the human soul. Plato's purpose, I suggest, in depicting soul in *Laws X* as being "the elder-born" was to enable him to show that soul was pre-eminently natural, existing before what the atheists regarded as the primary natural substances, e.g. fire or air,[590] and thus, as primary creations, did not derive later from art or reason. Having taken, therefore, the perspective of his opponents in order to win them over, it would have defeated Plato's purpose to depict soul as being divine or primal to start out with, as that was precisely what he was setting out to prove, namely, that the gods exist. Rather, by proving first that the soul exists and second, by way of being a self-moved mover, that it was divine, he could then argue, by way of identity, that the gods existed. Thus, after establishing that the soul by definition was the 'self-moved mover' in living things,[591] Plato then quickly asserts that the soul, in cleaving to divine reason, was itself a divinity or in cleaving to unreason, the opposite (sort).[592]

Plato rarely differentiates in *Laws X* between the different *kinds* of souls, i.e. between human souls, the primal souls, and heavenly souls, or between souls and gods, nor does he mention the World Soul, moving seamlessly from a discussion of souls that move the world to gods that also move the world. What this suggests is that Plato's primary concern in this work with regard to soul is its nature generally as the self-moved mover within all things that make up the universe. Only secondarily does he become interested in the individual powers of the primal souls, namely, the Beneficent and Non-beneficent Souls, as they relate to the world and to other souls. The brevity of Plato's discussions on the cause(s) of evil, which sporadically surface in this dialogue, although important and slightly more forthcoming than in the *Timaeus*, is in keeping with a similar stance or aversion that Plato has to all discussions of this kind (i.e. involving evil).

The second problem concerns the parameters or identity of 'soul' as it is understood in *Laws X* when compared to the *Timaeus*. My position is that a problem only arises if one attempts to carry the comparison of soul from *Laws X* with that of the *Timaeus* rigorously through. This cannot be done, chiefly because the considerations undertaken in *Laws X* never extend to a point outside of the generation of the Cosmos, whereas more than half of the *Timaeus* is concerned with either the pre-generative Cosmos or the process of generation. Thus, the Demiurge and Ananke understood as primary causes existing outside of the material universe in the *Timaeus* are indiscernible in the post-generative Cosmos of *Laws X*. What Plato is able to discern, however, and thus postulate in *Laws X*, is the existence of 'self-moved movers' or 'primal souls' within the Cosmos,

---

590. *Lg*. 892c2-7.
591. *Lg*. 895a5-b7.
592. *Lg*. 896e8-897b5.

which he explains are the "initial principle[s]" behind the sequences of movements that take place in the generated world.[593]

This apparent gap between the idea of 'primal souls' in *Laws X* and 'primal powers' in the *Timaeus*, or between that of the 'self-moved movers' and that of 'unmoved movers,' is effectively closed, I suggest, with one merging into the identity of the other, as the pre-generative perspective or aspect changes into the post-generative. These powers are essentially the same, although formally different. In the *Timaeus*, because it is a cosmology dealing with ancient beginnings the causal powers are understood as transcendent and impersonal with respect to the world they generate, their role in the world, if any, unrealized and hence, yet to be comprehended or recorded. In *Laws X*, however, these same powers viewed from within a world now long in existence, but often forgotten, misunderstood or denied by mankind as its source and unrecognized as causes of ongoing noetic or anoetic movement (Plato's understanding as it has developed since the *Timaeus*), are argued by Plato to be immanent and mindful of creation. Thus, at *Lg.* 896a5-9 the Athenian asks: "Have not we got ourselves a satisfactory proof that soul is identical with the original source of the generation and motion of all past, present and future things and their contraries?" Clinias agrees, replying that 'soul,' being the source of motion, is the most ancient thing there is. The Athenian's final word on this is that since soul has been proven to be prior to matter, then soul is the master and matter is its natural subject (*Lg.* 896b10-c3). Although, as pointed out earlier, Plato does not always distinguish clearly when he moves from a general discussion about 'soul' to a more specific account, as, for instance, when he discusses 'primal souls,' he does nonetheless occasionally do this, making a comparison with the primal powers of the *Timaeus* possible. Moreover, in the *Timaeus* neither the World Soul nor the human soul are ever described as "masters" of the generated world, rather, their description is always one of 'moved movers,' their movement having been organized by the Demiurge and in this sense 'moved.' I would suggest, therefore, that the concept of the 'primal soul' in *Laws X* as a "self-moved mover" (although it is said to move itself, like the primal powers in the *Timaeus* it is not moved by any *other* power) and "master" is closely linked to the idea of the 'primal power' in the *Timaeus* (understood as an "unmoved mover" and in the case of the Demiurge, as "generator") but not to the World Soul, which although not discussed can be assumed to still exist in *Laws X*, or to human souls.

What I suggest transpires, therefore, is that when the Demiurge and Ananke, although initially understood solely as the primal powers and unmoved movers within the metaphysical realm of the *Timaeus* and hence, the masters of generation, are later understood to also be active within the post-generative world of *Laws X*, they are transformed by Plato for that purpose into self-moved movers (as sources of noetic or anoetic movement) or 'primal souls.' Hence, in this way,

593. See *Lg.* 894e4-895a3.

by not only generating the Cosmos, but also by directly influencing the works of generation as well, these powers retain their status as its ultimate masters. A strong indication, however, that in addition to the primal souls now proposed Plato still retains his original idea of the generative unmoved movers, the Demiurge (divine reason or Nous) and Ananke (unreason), is the passage at *Lg.* 897b1-4 where he states:

> These are the instruments soul uses, whether it cleaves to divine reason (soul itself being, if the truth were told, a divinity), and guides everything to an appropriate and successful conclusion, or allies itself with unreason and produces completely opposite results.

With regard to the idea of the human soul "cleaving" to divine reason or unreason Plato explains that "a soul is allied with different bodies at different times, and perpetually undergoes all sorts of changes, either self-imposed or produced by some other soul."[594] With respect to these changes being self-imposed, Plato introduces here an element of 'will' operating within the human soul to effect its own changes, e.g. as it cleaves to 'reason' or 'unreason.' Thus, he states "Our King . . . [has] left it to the individual's acts of will to determine the *direction* of these changes" (*Lg.* 904a2-c4). He further writes: "Take a soul that becomes particularly full of vice or virtue as a result of its own acts of will and the powerful influence of social intercourse" (*Lg.* 904d4-6).

4.3.7 A call to vigilance

This leaves the question of the precise role of the primal soul open, in particular, the role of the Non-beneficent Soul as it relates to the human soul. If the human soul cannot choose what it believes to be evil, which is Plato's long-held position, how is it possible that it can be influenced by an 'evil' soul or by the 'unreason' of Ananke? If it has a 'will,' moreover, to act according to its own desires and these desires are for its own good, how is it possible that it can be coerced to the contrary? The answers, I suggest, to these questions lie in the nature of 'soul' itself. Soul, from the perspective of the *Timaeus*, is, as I have argued, begotten (generated) from its primal source, the Demiurge, and being generated by the Demiurge and therefore ordered, is completely good. Soul, from the perspective of *Laws X*, however, is described as a 'self-moved mover' which moves everything in the universe in either an organized or disorganized way, and thus, can be linked, in part, to the concept of Ananke, the force for disorder, as its progenitor. Plato never discusses 'soul' in the *Timaeus* as in any way emanating from Ananke, although he does talk of Ananke as being not completely subdued and therefore of remnants of disorder being left in the Cosmos.

---

594. *Lg.* 903d3-5.

What these diverse descriptions presume is not that the human soul was also generated from Ananke (as though containing an element of evil or some, as evil souls) and thus naturally contains disorder within itself—this is impossible, given that Ananke by definition is the disorganizing force and does not generate anything—rather, Plato, in *Laws X*, is prepared to admit that primal and powerful souls, whose principal and only movement is either order or disorder, have been in existence from the very outset of generation and have the power, by their rational or irrational movements, to influence the movement of human souls.

In the Timaeus Plato did not admit to the generation of souls other than the World Soul and human souls. Moreover, whereas in this earlier work the human soul as a 'moved mover'[595] exhibits little of the freedom that Plato now gives it in *Laws X* and in which latter work the World Soul is never discussed, Plato does not give any indication in *Laws X* that he has substantially changed his position with regard to the generation of these souls or that they are no longer 'moved movers' with respect to their general make-up. The developed notion in *Laws X* of the 'will' as a dynamic aspect of the human soul in choosing secondary ends, however, does make the move to a 'self-moved mover' an appropriate description of the human soul. Further, by admitting to the ongoing presence of non-generated *self-moved* movers, to wit, in the form of Beneficent and Nonbeneficent Souls, Plato now introduces into *Laws X* the idea of powerful souls capable of influencing human souls and impacting on the World Soul ('soul' as now generally understood), thereby accounting for perpetual change in the universe, which change results in both good and bad ends. In this way, by initially seeking for an explanation of how belief in a false opinion takes place, i.e. non or wrong belief in the gods—a decidedly bad end from Plato's perspective, Plato has moved from the ideal to the real and in so doing, has enriched his metaphysics with regard to the Cosmos, extending the notion of causal powers into the universe as causal souls.

Whether the actual number of "*divine* [italics my own] draughts-player[s]"[596] in the universe has changed is another issue. I would suggest that it has not—only the perspective has changed. When Plato uses the term 'divine,' he undoubtedly is referring to the Demiurge, the ungenerated gods, the generated gods or the Beneficent Soul. However, implicit to the idea of a 'draughts-player' is the idea of a game and thus, the *other* player and although left unacknowledged, is, I suggest, very much an active but silent player in the cosmic game. This "*undivine* draughts-player," therefore, taking the form of Ananke in the *Timaeus* and the Non-beneficent or 'Evil' Soul(s) in *Laws X*, has also been given a powerful role by Plato, namely, of maintaining disorder and effecting ongoing disor-

---

595. Moved by the very nature of its construction and with such movement as its purpose warrants.

596. See *Lg.* 903b4-e1.

der in the universe and thus, of challenging its fellow player(s) in the latter's desire for the triumph of virtue and defeat of vice.[597]

The question of exactly *how* this disorder is effected in the human soul is left unanswered, but there are clues as to the direction which Plato wishes to go. A soul becomes "full of vice" (or virtue)[598] by acts of its own will as an individual determines the direction of changes in its soul.[599] Plato also talks, as he nears the end of *Laws X*, of there being a good element as a constitutional part of the soul and a bad element. It is possible that Plato understands the human soul in terms of it being a receptacle for motion, whereby it receives the motions of another soul according to the direction of its will or alternatively, where it is able to copy the motions of the soul it wishes to imitate (i.e. of a 'good' or 'evil' soul) or again, where it picks up sympathetically the motions of a soul simply by being in proximity to it.

What is certain is that neither the 'Beneficent Soul' nor the 'Non-beneficent Soul' is understood to be passive. On the contrary "our King" wishes for the triumph of virtue, but in this game of virtue over vice, the bad outnumber the good, with Plato stating, "the battle we have on our hands is never finished, and demands tremendous vigilance."[600] The question remains of vigilance against what? There is only one answer according to Plato and that is the disordered motions of the Non-beneficent Soul, the successor, I argue, of Ananke. What ruins us, Plato states, is injustice and senseless aggression and what saves us is justice and sensible moderation. He hints that in possessing a 'will' (for good), we are able to learn from our mistakes, not wishing to be destroyed or our lives to be imbalanced. Such injustice and aggression result, Plato suggests, from the 'will' at times following untoward, naive movements of the soul, seeking good where there is none. And what are these mistakes? Plato cites 'excess' and 'acquisitiveness' as prime examples of the motions that destroy or imbalance and when left unchecked become disease in the body, plague in the land, injustice in society, and ultimately, 'evil' in man.[601]

597. See *Lg*. 904a6-c4.
598. *Lg*. 904d4-6.
599. *Lg*. 904b8-c4.
600. *Lg*. 906a2-6.
601. *Lg*. 906a7-c6.

APPENDIX

Summary Commentary

A. E. Taylor on νοῦς[602]

Timaeus	Greek	Page in Commentary/ Comments	Translation
29b6	μέτα νοῦ	74 / None.	'by thought'
30c2-31b3	νοῦς	82 / Reference is not actually being made to νοῦς itself, but rather, to its relation to νοητὰ ζῷ at 30c3 and 31a5 as held by the Neo-Platonists, whose interpretation distinguishes between the Demiurge and creator, i.e. νοῦς and the supreme God, i.e. νοητὰ ζῷ. From the latter or the One proceeds νοῦς and from νοῦς proceeds the Soul of the World, which three form the so-called 'trinity' of Plotinus. Taylor concedes that they are right enough. It is his next comment where he informs us of his understanding of νοῦς. He rejects the position that the νοητὰ ζῷ is a 'god' superior "to the mind which creates the οὐρανός," stating that this is a development for which the dialogues provide no support. Thus, in this passage νοῦς is understood as a mind which creates. Taylor further states that God in the dialogues is the / ἀρίστη	

602. Taylor, *A Commentary on Plato's Timaeus*, passim. I have limited my choice to key passages from this work.

Timaeus	**Greek**	**Page in Commentary/ Comments**	**Translation**
30c2-31b3 (cont.)	νοῦς	ἀρίστη ψυχή, whereas the νοητὸν ζῷν is not a ψυχή at all, but a system of εἴδη [forms]. Hence, "the Demiurge of the *Timaeus* is exactly the 'best ψυχή' which is said in the *Laws* to be the source of the great cosmic orderly movements, that is, he is God, and if we are to use the word God in the sense it has in Plato's natural theology, the only God there is." From this passage it can be ascertained that for Taylor νοῦς is not only the mind which creates, but as "the only God there is" can correctly be translated as 'Mind.'	
35b1-3	νοῦς	117 / This is another passage in which νοῦς does not appear itself, but where Taylor refers to it obliquely, appealing to Plutarch. In discussing ἀνάγκη, Plutarch is quoted as describing it "in plain words a disorderly and maleficent soul. This was what soul was by itself, but it received intelligence (νοῦς) and reason (λογισμός) and sane attunement (ἔμφρων ἁρμονία) that it might be the soul of an ordered world (κόσμος)." Although Taylor disagrees with Plutarch on other grounds, he does not quibble with his understanding of νοῦς as intelligence. Also, the distinction of λογισμός as 'reason' is valuable to note.[603]	
		123 / Νοῦς is again brought into Taylor's discussion, this time with respect to ψυχή, in which his understanding is further aided by the psychological acumen of Proclus. Thus, while referring to soul (ψυχή) and paraphrasing Proclus, and at times quoting him, Taylor writes "it is not a pure unity without plurality, like νοῦς (intelligence)." Further, what "he [Plato] means by the Indivisible [35a3] (Τῆσ ἀμερίστου) is an intelligence / or	

603. Plutarch's *Moralia*, "On the Creation of the Soul in the *Timaeus*," 1014d-e.

Timaeus	**Greek**	**Page in Commentary/ Comments**	**Translation**
35b1-3 (cont.)	νοῦς	or νοῦς. But he does not mean *any* νοῦς; he means *the* νοῦς which belongs to the κόσμος in particular . . . and what he [Plato] is saying is that the cosmic *soul* is the connecting link between the νοῦς or 'understanding' of the κόσμος and the organic life which the κόσμος enjoys. Now, understanding in its own nature is eternal and possesses truth once for all and all at once. But souls do not . . . So the soul may be said to be at once eternal—in so far as it enjoys intelligence—and a thing of time."[604] Taylor's method of using the exegesis of others is to let his paraphrase or quote of their words stand unless he explicitly states that he disagrees or cares to explain further. In this case he is in agreement with Proclus. Thus, νοῦς is to be understood as 'understanding' or 'intelligence,' which 'intelligence' is eternal and possessing truth "once for all and all at once" and is "a pure unity without plurality."	
36d1-2	τὸν νοῦν	153 / In this section Taylor is quoting Aristotle's criticism in *De Anima* (A 406b 26ff.) of 36b5-d3 in the *Timaeus*. Taylor argues that Aristotle falls foul here in his criticism of Plato and that as a lecturer he was "only talking for victory" feeling the need for "a little judicious levity" (154). Aristotle's interpretation of τὸν νοῦν as 'Intelligence' when he is trying to reason out Plato's thought, however, is not questioned.	'Intelligence'
36d8	κατὰ νοῦν	175 / "Simply, in the usual sense of the phrase." Taylor rejects as "an unfortunate fancy" Archer-Hind's[605] suggestion that "Probably, as in / *Phaedo* 97 D,	'to his mind' 'as he intended'

604. Proclus, *In Tim.* ii. 120 (Diehl).
605. A-H, *TP*, 115.

Timaeus	**Greek**	**Page in Commentary/ Comments**	**Translation**
36d8 (cont.)	κατὰ νοῦν	*Phaedo* 97D, there is a double meaning in these words 'to his mind' and 'according to reason'," since A-H offers no support for the 'probability' of the second meaning.	
37c2	νοῦς	182 / "νοῦς ἐπιστήμη τε is rather to be regarded as a single concept, 'knowledge,' as opposed to 'belief' or 'opinion,' which may or may not be true."	
37c3-5	νοῦς	183 / Again, this is a passage which does not contain νοῦς, but is one in which νοῦς figures predominantly in the discussion, providing an important clue to Taylor's understanding of it as 'Mind,' both in the context of this discussion and in the passages which follow. In this section Plato is talking about the ψυχή of the οὐρανός. Moreover, Timaeus throughout is thinking of the ψυχή as self-moving. With reference to two contrasted expressions, δόξαι καὶ πίστεις and νοῦς ἐπιστήμη τε, δόξα and ἐπιστήμη are both described as not being capable of existing outside a ψυχή, equivalent to the doctrine that thinking and knowledge are only to be found in a ψυχή, rendering, Taylor argues, such phrases as 'impersonal thought' or 'unconscious mind' nonsense.[606] He further states that although in Ionic νοῦς means 'Mind,' e.g. in Anaxagoras, it rarely, if ever, means that in classical Attic prose, where its meaning "is much more often the 'sense' of a statement, the 'purpose' or 'intention' of an act, or a plan of policy, or the like." Thus, so far as there is a classical Attic equivalent for 'the Mind,' when ψυχή itself is not used, the word is more commonly διάνοια. The point that Taylor is making is that when νοῦς is used in connection with ψυχή it means 'Mind,' i.e. at 37c2. / The	

606. See *Sophistes* 249a, *Philebus* 30c, *Ti.* 30b8.

Timaeus	**Greek**	**Page in Commentary/ Comments**	**Translation**
37c3-5 (cont.)	νοῦς	The implication of this inference is that even where the connection between νοῦς and ψυχή is less obvious, i.e. where νοῦς is being discussed and prior discussions of the ψυχή of the οὐρανός or its self-moving attribute are presumed, in these passages 'Mind' should be taken as the meaning of νοῦς and not the 'purpose' or 'intention' of an act, or 'reason' with a capital "R" as in Cornford.[607]	
41a7-d3	νοῦς	247 / Νοῦς does not figure in this passage, nor does Taylor directly talk about it. However, he does discuss the "drift of the Creator's speech," noting that it is the Creator's will, the same power which put them [the newly fashioned gods] together, which will also assure their continuance. Thus νοῦς, which Taylor has already connected very closely to both the 'understanding' of the κόσμος and the 'self-moving ψυχή' of the οὐρανός, by its proximity to the Creator also acquires the nuance of 'will.' What is understood as 'will' by Plato is this power that the Creator, the "best ψυχή," uses to create and sustain (and a 'wicked one,' to dissolve)[608] the Cosmos. However, whereas the idea of 'will' is easily incorporated into the notion of 'Mind,' since the idea of mind is closely associated with its power to act, such is not the case with either 'Reason' or 'Intelligence,' which are not so associated. 'Reason' of itself is not a power, e.g. a person can, and often does, think and act *un*reasonably. Rather, it is a mode of understanding whereby events or existents are connected logically one to another by the human mind, whereupon the mind reflects and if it so chooses, acts in relation to these. 'Intelligence,' on the other hand, although closer in meaning to 'Mind' than 'Reason' (we can speak of / 'an	

607. Cornford, *PC*, 160.
608. See Bury, 89.

Timaeus	**Greek**	**Page in Commentary/ Comments**	**Translation**
41a7-d3	νοῦς	'an intelligence' as having being), must be rejected on similar grounds. Wherever there is human life there is 'intelligence,' but the possibility always remains implicit to this notion that there is intelligence other than human, since primarily it is an indicator of design, whether of a rational or creative nature, or issues directly or indirectly (bees build hives) from its source. The fact that 'an intelligence' could be less than human does not accord well with the notion of the 'best ψυχή' possible. The idea of the power to sustain, moreover, is altogether missing, with the end result that it is often used simply as a synonym for 'reason.' Thus, because of the idea of power which is associated with 'will' and the close association of νοῦς with the 'self-moving ψυχή,' whose 'will' generated the Cosmos, 'Mind' is a better choice for νοῦς over 'Reason' or 'Intelligence.'	
46d5-6	νοῦς	292–293 /	'mind'
		Like several of the above passages, νοῦς figures in the discussion, but does not occur in the passage itself. It is worth quoting Taylor in full here: "This point that ψυχή, 'the self-moving,' is the *only* thing which exhibits νοῦς is repeated because of its philosophical and theological significance. It is meant to exclude what we call 'pantheism,' 'de facto teleology,' 'unconscious purpose.' Plato, no less than Timaeus, holds it fundamental that where there is regular orderly motion subserving a 'good' purpose, there is also real and actual mind. The most important proposition about ψυχή in the Platonic philosophy is precisely the 'synthetic *a priori*' proposition that the 'self-moved' . . . is the only purposive agent, the only thing which exhibits νοῦς."	

BIBLIOGRAPHY

Ackrill, J. L. *Essays on Plato to Aristotle*, Oxford: Clarendon Press, 1997.

Allen, R. E., ed. *Studies in Plato's Metaphysics*. London: Routledge & Kegan Paul, 1965.

Archer-Hind, R. D. *The "Timaeus" of Plato*. London: Macmillan, 1888.

Aristotle. *De Caelo*. Oxford: OCT, 1936.

———. *De Generatione et Corruptione*. Edited by H. H. Joachim. Oxford: Clarendon Press, 1922.

———. *Physics*. W. D. Ross, ed. Oxford: Clarendon Press, 1936.

———. *The Metaphysics*. H. Lawson-Tancred, trans. London: Penguin, 1998.

Barnes, J. *Early Greek Philosophy*. London: Penguin, 1987.

Brisson, L. *Le Même et l'Autre dans la Structure Ontologique du Timée de Platon*. Paris: Éditions Klincksieck, 1974.

Burnet, J., ed. *Platonis Opera*. Vol. IV, *Timaeus*. Oxford: Oxford University Press, 1903.

———. *Platonism*. Berkeley: University of California Press, 1928.

Bury, Rev. R. G. *Plato* with an English Translation. Book VII. Cambridge: Harvard University Press, 1952.

Calvo, T. and L. Brisson, eds. *Interpreting the Timaeus-Critias*. IPS 9. Sankt Augustin: Academia-Verl., 1997.

Cherniss, H. "The Relation of the *Timaeus* to Plato's Later Dialogues (1957)." In *Studies in Plato's Metaphysics*, edited by R. Allen, 339–78. London: Routledge & Kegan Paul, 1965.

———. "The Source of Evil According to Plato." *PAPS* 98 (1954): 23–30.

Cleary, J. J. "Plato's Teleological Atomism." In *Interpreting the Timaeus-Critias*, edited by T. Calvo and L. Brisson, 239–48. Sankt Augustin: Academia-Verl., 1997.

Cornford, F. M. *Plato's Cosmology: The Timaeus of Plato*. 1935. Reprint, London: Routledge, 1997.

Diels, H. and W. Kranz, eds. *Die Fragmente der Vorsokratiker*. 3 vols. 6th ed. Berlin: Weidmann, 1951-52.

Dihle, A. *The Theory of Will in Classical Antiquity*. Berkeley: University of California Press, 1982.

England, E. B. *The Laws of Plato*. Vol. II, bk. XII. Manchester: The University Press (Longmans, Green & Co.), 1921.

Hackforth, R. "Plato's Theism." In *Studies in Plato's Metaphysics*, edited by R. Allen, 439–47. London: Routledge & Kegan Paul, 1965.

Hamilton, E. H. and H. Cairns, eds. *Plato: Collected Dialogues*. 1961. Reprint, Princeton: Princeton University Press, 1980.

Jowett, B., trans. *The Dialogues of Plato*. Vol. III, *Timaeus*. London: Oxford University Press, 1953.

Lee, H. D. P., trans. and ed. *Plato: "Timaeus."* Baltimore: Penguin Books, 1965.

Liddell, H. G., and R. Scott. *Greek-English Lexicon*. Edited by H. S. Jones. With Suppl. Oxford: The Clarendon Press, 1968.

Lutoslawski, W. *The Origin and Growth of Plato's Logic*. London: Longmans, Green, and Co., 1897.

Mason, A. S. 1990. Reason and Necessity in Plato's *Timaeus*, D. Phil. thesis, Braesnose College, Oxford.

Mohr, R. D. *The Platonic Cosmology*. Leiden: E. J. Brill, 1985.

Morrow, G. R. "Necessity and Persuasion in Plato's *Timaeus*." In *Studies in Plato's Metaphysics*, edited by R. Allen, 421–37. London: Routledge & Kegan Paul, 1965.

Owen, G. E. L. "The Place of the *Timaeus* in Plato's Dialogues (1953)." In *Studies in Plato's Metaphysics*, edited by R. Allen, 313–38. London: Routledge & Kegan Paul, 1965.

Pangle, T. L. *The Laws of Plato*. Chicago: University of Chicago Press (Basic Books, Inc.), 1980.

Ritter, C. *The Essence of Plato's Philosophy*. London: George Allen & Unwin Ltd, 1933.

Rivaud, A. *Plato*, Oeuvres Complètes, Tome X, *Timée-Critias*. Paris: Collection des Universités de France, 1925.

Robinson, T. M. *Plato's Psychology*. Toronto: University of Toronto Press, 1970.

———. "The Argument of the *Timaeus* 27d ff." *Phronesis* 24 (1979): 105–9.

Saunders, T. J., trans. *Plato: "The Laws."* London: Penguin Books, 1970.

Sorabji, R. *Necessity, Cause and Blame*. London: Gerald Duckworth & Co. Ltd., 1980.

Taylor, A. E. *A Commentary on Plato's "Timaeus."* Oxford: The Clarendon Press, 1928.

———. *The Laws of Plato*. London: J. M. Dent & Sons Ltd., 1934.

Teloh, H. *The Development of Plato's Metaphysics*. University Park; London: Pennsylvania State University Press, 1981.

Vallejo, A. "No, It's Not a Fiction." In *Interpreting the Timaeus-Critias*, edited by T. Calvo and L. Brisson, 141–48. Sankt Augustin: Academia-Verl., 1997.

Vlastos, G. "Creation in the *Timaeus*: Is it a Fiction?" In *Studies in Plato's Metaphysics*, edited by R. Allen, 401–19. London: Routledge & Kegan Paul, 1965.

———. "Degrees of Reality in Plato." In *New Essays on Plato and Aristotle*, edited by R. Bambrough, 1–19. London: Routledge & Kegan Paul, 1965.

———. "The Disorderly Motion in the *Timaeus*." *Classical Quarterly* 33 (1939): 71–83. Reprinted in *Studies in Plato's Metaphysics*, edited by R. E. Allen, 379–99. London: Routledge & Kegan Paul, 1965.

———. *Plato's Universe*. Oxford: Clarendon Press, 1975.

Wright, M. R., ed. *Reason and Necessity*. London: Gerald Duckwork & Co., 2000.

Zeyl, D. J., trans. *Timaeus*. In *Plato: Complete Works*, edited by John M. Cooper. Indianapolis: Hackett Publishing Company, Inc., 1997.

INDEX I

Word Occurrence/Usage in the *Timaeus* συνίστημι / συνίσταμαι συστῆναι (and noun formations)

A. συνίστημι / συνίσταμαι

συνέστη (3sg2aor)	25a6	there existed a confederation / a dynasty had arisen [Atlantis myth]
συνιστὰς	29e1	He that constructed (it) [the universe]
συνέστησεν.		constructed [Becoming and the All]
συνιστὰς	30b5	constructed [reason within soul]
συνιστὰς (mn)	30c3	the Constructor [of the Cosmos]
συνέστησεν.		(did) construct (it) [Cosmos]
συνέστηκεν	30d1	that have been fashioned [living creatures]
συνέστησε.	31a1	constructed (it) [cosmos as a living creature]
συνιστάναι	31b7	to construct [the body of All]
συνίστασθαι	31b8	should be conjoined [two things alone]
συνέστησεν	32c7	had constructed it [the Cosmos]
συνιστάς,		the Constructor of it [its Constructor]
συνεστήσατο	35a1	(he) made [her = Soul]
συνέστησεν	35a5	(he) compounded it [intermediate existence]
συνιστάντι	36d8	of its Constructor
συνισταμένῳ	37e2	the construction [of the Heaven]
συνέστη	45c5	it [the stream of vision] forms
συνίστατο	**48a5**	**was wrought [the universe]**
συνιστάναι	53b6	constructed them [the four elements]
συνέστηκεν.	53c8	is composed [of triangles]

A. συνίστημι/συνίσταμαι (cont.)

συνέστηκεν.	54a7	is constructed [as a third]
συνιστάμενα	54e3	are combined [four equilateral triangles]
συνίσταται (3sg/mp)	55a4	is constructed [the first solid figure]
συνιστάμενον,	55b6	combined [four]
συνεστὸς	56b2	(it) is composed
συνίστασθον	56e5	(unite to) make [two corpuscles of fire]
συνίστασθαι	57b1/2	to be re-compounded [a few of the smaller corpuscles]
συνίσταται	60c5	is that composed [fairer sort of stone]
συνεστήσατο,	69c1	(He) constructed [this present universe]
συνήσειν	71a3	(συνίημι = to understand)
συνέστησε	71b1	(God) constructed the form of [the liver]
συνίστασθαι	72e3	its construction [the remainder of the body]
συνίστησιν	73e1	he compounded [the bone]
συνέστησεν·	74d1/2	molded [flesh]
συνέστησεν,	75a6	he fashioned [the flesh]
συνιστάντες	76e1	those who were constructing (us) [the Gods]
συνίσταται	78a2	(all bodies) composed (of smaller particles)
συνίσταται	78a5	(all things) composed (thereof)
συνεστήσατο	78c1	(He) constructed [inward parts of the veil/body irrigation]
συνεστήκῃ (ptcpf)	78e1	preserves its structure [the mortal living creature]
συνίσταται,	81e6	the origin/creation [= putting together of disease]
συνεστηκυιῶν,	82b8	(in order) of construction [structures naturally secondary]
συνισταμένου	83d7	is formed [whey of phlegm = sweat and tears]
συνίσταται	89c2	(the triangles of each creature) are constructed

B. συστῆναι (and noun formations)

σύστασις. (fn)	32c6	the construction [of the Cosmos]
συστάτῳ	33a3	composite [surround a . . . body]
σύστασιν	36b7	(this) that he had put together
σύστασις	36d9	the construction [of the Soul]
συστήσας	41d9	(and when) he had compounded [the whole]
συστάσεως	**48a2**	**a battling/striving together [of both Might and Mind]**
σύστασιν,	54a5	the construction [of these bodies]
συστήσεται,	54c7	will form themselves [many small bodies]
συστάντων,	55a5	(second solid) [is] constructed [out of]
συστάντος	55c3	(thus) constructed [the shape of the body]
συστάσεως	**55c4**	**(one other) compound [figure]**
συστάντα	56d7	(becoming) a compound [of one corpuscle]
συστὰν	57a2	has been re-composed [any of the other Kinds]
σύστασιν,	57c9	is due to the construction [of each two elemental triangles]
συστάσει	58b2	in their construction [Kinds with larger constituent parts]
συσταθεὶς	59c2	being compounded [bright and solid waters = bronze]
συστάν	60d1	in its composition [fairer sort of stone]
συστάσεως	**60e4**	**(the interstices of) its structure/texture [Lee, 86]**
ἀσύστατον	61a1	(earth) not condensed
συστάντα	61a6	(earth) is condensed
σύστασις	66c2	the composition [of the particles]
συστήσαντες	71d5	[they who] constructed [us]
σύστασις	72c2	the structure [of the organ]
σύστασις	75b3	the structure [of the head]
συστάσεως	**78b2**	**its (own) structure [the belly's]**
συστάσεων	82b8	(in) the structures [of diseases]
συστασῶν	83d2	(bubbles) are formed [serum]
συστάσεων	89a6	renovations [of the body]
σύστασις	89b5	(in its) structure [of every disease]
συστάσεως·	**89c5**	**(with respect to) the structure [of diseases]**
συστήσαντες	91a3	(the gods) by constructing [an animate creature/men/women]

INDEX II

Word Occurrence/Usage in the *Timaeus*

τὸ αἴτιον/αἴτιος, α, ον
ἡ αἰτία

A. τὸ αἴτιον/αἴτιος, α, ον
B. ἡ αἰτία

αἰτίαν	18e3		was due to [i.e. was the cause of] chance [mating]
αἴτιον	22c1	τὸ	And this is the cause [lack of historical knowledge]
αἰτίας	22e4	δι᾽ ἅς	for these reasons [facts of history]
αἰτίου	28a4	ὑπ	owing to some Cause [everything comes to be]
αἰτίου	28a5		without a Cause [nothing can become]
αἰτίαν	29d7		state the cause [the reason why] He constructed Becoming and the All
αἰτίαν	33a6	τὴν	For this reason [the world was fashioned a single whole]
αἰτίας	38d7	δι᾽ ἅς	And for what reasons [placement of heavenly bodies]
αἰτίας	40b4	τῆς	From this cause (origin) [came into being the fixed stars]
αἴτιον.	42e3		a cause of evil to itself [the mortal creature]
αἰτίας	44c7	δι᾽ ἅς τε	the causes and divine counsels [existence of soul]
συναιτίων	46c7	τῶν	the auxiliary causes

A. τὸ αἴτιον/αἴτιος, α, ον
B. ἡ αἰτία (cont.)

συναίτια	46d1		(supposed to be not) auxiliary causes,
αἴτια	46d2		primary Causes [of all things]
αἰτίας	46d8	τὰς	the Causes [which belong to intelligent nature]
αἰτιῶν	46e3	τῶν	(must declare both kinds) of Causes
συμμεταίτια	46e6		accessory causes [which give eyes their power]
αἰτία	47a2		(vision) . . . is the cause [of the greatest benefit to us]
αἰτία	47b6		the Cause (and purpose of that best good)
αἰτίας,	48a7		(the form of) the Wandering/Errant Cause
αἰτιῶν	57c8		(Such are the) causes [of the pure primary bodies]
αἰτιατέον	57c9		is due to [the construction of two elemental triangles]
αἰτία	61b6		they are the sole causes [i.e. portions of fire]
αἰτίας	61c5		the causes which [account for qualities of Four Kinds]
αἰτίας	65b7	τὰς	the agents which cause them [the bodily affections]
αἴτιον	66b6	τὸν	the cause (of these processes is termed "acid")
αἰτίας	67b1	δι' ἃς	the causes [whereby affections of hearing are produced]
αἰτίων	67c8	τῶν	the causes which produce it [the visual stream]
αἰτίας.	67e4	τὰς	(appear different) for reasons [causes] stated (but are really identical affections)
αἰτίαις	68e4	(fem.)	and their inherent properties, he (the Artificer) used as subservient [αἰτίαις ὑπηρετούσαις] causes; 68e1-69a5 is an important summary passage in the *Timaeus* re the causes and is discussed at length within chapter 5 of Part I.

A. τὸ αἴτιον/αἴτιος, α, ον
B. ἡ αἰτία (cont.)

αἰτίας	68e6	διὸ (fem.)	two kinds of causes [the ἀναγκαῖον and the θαῖον, the necessary and the divine]
αἰτίων	69a7	τῶν	the various kinds of causes
αἰτίοις	76c6		of the causes mentioned [i.e. processes re growing of hair]
συναιτίοις	76d6		(whereas these were) the auxiliary causes
αἰτιωτάτῃ	76d7	τῇ	(it was wrought by) the greatest of Causes (divine Purpose)
αἰτίαις	79a6		the causes [concerning the process of respiration]
αἰτιατέον	87b4		(all of us who are wicked, become wicked owing to two quite involuntary) causes.
αἰτίαις	87c2		and the causes which serve this [remedial treatment of body and mind]
αἰτιᾶσθαι	88a7		(and makes them [the doctors] ascribe the malady) to the (wrong) cause

Author Index

Subject Index

About the Author

S. R. Charles is a philosopher and metaphysician specializing in Ancient Philosophy, particularly late Platonic scholarship, as well as ethics. Dr Charles is the recipient of the Archimedes Award from the University of Toronto (1981) and the Bruce of Grangehill Prize from the University of Edinburgh (2000), having studied and taught at both institutions. Publications include '*Plato's Devil*,' Chapter 7 of *Earth's Abominations: Philosophical Studies of Evil*, Daniel M. Haybron, (ed.), 89–104 (Amsterdam: Rodopi VIBS, 2002). Dr. Charles is also an accomplished composer and harpist.

www.ingramcontent.com/pod-product-compliance
Lightning Source LLC
LaVergne TN
LVHW101445200926
848483LV00001B/51